PHILOSOPHY AND MATHEMATICS

From Plato to the Present

Titles in Philosophy

Ayer, Sir Alfred J. • Oxford University • *The Origins of Pragmatism*—Studies in the Philosophy of Charles Sanders Peirce and William James

Ayer, Sir Alfred J. • *Metaphysics and Common Sense*

Baum, Robert J. • Rensselaer Polytechnic Institute • *Philosophy and Mathematics*—From Plato to the Present

De Lucca, John • Queen's University • *Reason and Experience*—Dialogues in Modern Philosophy

Hanson, Norwood Russell • *Perception and Discovery*—An Introduction to Scientific Inquiry

Hudson, W. D. • Exeter University • *Reason and Right*—A Critical Examination of Richard Price's Moral Philosophy

Humphreys, Willard C. • Evergreen College • *Anomalies and Scientific Theories*

Ross, Stephen David • State University of New York, Binghamton • *Moral Decision*—An Introduction to Ethics

Ross, Stephen David • *In Pursuit of Moral Value*

PHILOSOPHY AND MATHEMATICS

FROM PLATO TO THE PRESENT

ROBERT J. BAUM

Rensselaer
Polytechnic
Institute

FREEMAN, COOPER & CO.

1736 Stockton
SAN FRANCISCO · 94133

Acknowledgments

We gratefully acknowledge permissions to quote from the following sources.

The Dialogues of Plato, translated by Benjamin Jowett, 3rd edition, 1892, The Clarendon Press, Oxford. *The Oxford Translation of Aristotle*, edited by W. D. Ross, The Clarendon Press, Oxford. *Descartes: Philosophical Letters*, translated and edited by Anthony Kenny, 1970, The Clarendon Press, Oxford. *The Philosophical Works of Descartes*, edited by E. S. Haldane and G. R. T. Ross, Cambridge University Press. *The Unpublished Scientific Papers of Isaac Newton*, edited and translated by A. Rupert Hall and Marie Boas Hall, Cambridge University Press. *Critique of Pure Reason*, by Immanuel Kant, translated by Norman Kemp Smith, St. Martin's Press, Inc., Macmillan and Co. Ltd., London, and The Macmillan Company of Canada Limited. "Le Nombre Entier," by Gottlob Frege, *Révue de Métaphysique et de Morale* 3, 1895, pp. 73–78, translated by V. H. Dudman, Macquarie University. "Are True Numerical Statements Analytic or Synthetic?" by Erik Stenius, *The Philosophical Review.* "Mathematics as an Experimental Science," by Sidney Axinn, *Philosophia Mathematica*, Vol. 5, June–Dec. 1968 (No. 1–2). "The Elusiveness of Sets," by Max Black, *The Review of Metaphysics*, Vol. XXIV, No. 4, pp. 614–636.

Printed in the United States of America

Library of Congress Catalog Card Number 73–84704

International Standard Book Numbers 0–87735–513–4
0–87735–514–2

Contents

Preface vii

General Introduction 1
1. Plato • Introduction 15
Selections 18
2. Aristotle • Introduction 39
Selections 44
3. Descartes • Introduction 78
Selections 81
4. Hobbes • Introduction 101
Selections 104
5. Locke • Introduction 116
Selections 119
6. Newton • Introduction 135
Selections 138
7. Leibniz • Introduction 149
Selections 152
8. Berkeley • Introduction 172
Selections 175
9. Hume • Introduction 193
Selections 196
10. Kant • Introduction 212
Selections 215
11. Mill • Introduction 235
Selections 238
12. Frege • Introduction 263
Selections 264
13. Three Contemporary Views 269
Stenius • Introduction and Selection 270
Axinn • Introduction and Selection 283
Black • Introduction and Selection 292

Bibliographies 310
Index 317

Preface

This book has been designed to fill several major gaps in the available literature on the philosophy of mathematics. The philosophy of mathematics as a sub-discipline of philosophy includes discussions of the epistemological status of the axioms of mathematics, the ontological status of the objects of mathematical knowledge (if there are any such objects), and the relation between mathematics as a subject of human study and other areas of human study and interest, including the physical sciences, ethics, religion, etc. And these questions can be traced back to the beginnings of the Western philosophical tradition.

Although there are a number of excellent books on the philosophy of mathematics presently available—both original texts and anthologies—most of them are devoted to the literature as it has developed out of the work of mathematicians, logicians, and philosophers such as Frege, Hilbert, Peano, Brouwer, and Russell at the beginning of this century. Much of this work is focused on a sub-area of the philosophy of mathematics known as the *foundations* of mathematics. Foundational studies are primarily concerned with the problems surrounding the establishment of the fundamental axioms of mathematics, and the questions of the consistency, completeness, etc. of alternative sets of axioms. Although foundational studies during the last hundred years have produced results of significant value to philosophers, as well as to mathematicians and logicians, their connections to general philosophical problems have not always been explicitly drawn out, and when they have been it has too often been in a highly technical manner unintelligible to non-specialists. The material included in this book has been selected in such a way that it focuses on the *general* epistemological and ontological issues underlying the recent work in the foundations of mathematics so that these issues can be understood by the general student of philosophy.

Many recent discussions of the *philosophy* of mathematics do not make any distinction between this general subject and its sub-area of foundational studies, and thus they give the impression that the study of the philosophy of mathematics also began with Frege, with perhaps some recognition of certain historical roots running back to Kant and Leibniz. Even those who recognize the full scope of the historical background of present-day work in philosophy of mathematics have little to turn to in the way of source materials. There has been too little work in the history of the philosophy of mathematics (a partial bibliography on this topic is included in this volume), and the original sources are in many cases inaccessible to any but the most determined persons. A study of the development of the philosophy of mathematics similar to the Kneales' *The Development of Logic* has yet to be made, and if and when it is attempted it will undoubtedly require a number of volumes to provide adequately for all thinkers who deserve attention.

Many of the original source materials have not been made available previously in English in a convenient form because it could not be done by simply collecting a number of essays or chapters from longer works of the pre-20th century philosophers. None of these philosophers wrote any essays or even chapters of books completely devoted to the nature of mathematics. Rather, their discussions of these issues permeated their writing. Thus, in order to present the philosophies of mathematics of a selected group of these philosophers, it has been necessary to go through their collected works and to extract those passages (sometimes only a few sentences at a time) which bear on the basic questions of the philosophy of mathematics. These excerpts are here arranged to present each philosopher's theory in a logical way. There is always a risk involved in pulling items out of context in such a manner, but great pains have been taken to assure that the following selections present each individual's theory in the way that—in the judgment of the editor—he would have presented it himself if he had been asked to write an essay on the nature of mathematics.

The originality of this project, and the lack of material available in print on the history of the philosophy of mathematics, offered many possibilities for editorial comments. Careful consideration of the needs and interests of the students and teachers who will be using this book, as well as of space limitations, indicated that it would be best to concentrate on only a few topics in the editorial material. The general introduction focuses on the philosophy of mathematics as an integral part of the traditional discussions of the questions of epistemology and ontology, and gives an outline of some of the problems involved in evaluating competing theories about the nature of mathematics. The introduction to each section focuses on three points of interest about the particular philosopher represented in that section:

- It provides relevant biographical information, including a statement of his training in and contributions to mathematics proper.
- It places him in his historical context, relating his work to that of both his predecessors and his successors—including 20th century philosophies of mathematics.
- It relates his discussions of the nature of mathematics to the larger context

of his general theories of epistemology and ontology, as well as to other relevant aspects of his philosophy.

There will undoubtedly be some differences of opinion about the list of readings, both with regard to the individual philosophers who have been included and excluded, and with regard to the specific passages selected to represent the theories of each individual. The choice was between leaving out many interesting theories of influential philosophers or expanding this project to a multi-volume work. Philosophers such as Plotinus, Augustine, Aquinas, Scotus, Occam, Bacon, Spinoza, etc. were regretfully excluded for the reason that such a comprehensive collection would be of little use except as a reference work for scholars. The philosophies of mathematics of many significant mathematicians were not included because in most cases their philosophical assumptions are implicitly rather than explicitly stated in their writings; and even where they discuss these issues explicitly, their theories are generally not particularly original, and their basic ideas are better expressed by one or more of the philosophers included in this book. It is the editor's belief that the theories presented in the first twelve sections of this volume are representative of the philosophies of mathematics which were held by most if not all pre-20th century mathematicians, and that they also include most if not all of the ideas which might be considered anticipatory of present-day theories.

With regard to 20th century writings on the philosophy of mathematics, nothing has been included in this volume which is readily available elsewhere. Thus this book can be used in conjunction with any of the other anthologies in print today (e.g. Hintikka, Benacerraf and Putnam, Lakatos, van Heijenoort, or the Bobbs-Merrill reprints) without fear of overlap. The contemporary essays have been selected in part on the basis of their diversity, their readability (including their lack of highly technical terminology or symbolism), and their focus on some of the traditional epistemological and ontological aspects of the philosophy of mathematics. These essays not only complement those available in other anthologies, but also go beyond the arguments presented in other texts (such as Barker, Beth, etc.). The historical materials complement *any* other text (original or anthological), although this book can also be used by itself either for an introductory course on the philosophy of mathematics or for an advanced seminar on the history of the philosophy of mathematics. These readings would also provide valuable supplementary material for courses on the history of mathematics or even the history of science in general. It has been suggested that these readings might be of use in a course on the Introduction to Philosophy, where the focus could be placed on the problem of certainty.

With regard to the conception, writing, and production of this book, I am indebted to so many persons in so many ways that I simply could not begin to name them all here. I hope that my many teachers and colleagues will recognize their influence on my thinking as evidenced in specific portions of the following work and/or in its overall concept and execution; I also hope they will understand my thanking them in this wholly inadequate way. There are several individuals who must be singled out for their special contributions. Professors Paul Olscamp, Robert Turnbull, and Charles Kielkopf exerted major influences on the conception and execution of this book through their encouragement of and assistance with my studies of the history of the philosophy of mathematics. Professors Stephen Ross, Louis Hammer, and David Wieck made numerous helpful criticisms of earlier drafts of the manuscript, and many of their suggestions have been incorporated into the present text. Professor Michael Scriven provided the encouragement and support needed to get the book on its way, in addition to offering several helpful general suggestions about the content and format. Christine Korin and Elizabeth Wortham did a superb job of typing and retyping hundreds of pages of manuscript, a job which went far beyond their required secretarial duties. William H. Freeman has patiently imposed his personal standards of excellence on every step in the writing and production of the book. The reader owes special thanks to Margaret Freeman, whose editorial talents have made many portions of the book much more readable than the author ever could have made them. I cannot begin to list the ways in which my wife, Gail, has aided and abetted my work on this book. The assistance which she so willingly contributed after long days at her own demanding profession can only be acknowledged here. My thanks will have to be conveyed in other ways.

R. J. B.

Troy, New York
June, 1973

PHILOSOPHY AND MATHEMATICS

FROM PLATO TO THE PRESENT

General Introduction

Contrary to a widely held opinion, the philosophy of mathematics is *not* an esoteric subject-matter studied only by a few specialists at the post-graduate level. Every student is exposed to certain elements of this subject in the first grade when he or she is taught that 2 + 2 *must* equal 4, and in high school when taught that the theorems in elementary geometry are *not* about the particular figures drawn in the text book or on the blackboard. Unfortunately, few teachers or students are aware of the philosophical significance of these apparently straightforward mathematical assertions, and students are seldom encouraged to challenge their truth or accuracy. It would not be appropriate to the task at hand to probe into the reasons for this educational situation; the important point to be made is that students *are* taught a few very naive and inadequate philosophical assumptions about the nature of mathematics in even the most elementary mathematics courses. The readings in this book are intended to provide an opportunity to discover some of the more sophisticated and significant philosophies of mathematics which have been formulated during the last two millenia. They are also offered as evidence that the study of this subject is not impossible for the average student, and that a careful study of some of the problems and the theories offered as solutions to them will deepen one's understanding not only of mathematics and philosophy, but also of other areas of interest and concern, including ethics and religion.

An examination of the recent literature on the philosophy of mathematics provides ample justification for the general opinion that the subject is accessible only to specialists proficient in symbolic logic and higher mathematics. There are indeed certain problems in the philosophy of mathematics which can be understood only by such a small group of specialists. But as many of these specialists have pointed out, the *fundamental* issues are more appropriately analysed from a general rather than a highly technical and esoteric point of view, and are accessible not only to the trained mathematician but also to the general philosopher and even to the interested layman.

The recent literature might also lead one to conclude that the study of the philosophy of mathematics began at the end of the nineteenth century with the work of such thinkers as Frege, Peano, and Russell. Such a belief concerning the history of the subject would be even more mistaken than that concerning its inaccessibility. Even though it is all that has been readily available in recent years, the modern literature is not the whole literature. The philosophical study of mathematics did not begin suddenly one hundred years ago; indeed, it is one of the earliest recorded subjects of human inquiry. The readings in this book have been selected not only to provide an introduction to the most basic problems faced by the philosopher of mathematics and to the variety of proposed solutions; they have also been chosen to illustrate the long history of man's concern with these problems and to emphasize the relevance of this special subject-matter not only to all other areas of philosophical inquiry, but to every manifestation of man's perennial search for certain knowledge.

From the earliest times, man has searched for the answers to a multitude of questions. Some are quite specific and concrete: When will the next flooding of the Nile occur? What was the cause of this child's death? Why did the sky suddenly go black? Others are more general and abstract: What is justice? Is there life after death? What are the ultimate constituents of the universe? Although the particular concrete questions are of more immediate concern in the everyday contexts in which they normally arise, the more general abstract questions have been considered by many to be ultimately of more importance and greater interest. An adequate answer to the question about the sky going black requires reference to more abstract notions such as those of the eclipse of the sun, and to the general principles of planetary motion. The real utility of abstract general knowledge is that just a few general principles are sufficient for answering innumerable concrete questions.

But abstract general principles alone are not sufficient for providing adequate answers to our questions. Answers have always been available—too many answers. The Greek Sophists went so far as to claim that equally convincing arguments can be given in support of every logically possible answer to any question. The question thus arises: Which answer, if any, is the *true* answer? Traditionally the demand was often for not merely the most probable answer, but rather for that answer which is absolutely certain. The evidence was required not merely to remove any reasonable doubts, but to establish the truth of the statement beyond the shadow of any doubt. René Descartes echoed this ancient demand, in his *Meditations*:

> I shall continue . . . until I have found something certain, or at least, if I can do nothing else, until I have learned with certainty that there is nothing certain in this world. Archimedes, to move the earth from its orbit and place it in a new position, demanded nothing more than a fixed and immovable fulcrum; in a similar manner I shall have the right to entertain high hopes if I am fortunate enough to find a single truth which is certain and indubitable.

Although the passages in this book may be said to form a part of the literature of the special area of philosophy known as the "philosophy of mathematics," they are equally significant as a part of the chronicle of Western man's quest for certainty. With few exceptions, the authors of the passages written before 1900 studied the nature of mathematical knowledge not for its own sake, but rather for the insights that such a study might provide into the nature of knowledge in general. Their concern was with *general* questions such as "Is certain knowledge possible?" and "What makes knowledge certain?". (It should be noted that many philosophers, particularly those before 1900, would have considered this wording redundant; for them "knowledge" meant "*certain* knowledge," and "uncertain knowledge" or "probable knowledge" involved an internal inconsistency as in "square circle." In present-day discussions the concepts of knowledge and certainty are usually defined independently.) Despite possible differences in motivation and perspective, the "traditional" philosophers arrived at conclusions which provide the foundations of and starting points for much of the work of today's philosophers of mathematics. A careful study of the traditional analyses of mathematical knowledge will provide an overview of Western man's search for certainty and also a feeling for the basic problems with which the philosopher of mathematics is still grappling today.

The quest for certain knowledge has proved to be difficult, and many philosphers now believe that it isutile. One of the most serious difficulties encountered in this search has been the vagueness and ambiguity inherent in the term "certainty" itself, both as it is used in everyday discourse and as it has been used by many philosophers (including most of those represented in this volume). The term is frequently used to refer to the *feelings* of certainty which everyone has experienced in conjunction with various beliefs. Most people have learned quite early in life that such feelings are not always a reliable guide, and thus distinctions are drawn between feeling "sort of" certain, "not too" certain, "extremely" certain, and various intermediate degrees of certainty. Most people will allow that even the strongest feeling that a claim is certain does not seem to be adequate to guarantee that no one will ever have doubts about it in the future, or even that it is true. Thus the term "certainty" has a stronger sense, one which requires going beyond the subjective feelings of certainty to some sort of objective criterion, and one which has often been used as the foundation of claims of knowledge. It is this absolute, infallible, objective kind of certainty which many philosophers have sought, although their failure to identify explicitly the ambiguity in the term "certainty" led others to devote much of their attention to the mere subjective feeling of certainty. But this does not mean that their efforts were futile or wasted, for, as the selections in this book show, many significant insights were gained into a variety of related issues, and the various senses of "certainty" were gradually identified and sorted out as a direct result of these earlier "misguided" efforts. Contemporary philosophers would not even be aware of the vagueness and ambiguity of the term, let alone have any deeper understanding of its different senses, had it not been for the contributions of these earlier philosophers.

The philosophical study of the nature of mathematical knowledge has played a variety of roles in the quest for certainty (in both the subjective and objective senses of the term as well as in its more confused senses). From at least as early as the fourth century B.C., mathematical knowledge was widely accepted in Western cultures as a prime example of knowledge which possesses the highest possible degree of certainty. (The expression "I am as certain of that as I am certain that 2 and 2 equals 4," is still one of the strongest assertions of certainty.) Thus, it is not surprising that many (but certainly not all) philosophers have begun their quest for certain knowledge with the assumption that mathematical knowledge is absolutely certain, and then have proceeded with an analysis of the nature of mathematics. If one can discover what exactly enables man to acquire absolutely certain knowledge of mathematical truths, then, it has been hoped, one will be able to use this same method to arrive at answers to many other questions with a similar degree of certainty, whether these questions be about the existence of God, the date of the next eclipse of the sun, or the meaning of "justice." If this hope proved to be overly optimistic, the philosopher could at least have a better understanding of why it is impossible to know the answers to some questions with certainty by identifying the basic differences between these questions and mathematical ones.

Other philosophers have begun their search with the analysis of a paradigm other than mathematics, or have formulated their theory of the nature of certain knowledge by some other means. But they could not ignore mathematical knowledge, and many have relied on it as a touchstone for evaluating the adequacy of their general theory. Even if a theory seemed plausible at first sight, if it could not account for the apparent certainty of mathematical knowledge it was often considered unacceptable, and had to be either revised or rejected. (Of course, it is possible that one might have even decided that this indicates that the original paradigm was not truly an example of certain knowledge, and then proceeded to find a new paradigm.)

A third and smaller group of traditional philosophers arrived at a more negative conclusion—that mathematics does not provide certain knowledge in any sense at all. These philosophers, in going against the "common sense" *and* the majority philosophical position of their day, had to explain not only what kind of knowledge mathematics does provide, but also why so many people held the mistaken belief that mathematical knowledge is certain (particularly in the objective sense of the term). This negative view has been defended in a variety of ways; some of the most interesting and significant arguments in support of it are included in this book. One strategy is to assert that only knowledge of directly perceived phenomena can be certain, and then to argue that mathematical knowledge is general and abstract and thus cannot be certain. Another approach involves the general assertion that *no* knowledge can be absolutely certain. This approach requires separating the concept of certainty from the concept of knowledge, or else admitting that there is no knowledge at all (i.e., that at best only highly probable opinions or beliefs are possible). Still another theory maintains that mathematics is merely an abstract game, involving the manipulation of meaningless symbols, and that it is a mistake to speak of mathematical *knowledge* at all since there is no object of knowledge. Even though the burden of proof was placed on these "negativists" for hundreds of years, the general consensus today (at least among philosophers and mathematicians) is that some of their theories are among the most plausible of those available. The arguments given in support of these theories today are very similar to those given in previous centuries, and it would be interesting to investigate the reasons why they were ridiculed at that time while they are so widely accepted now. Unfortunately, that is beyond the scope of this book.

The basic questions concerning certain knowledge belong to the sub-discipline of philosophy known as *"epistemology"* or the theory of knowledge. Although they differ considerably in their analyses of the nature of mathematical knowledge, most of the philosophers represented in this book would agree on the basic assumption that knowledge of any kind involves some kind of relation between a knower and that which is known. It is almost tautological that there can be no knowledge if there is no knower, or if there is nothing to be known; it is much more interesting and informative to talk about the *nature* of the knower or the thing known. Thus, in dealing with the epistemological questions, such as "What is the nature of mathematical knowledge and how is it acquired?", most philosophers have found it necessary or worthwhile to deal with a number of other questions, such as "What is the nature of the objects of mathematical knowledge?", "What is the nature of the knower?", and "What is the relation between the mathematician and the objects of mathematical knowledge?" Questions such as these overlap a second sub-discipline of philosophy known as *"ontology"* (the science of being, or the study of what exists). A brief outline of some of the basic ontological theories will therefore be of considerable relevance to the understanding of the subsequent selections.

Before discussing some of the different ontological theories, we should observe that the need for such a discussion also arises when a more linguistically oriented approach is taken to the basic epistemological questions. Some philosophers have focussed their attention on criteria of the truth or falsity of statements, and have reformulated the question of mathematical certainty to read "Under what conditions, if any, can mathematical propositions be asserted to be true?" (Ontological questions can be similarly reformulated; for example, "Under what conditions can something be said truly to exist?") It might seem at first glance that such questions are concerned simply with the nature of language or with the meanings of certain words. However, the extent to which epistemological and ontological questions can be reduced to linguistic questions is dependent on the theory of meaning which one is using. Granted the assumption that only *meaningful* propositions can be true or false, sequences of marks such as "aQ*7y &r-Z!3/" are neither true nor false since they are meaningless (in the context of ordinary English discourse). Most traditional philosophers (as well as many contemporary ones) have adhered to one version or another of what has been called the "correspondence theory" of meaning, which stipulates that the meaning of any word is the object, event, or relation in the 'real' world, or the idea in the mind, which is denoted or signified by that word. Under this theory, a particular statement is true if and only if the entities named by the words in the statement *actually* stand to one another in the relation indicated in the statement. Thus if one accepts some version of the correspondence theory, the answer to the question "Under what conditions can statement X be asserted to be true?" requires an analysis of the kinds of things which exist. There are some serious difficulties with the correspondence theory, but the alternatives which have been formulated to replace it are all quite complex and there is no consensus among philosophers at present in support of any one of them. Some of these alternative theories of meaning are discussed in various selections in this book, and are commented on in the introductions to the appropriate sections.

The questions about the natures of the knowers and things-known were traditionally answered in the context of discussions about the kinds of fundamental "building blocks" or basic constituents of the universe. Some of the terms which have been used in such discussions are "mind," "soul," "body," "matter," "atom," "monad," "God," "idea," "universal," "sense-data," and "Form." Many of these terms have been used by more than one philosopher, but few if any of them have been used in exactly the same way or with the same meaning by any two thinkers. This is well illustrated by the term "substance," the widespread use of which has tended to conceal significant differences among the theories in which it is used. In some philosphical systems it functions as a technical term possessing special connotations, while in other contexts it has been used quite casually and even carelessly with no clearly stated definition. Some philosophers have defined substance as that whose essential characteristic is its lack of dependence on anything else for its existence. Other philosophers have asserted that there are no such things as substances, or have rejected the term "substance" as being meaningless—although neither of these groups has categorically denied the existence of every kind of basic entity. Still other philosophers have raised questions concerning the meaning of the term "exists," and they have suggested that the philosophy of language is in a significant way more basic than ontology. This brief sketch is far from complete, but it is sufficient to suggest *some* of the terminological and other complexities which have appeared in over two thousand years of discussions concerning the ultimate natures of knowers and objects of knowledge.

Not only have many different kinds of fundamental entity been postulated by various philosophers, but these different kinds have been combined in a variety of ways. There is an almost limitless number of ways in which the various kinds of basic entity which have been considered by philosophers *can* be combined. It is impossible to discuss all of these possibilities here, but the most important elements of such combinations can be illustrated by a brief discussion of the kinds of ontological theories which can be constructed using just four kinds of "building blocks"—minds, bodies, ideas, and God. It should not be assumed that these terms are being used in the context of this discussion to refer to concepts drawn from the theory of any particular philosopher, although they do in fact approximate to varying degrees the concepts used by many of the philosophers whose theories are included in this book. The four terms will be used here as non-technical terms bearing meanings similar to those which they possess in ordinary discourse. Thus, "body" will be interpreted as referring to any object which can be perceived by means of the five external senses, that is, those objects which possess qualities such as shape, spatial location, motion, etc. (Philosophers have disagreed sharply as to exactly which properties actually belong to bodies. We will duck this question here, since it will be discussed in considerable detail in later sections of this book.) "Mind" will be assumed to refer to those things which possess 'mental' properties such as thoughts and feelings. It is not necessary to answer here the widely debated question concerning the immortality or indestructibility of minds; again, this is an issue which is discussed in some of the later sections of this book. Although many people ordinarily assume that what they refer to as "ideas" exist only in minds, the term can be defined in such a way as to make ideas theoretically independent of minds without making it incomprehensible; such an interpretation is being assumed here. "God" will be taken to refer to the originating and/or sustaining cause of all other existing things. This Supreme Being can be assumed to possess properties such as eternality, omniscience, omnipotence, infinitude, etc., although it should not be identified with the deity of any particular religion.

Given even this limited set of four types of basic entity, it is still possible to construct a large number of different ontological theories. The four terms can be combined in fifteen different ways, in so far as one assumes the actual existence of all four kinds, or any three or two kinds, or only one of the four kinds. A theory could also assert that none of these four kinds of entity really exists, but then it would have to postulate the existence of some other kind of being, since a theory which maintained that *nothing* exists would seem to involve a factual absurdity, if not a logical contradiction. The theories which would appear to agree with most people's common-sense assumptions are those which assert the existence of *at least* minds and bodies, with the possible addition of God and/or ideas. Although such theories may appeal to common sense, and may have other attractive features, they also have several serious weaknesses. For example, in defining bodies and minds it is often considered necessary to deny explicitly that they have anything in common with each other (e.g., the former fills space and the latter has no spatial characteristics at all), because if they did have anything in common then it might be argued that they are both composed out of a single different and 'more basic' kind of being. But if it is assumed that non-physical minds know non-mental bodies, it would seem that the two *must* have something in common which would permit the apparent interaction between the knower and the known. There are various strategies which can be used for surmounting such difficulties, but they generally introduce additional problems into the dualist or pluralist theories. Some of these will be discussed in greater detail later in this book.

The difficulties surrounding the dualist and pluralist theories are sufficiently serious to possibly outweigh any initial implausibility that may be associated with monist theories due to their apparent conflict with common sense. Such theories need not violate the demand that there be both a knower and a known, since it is clearly possible that a being can know another being of the same kind or that it can know itself. The most popular monist theories have traditionally been those which assert the existence of only bodies ("materialism") or only minds ("idealism"). Although such theories are indeed more plausible than they may at first appear, the various versions of them which have been formulated have their weak points. Some of the more significant problems of both materialism and idealism involve their respective accounts of the nature of ideas. It is widely accepted that ideas have few if any of the characteristics of bodies; in particular, they have neither mass nor spatial location, and it is difficult indeed to talk of them in purely materialistic terms. On the other hand, idealists (and also many dualists and pluralists) have also encountered difficulties in explaining the relation between minds and ideas. The idea of a circle would seem to be something quite distinct from the mind which knows it or in which it might be said to exist. Some idealists would assert that the idea is a *property* of that mind, but such an assertion requires much elaboration as well as a carefully worked out argument in defense of its quite unorthodox use of the concept of being a property. The other two monist theories have been advocated less often, and they involve even greater difficulties. Pantheism is the theory which holds that God is the only being that really exists and that everything else is a "mode" or property of God—thus God *is* the world. A theory which asserts that only ideas exist might be called "idea-ism." It is difficult to determine conclusively whether any philosopher has held such a theory, although a possible candidate for such classification is David Hume, whose theory is included in this volume.

Some of the variations of the theories based on the fifteen combinations of the four basic types of entity just discussed are elaborated on in the following selections, and additional criticisms are also provided. Examples are also given of theories which assert the existence of quite different kinds of fundamental entities (e.g., Leibniz' monads) and which approach the whole problem of the nature of the knower and the objects of knowledge from other perspectives (e.g., Kant's critical philosophy). And the theories included in this volume differ from one another in other significant ways as well.

Although it may seem that more than enough possibilities for consideration in a theory of mathematical knowledge have already been brought up in the discussion of the different types of potential knowers and objects of knowledge, there are additional complicating factors yet to be considered. It has already been indicated that the knower in a knowing-situation may be composed of any of several kinds of basic entity; the additional factor that must be taken into account is the *role* of the knower in the knowing-situation. In some theories, the knower is considered as totally passive—the knowledge is produced simply by the object-known actively impressing itself upon the knower—and in these cases the differences between the philosophies of mathematics is entirely dependent on differences between the things-known and/or the knowers. Many theories, however, maintain that the knower plays an essential *active* role in the knowing-situation. It is thus possible that two or more philosophers can be in complete agreement regarding the ultimate types of being which exist, as well as about the nature of the knower and the thing-known, and yet they can still hold different theories of the nature of mathematical (or any other kind of) knowledge. This aspect of the theories of mathematical knowledge is too complex to be described in detail here, but it will be elaborated on in the relevant section introductions.

The question of the role of the mind in the acquisition of mathematical knowledge is one of the central topics of debate among present-day philosophers of mathematics. The basic disagreement on this matter is over two issues—whether the *objects* of mathematical knowledge exist independently of the mind prior to their being known or are in some way created by the mind, and whether mathematical *theories* are discovered or created by the mind. The terminological distinction between "discovery" and "creation" is of value for identifying certain prejudices not only of philosophers but also of some mathematicians whose writings appear to be of a purely mathematical nature. If a mathematician (or a philosopher) uses terms such as "invent," "create," or "construct" frequently and consistently in his work, it would seem justifiable to conclude that he probably believes (tacitly at least) that either the objects of mathematics or the mathematical theories (or both) are creations of the human intellect. On the other hand, if he consistently uses terms such as "discover" throughout his work, he probably believes that the objects of mathematical knowledge are independently existing entities or properties of such entities and that the theories concerning such entities also exist independently of the particular knower. If he uses both terms (and many philosophers and mathematicians do), it may indicate one of several things: he may be naive and totally unaware of the full implications of these terms; he may be simply confused and/or uncertain as to which of the alternative theories is the correct one; or he may hold that the objects of mathematics are created and the theories about them are discovered, or vice versa. Other philosophers have argued that none of these alternatives is correct and that the application of either of the terms "invent" and "discover" to mathematical knowledge is fundamentally misguided; however, the terms are still widely used in the literature and it is doubtful that such usage will cease too quickly. (The use of the terms "invent," "create," and "construct" in the section introductions is intentional, and reflects the prejudices of the editor on these issues.)

Various individuals have focused their attention on a number of special problems in the philosophy of mathematics, in addition to the general ontological and epistemological problems outlined above. One of the most interesting and important sets of questions discussed in many of the following selections concerns the relation between the two basic branches of mathematics—arithmetic and geometry. The most obvious question concerning this distinction is that of the ontological status of the objects of the two disciplines, but questions have also been raised concerning their methodologies and the meanings of their basic terms as well. A surprising number of different theories have been formulated during the last 2500 years about the similarities and differences between arithmetic and geometry, some of the most important of which are included among the following selections. These questions are discussed further in the appropriate section introductions.

Another important problem discussed by philosophers of mathematics is that of the concept of infinity. Philosophers have frequently discussed this concept in other contexts—such as in discussions of the size or age of the universe, and in proofs of the existence of God—as well as with regard to its use in mathematics. In mathematical contexts it has generated a variety of difficulties—both in arithmetic (i.e., with regard to number) and in geometry (i.e., with regard to space). Different problems have also been encountered in dealing with the infinitely large and infinitely small. Some of the most significant debates among 20th century philosophers have revolved about questions concerning the concept of infinity, and it is their conflicting theories of infinity which distinguish two of the most important schools of philosophy of mathematics. Indeed, so many complex and important problems have appeared in so many contexts concerning the concept of infinity that it would be futile to try even to outline all of them in this volume. A number of interesting and significant treatments of a few of the problems of infinity have been incorporated into the following selections, but it must be noted that these discussions do not touch on all of the problems nor do they present more than a few of the solutions which have been proposed. Unfortunately, some of the most important discussions of infinity (e.g., Kant's) could not be included in this book for several reasons, not the least important of which is the fact that an adequate collection of such material would require at least several complete volumes.

Although the philosophy of mathematics (especially as defined in recent years) involves the examination of a number of other concepts—including completeness, consistency, independence of axioms, etc.—only one more problem can be outlined here. This is the basic question—"Why should anyone do mathematics?" or "What is the value or use of mathematics?" This problem is at least as complex as the problem of infinity. It is of immediate interest to both the student and the professional mathematician who might be concerned with determining the wisest or best thing to do with the time available in this life. The question also involves the important problem of the distinction (if there is one) between pure and applied mathematics, which in turn is related to some points of fundamental disagreement among philosophers of mathematics. For example, one group of philosophers (who are known as "instrumentalists" or "pragmatists") have argued that the value (and even the meaning) of mathematics is completely dependent on the uses to which it can be put, while many others have asserted that mathematics is intrinsically valuable. The following selections include a variety of statements concerning the value and use of mathematics, as well as of some related issues, including the problem of the distinction between pure and applied mathematics.

It was mentioned earlier that the study of the nature of mathematics played a significant role in the general systems of many philosophers. A general overview of some of the basic concepts and theories of epistemology and ontology has been given, and some of the interconnections among these concepts and theories and the central issues of the philosophy of mathematics have been outlined. It remains to be noted that many discussions of the philosophy of mathematics have touched directly on most of the other areas of philosophical inquiry, including ethics, philosophy of language, aesthetics, logic, political philosophy, philosophy of religion, and philosophy of science. In addition, since each of these areas is generally in close contact with the central areas of epistemology and ontology, it follows that there are at least indirect connections among most of them. The explicitly asserted direct connections between the philosophy of mathematics of a particular philosopher and any of these other aspects of his work are included in the passages comprising the various sections. The most important of the indirect and/or implicitly asserted connections are described in the section introductions.

If one attempts to establish criteria for determining with certainty whether or not mathematical knowledge is certain, as many philosophers have in fact tried to do, there is nothing to prevent one from asking for criteria for establishing the certainty of theories about the certainty of the theories about the certainty of mathematical knowledge, and so on *ad infinitum.* One strategy for avoiding such an infinite regress of certainty-criteria is to find some theory of certainty which is applicable to itself. Such a theory would be blatantly, but not necessarily viciously, circular. The most serious problem arises from the fact that it is possible to formulate a number of different theories of certainty, each of which is applicable to itself, and thus additional criteria are still necessary to determine which of these theories is the correct one.

In recent years, philosophers have turned their attention away from the quest for certainty, and have instead sought criteria for determining the truth or correctness of particular theories, or for selecting the best of several competing theories. They have suggested a number of criteria which are supposed to enable one to determine clearly and definitely which of a given group of theories is the 'true' or 'correct' theory. All of them have proven to be inadequate for a variety of reasons. One of the most obvious criteria for a correct or true theory is that the theory must "fit all the relevant facts." This criterion is deceptively simple and straightforward; the real difficulties do not appear until one tries to apply it to the evaluation of a particular theory. What exactly is a "fact?" How can one determine which facts are "relevant" to a particular theory? Even if agreement can be reached as to what is to constitute the set of relevant facts, it turns out on close examination that most, if not all, widely accepted theories, even in the physical sciences, cannot account for *every* relevant fact. In addition, there is much uncertainty about what exactly is meant when it is asserted that a theory must "fit" the facts. It clearly cannot "fit the facts" as a suit of clothes might "fit a man," but it is not clear what the relation between a theory and the facts may be. Also, there is no general agreement as to what the facts are in mathematics. Thus this criterion is not very helpful for identifying the correct theory in the philosophy of mathematics.

Another frequently suggested criterion asserts that any theory, to be true, must be consistent; but this assertion is ambiguous. The most plausible interpretation would seem to be that the theory must be *internally* consistent; that is, it must contain no statements which contradict other statements within the same theory. However, it has been demonstrated that for any given set of facts, there are an infinite number of internally consistent theories which can be constructed which "fit" them equally well on whatever particular account of "fitting" one may use. Thus, internal consistency is at best a necessary but not sufficient condition of the truth of a theory, and additional criteria must still be found to narrow the field of alternatives to one. If the consistency-criterion is interpreted in a broader sense as requiring that the true theory be consistent with other theories as well as itself, additional problems are generated, not the least of which is that of determining which other theories it must be consistent with.

Even though there are an infinite number of internally consistent theories which can be constructed to "fit" any given set of facts, most of these can be rejected out of hand because they are extremely complex. This suggests that a 'simplicity criterion' would be sufficient for selecting the true theory from among those which have been established as fitting the facts (however this might be done) and as satisfying the consistency criterion (in either the broader or narrower sense)—the *simplest* theory is the true one. A serious problem with this criterion is that there does not seem to be any adequate reason which can be given as to *why* the simplest theory should be the true one. (Many scientists and philosophers in the 17th and 18th centuries argued that since the universe was created by God, and since God would not do anything unnecessary, everything in the universe must be arranged in the simplest possible way; there is no need to spell out the criticisms to which such an argument might be subjected.) Another problem with this criterion is that serious difficulties have been encountered in attempting to establish a rigorous definition of "simplicity," and thus far no criteria have been formulated which are adequate for distinguishing the "simplest" theory from among certain sets of alternatives.

Although the "fact-fitting," "consistency," and "simplicity" criteria are inadequate for determining whether a particular theory is true or not, they have been appealed to (explicitly or implicitly) for over two thousand years by both philosophers and scientists who have been concerned with defending or attacking all kinds of theories. The widespread use of the simplicity criterion is reflected, for example, in the fact that few philosophers have ever formulated ontological theories which postulate the existence of more than four kinds of basic entity. Likewise, most philosophers and scientists at least *believed* (even though many were mistaken in their beliefs) their theories to be internally consistent and to "fit the facts" in some way. It would seem that in practice philosophers and scientists have never been as concerned with finding absolutely certain theories as determined by some set of rigorously formulated criteria, as they have been interested in identifying those theories which they believed to be best suited for performing particular tasks at hand. It also appears that if and when they became convinced that another theory would do the job better than their present theory, they would replace the one with the other. If this is indeed an accurate characterization of the practice of philosophers and scientists up to this point in history, there would seem to be some justification for continuing in this manner. Thus instead of giving up in frustration at not being able to find *the* true, correct, and absolutely certain theory of the nature of mathematics (or any other theory), one can be satisfied with simply ascertaining which of the available theories is most adequate for performing a particular task, and then stick with that theory until he can find one that does the job better.

This is still not an easy task. It has been only during the last several decades that philosophers have begun to recognize the multiplicity of functions which most theories must be able to perform. They must not only provide explanations for various phenomena, and produce accurate predictions of future events (in the physical sciences at least); they must also provide a degree of psychological satisfaction, they must be aesthetically pleasing, and they must be capable of generating new research and new discoveries, among other things.

As its theories have become more technical and abstract, the philosophy of mathematics has tended to become an isolated and independent discipline; and as it has become more independent and isolated, its theories have been required to do less and less in a more and more rigorously defined manner. This has resulted in some important and valuable developments in recent years, and it certainly cannot be criticized. However, too many philosophers have lost sight of the fact that the study of the nature of mathematics has traditionally been, and still can be, carried out in a less isolated context, that it can deal with a much broader variety of problems, and that it has a relevance to many other areas of human interest and experience. It is not implausible that theories possessing different degrees of technical sophistication should be necessary for satisfying the needs of the mathematician who is concerned with certain technical problems in his discipline, and the needs of the philosopher or layman (or even mathematician) who is interested in such issues as the certainty of ethical judgments or the formulation of some sort of general world view which can be used for sorting out and relating the various elements of experience. It is the viewing of the philosophy of mathematics in such broader contexts which provides a common thread that runs through the following selections.

Although there are points of similarity among the fifteen theories included in this book, each one when taken as a whole is quite distinctive. A number of these theories imply that mathematical knowledge is acquired by a process of discovery rather than of creation; but a careful study of each of these theories will reveal that there are very significant differences among them with regard to their analyses of the nature of mathematical entities, the certainty of mathematical knowledge, the similarities and differences between mathematical and other kinds of knowledge, and numerous other points. And although these fifteen theories have been selected because among them they provide statements of most of the basic positions on the issues of the philosophy of mathematics, they do not begin to represent all of the combinations and permutations of these basic positions which have been put forward by philosophers of mathematics during the past 2500 years.

The following selections should be viewed as merely providing an introduction to the basic problems of the *philosophy* of mathematics and a few of the possible solutions to them. They are intended to whet the reader's intellectual appetite and to propel him or her into investigating other alternatives before making any decisions as to which best satisfies his or her own needs or most closely approaches being the 'true' theory. The evaluation of such theories is not easy, and the more alternatives which are taken into consideration, the more difficult it becomes. But it is also true that the greater the number of choices, the more interesting, exciting, and rewarding the search for the best theory becomes.

1

PLATO · Introduction

Although the history of Western philosophy can be traced back to at least Thales of Miletus, who lived around 600 B.C., Plato (428–345 B.C.) was the earliest philosopher whose writings have survived until today in significant quantity. Nevertheless, the remaining fragments of the writings of the pre-Platonic philosophers and second-hand reports of their theories indicate that even the earliest philosophers were concerned about the nature of mathematical knowledge. Some of these early philosophers, including Thales, Pythagoras, and Zeno of Elea, made important contributions to the development of mathematics as well as to philosophy; in point of fact, it is doubtful that any of the early Greeks ever distinguished mathematics and philosophy as two different subject matters.

Plato was familiar with the work of his predecessors, and it influenced his own thinking in a variety of ways. Thus many of the ideas which comprise his philosophy were the products of a process which had been going on for at least two hundred years. He was no exception to the rule that the ancient Greek intellectuals were mathematicians as well as philosophers. Although he did not make any significant original contributions to mathematics, he did make mathematics a central part of the course of study at his Academy, and several of his students became notable figures in the history of mathematics.

Plato argued that there is a clear distinction between knowing and believing or "opining" (he always meant *certain* knowledge when he spoke of knowledge). He gave different accounts of the ground of the distinction, often using sense experience as the paradigm of opinion and mathematics as an example of knowledge. Sometimes he emphasized the psychological states of the mind or soul. In other passages he emphasized the differences between the objects of knowledge and opinion; he asserted that the objects of opinion belong to the "world of becoming," whereas the objects of knowledge belong to the realm of the "eternal Forms." Having assumed that knowledge is eternal and unchanging, Plato argued that the objects of knowledge, the Forms, must be eternal and unchanging and exist independently of individual knowers. Similarly, he concluded that the objects of uncertain and changing opinion must be changeable, perishable, and in some sense 'unreal.' Plato sometimes spoke of this "world of becoming" as having a status "between being and nothingness," and he equated it with the world of sense experience. Thus no *knowledge,* whether of mathematics or anything else, can be derived from or directly related to sense experience.

Plato also made additional distinctions within the two categories of opinion and knowledge with his metaphor of the divided line (see 1.6). He believed that mathematical knowledge is the lowest or most easily grasped type of knowledge and thus it provides the natural point of transition from opinion to knowledge. Knowledge of the basic principles of ethics, aesthetics, and politics (which were Plato's main interests) is on a somewhat higher level of abstractness and is more difficult to acquire.

Plato gave a sophisticated account of the tri-partite nature of the soul which anticipated elements of Freud's theories by over two thousand years. He offered several arguments to prove the immortality of the individual soul. The argument concerning the slave-boy's pre-knowledge and "recollection" of the geometrical theorem (see 1.10) is part of a proof of the preexistence of the soul as well as of the existence of the Forms. (To prove the *immortality* of the soul, additional arguments are of course necessary to demonstrate that the soul will not cease to exist in the future.) This is related to the general problem of *how* one can acquire knowledge other than through the senses, if it is assumed that there is certain knowledge that the senses cannot provide. Plato's theory of the pre-existence of the soul and its direct (and essentially passive) perception of the Forms prior to its "entrapment" in a physical body was not accepted by many of his successors, but the theories offered as replacements for it (particularly the theories of innate ideas in the 17th and 18th centuries) were little, if any, better.

In the *Timaeus*, one of his last and most highly metaphorical dialogues, he spoke of a "demiurge," a god-like being which originally created order out of a previously existing chaos. Although scholars disagree as to how this discussion is to be interpreted, it seems likely that this being (if it is to be taken seriously at all) did not create the Forms, but at most used the already existing Forms as a model for creating order out of the primeval chaos. Thus, Plato's philosophy of mathematics requires only the Forms and souls; the objects of mathematical knowledge and the knower are in no way affected by anything else.

Despite the many criticisms to which Plato's theory of independently existing mathematical entities has been subjected over the past two millenia, it has shown surprising staying-power, and some of the most respected philosophers of the 20th century have incorporated variations of it into their own philosophies of mathematics. This is not to imply that they necessarily accepted the entire theory of the Forms, or the pre-existence of the soul, or even the existence of a soul. Nor did they always accept Plato's characterization of sense-perception as wholly unreliable and incapable of providing more than "opinion." For a substantial part of his career, Bertrand Russell held the belief that the objects of mathematical knowledge (as well as the objects of several other kinds of knowledge, including ethics) exist independently of any minds, and that they are known or 'perceived' by some means other than with the five external senses. (A clear statement of this position is made in his 1912 book entitled *The Problems of Philosophy.*) He considered such knowledge to be eternal and unchanging, but he also asserted that some knowledge derived from the senses is also absolutely certain. The basic difference between the two kinds of knowledge is that the former is universal or general (e.g., the Pythagorean theorem applies to *all* right triangles, not just the particular one that I am looking at here and now), whereas the latter is only about particular experiences or sensations. Russell ultimately modified his theory of universal knowledge, but the fact that such a brilliant mind believed for so long that any adequate account of the nature of mathematics requires independently existing non-sensually perceived entities indicates the depth of Plato's original insight.

In summary, it must be recognized that Plato's philosophy of mathematics reflects his whole ontology and to a certain extent provides the foundation of much of his general philosophy. Even his study of moral and political philosophy was essentially dependent on his analysis of the nature of mathematics and of man's knowledge of it. Plato thus proved to be a precursor not only of the many philosophers who included independently existing abstract mathematical entities in their systems; he led the way for the subsequent philosophers in the Western tradition who, almost without exception, found it helpful (and usually necessary) to explore the nature of mathematics in order to adequately pursue their primary interests—whether they were in metaphysics, ethics, aesthetics, political philosophy, or any other area.

Sources

All selections are from Benjamin Jowett's translation of the works of Plato, third edition, 1892, Oxford University Press. The numerals in parentheses refer to the standard marginal numerations of Plato's works.

1.1 *Phaedrus* (246)
1.2 *The Republic,* V (476–479)
1.3 *Timaeus* (51–52)
1.4 *Phaedrus* 247)
1.5 *Theaetetus* (185)
1.6 *The Republic,* VI (508–511)
1.7 *The Republic,* VII (533–534)
1.8 *The Republic,* VII (514–517)
1.9 *The Republic,* VII (525–527)
1.10 *Meno* (80–87)

PLATO · Selections

The nature of the knower—the soul.

1.1 [Socrates asserts:] Of the nature of the soul, though her true form be ever a theme of large and more than mortal discourse, let me speak briefly, and in a figure. And let the figure be composite—a pair of winged horses and a charioteer. Now the winged horses and the charioteers of the gods are all of them noble and of noble descent, but those of other races are mixed; the human charioteer drives his in a pair; and one of them is noble and of noble breed, and the other is ignoble and of ignoble breed; and the driving of them of necessity gives a great deal of trouble to him. I will endeavour to explain to you in what way the mortal differs from the immortal creature. The soul in her totality has the care of inanimate being everywhere, and traverses the whole heaven in divers forms appearing:—when perfect and fully winged she soars upward, and orders the whole world; whereas the imperfect soul, losing her wings and drooping in her flight at last settles on the solid ground—there, finding a home, she receives an earthly frame which appears to be self-moved, but is really moved by her power; and this composition of soul and body is called a living and mortal creature. For immortal no such union can be reasonably believed to be; although fancy, not having seen nor surely known the nature of God, may imagine an immortal creature having both a body and also a soul which are united throughout all time. Let that, however, be as God wills, and be spoken of acceptably to him.

Knowledge and opinion have different objects.

1.2 [Socrates is now recounting a conversation with Glaucon:] Does he who has knowledge know something or nothing? (You must answer for him.)

I answer that he knows something.

Something that is or is not?

Something that is; for how can that which is not ever be known?

And are we assured, after looking at the matter from many points of view, that absolute being is or may be absolutely known, but that the utterly non-existent is utterly unknown?

Nothing can be more certain.

Good. But if there be anything which is of such a nature as to be and not to be, that will have a place intermediate between pure being and the absolute negation of being?

Yes, between them.

And, as knowledge corresponded to being and ignorance of necessity to not-being, for that intermediate between being and not-being there has to be discovered a corresponding intermediate between ignorance and knowledge, if there be such?

Certainly.

Do we admit the existence of opinion?

Undoubtedly.

As being the same with knowledge, or another faculty?

Another faculty.

Then opinion and knowledge have to do with different kinds of matter corresponding to this difference of faculties?

Yes.

And knowledge is relative to being and knows being. But before I proceed further I will make a division.

What division?

I will begin by placing faculties in a class by themselves: they are powers in us, and in all other things, by which we do as we do. Sight and hearing, for example, I should call faculties. Have I clearly explained the class which I mean?

Yes, I quite understand.

Then let me tell you my view about them. I do not see them, and therefore the distinctions of figure, color, and the like, which enable me to discern the differences of some things, do not apply to them. In speaking of a faculty I think only of its sphere and its result; and that which has the same sphere and the same result I call the same faculty, but that which has another sphere and another result I call different. Would that be your way of speaking?

Yes.

And will you be so very good as to answer one more question? Would you say that knowledge is a faculty, or in what class would you place it?

Certainly knowledge is a faculty, and the mightiest of all faculties.

And is opinion also a faculty?

Certainly, he said; for opinion is that with which we are able to form an opinion.

And yet you were acknowledging a little while ago that knowledge is not the same as opinion?

Why, yes, he said: how can any reasonable being ever identify that which is infallible with that which errs?

An excellent answer, proving, I said, that we are quite conscious of a distinction between them.

Yes.

Then knowledge and opinion having distinct powers have also distinct spheres or subject-matters?

That is certain.

Being is the sphere or subject-matter of knowledge, and knowledge is to know the nature of being?

Yes.

And opinion is to have an opinion?

Yes.

And do we know what we opine? or is the subject-matter of opinion the same as the subject-matter of knowledge?

Nay, he replied, that has been already disproven; if difference in faculty implies difference in the sphere or subject-matter, and if, as we were saying, opinion and knowledge are distinct faculties, then the sphere of knowledge and of opinion cannot be the same.

Then if being is the subject-matter of knowledge, something else must be the subject-matter of opinion?

Yes, something else.

Well then, is not-being the subject-matter of opinion? or, rather, how can there be an opinion at all about not-being? Reflect: when a man has an opinion, has he not an opinion about something? Can he have an opinion which is an opinion about nothing?

Impossible.

He who has an opinion has an opinion about some one thing?

Yes.

And not-being is not one thing but, properly speaking, nothing?

True.

Of not-being, ignorance was assumed to be the necessary correlative; of being, knowledge?

True, he said.

Then opinion is not concerned either with being or with not-being?

Not with either.

And can therefore neither be ignorance nor knowledge?

That seems to be true.

But is opinion to be sought without and beyond either of them, in a greater clearness than knowledge, or in a greater darkness than ignorance?

In neither.

Then I suppose that opinion appears to you to be darker than knowledge, but lighter than ignorance?

Both; and in no small degree.

And also to be within and between them?

Yes.

Then you would infer that opinion is intermediate?

No question.

But were we not saying before, that if anything appeared to be of a sort which is and is not at the same time, that sort of thing would appear also to lie in the interval between pure being and absolute not-being; and that the corresponding faculty is neither knowledge nor ignorance, but will be found in the interval between them?

True.

And in that interval there has now been discovered something which we call opinion?

There has.

Then what remains to be discovered is the object which partakes equally of the nature of being and not-being, and cannot rightly be termed either, pure and simple; this unknown term, when discovered, we may truly call the subject of opinion, and assign each to their proper faculty—the extremes to the faculties of the extremes and the mean to the faculty of the mean.

True.

The nature of the objects of knowledge—the eternal Forms.

1.3 Thus I state my view:—If mind and true opinion are two distinct classes, then I say that there certainly are these self-existent ideas unperceived by sense, and apprehended only by the mind; if, however, as some say, true opinion differs in no respect from mind, then everything that we perceive through the body is to be regarded as most real and certain. But we must affirm them to be distinct, for they have a distinct origin and are of a different nature; the one is implanted in us by instruction, the other by persuasion; the one is always accompanied by true reason, the other is without reason; the one cannot be overcome by persuasion, but the other can: and lastly, every man may be said to share in true opinion, but mind is the attribute of the gods and of very few men. Wherefore also we must acknowledge that there is one kind of being which is always the same, uncreated and indestructible, never receiving anything into itself from without, nor itself going out to any other, but invisible and imperceptible by any sense, and of which the contemplation is granted to intelligence only. And there is another nature of the same name with it, and like to it, perceived by sense, created, always in motion, becoming in place and again vanishing out of place, which is apprehended by opinion and sense.

1.4 But of the heaven which is above the heavens, what earthly poet ever did or ever will sing worthily? It is such as I will describe; for I must dare to speak the truth, when truth is my theme. There abides the very being with which true knowledge is concerned; the colourless, formless, intangible essence, visible only to mind, the pilot of the soul. The divine intelligence, being nurtured upon mind and pure knowledge, and the intelligence of every soul which is capable of receiving the food proper to it, rejoices at beholding reality, and once more gazing upon truth, is replenished and made glad, until the revolution of the worlds brings her round again to the same place. In the revolution she beholds justice, and temperance, and knowledge absolute, not in the form of generation or of relation, which men call existence, but knowledge absolute in existence absolute; and beholding the other true existences in like manner, and feasting upon them, she passes down into the interior of the heavens and returns home. . . .

The Forms are not known by means of the senses.

1.5 [Socrates asks:] Very good; and now tell me what is the power which discerns, not only in sensible objects, but in all things, universal notions, such as those which are called being and not-being, and those others about which we were just asking—what organs will you assign for the perception of these notions?

[Theaetetus replies:] You are thinking of being and not-being, likeness and unlikeness, sameness and difference, and also of unity and other numbers which are applied to objects of sense; and you mean to ask, through what bodily organ the soul perceives odd and even numbers and other arithmetical conceptions.

You follow me excellently, Theaetetus; that is precisely what I am asking.

Indeed, Socrates, I cannot answer; my only notion is, that these, unlike objects of sense, have no separate organ, but that the mind, by a power of her own, contemplates the universals in all things.

You are a beauty, Theaetetus, and not ugly, as Theodorus was saying; for he who utters the beautiful is himself beautiful and good. And besides being beautiful, you have done me a kindness in releasing me from a very long discussion, if you are clear that the soul views some things by herself and others through the bodily organs. For that was my own opinion, and I wanted you to agree with me.

The degrees of knowledge.

1.6 [Socrates is again recalling his conversation with Glaucon:] And which, I said, of the gods in heaven would you say was the lord of this element? Whose is that light which makes the eye to see perfectly and the visible to appear?

You mean the sun, as you and all mankind say.

May not the relation of sight to this deity be described as follows?

How?

Neither sight nor the eye in which sight resides is the sun?

No.

Yet of all the organs of sense the eye is the most like the sun?

By far the most like.

And the power which the eye possesses is a sort of effluence which is dispensed from the sun?

Exactly.

Then the sun is not sight, but the author of sight who is recognized by sight.

True, he said.

And this is he whom I call the child of the good, whom the good begat in his own likeness, to be in the visible world, in relation to sight and the things of sight, what the good is in the intellectual world in relation to mind and the things of mind.

Will you be a little more explicit? he said.

Why, you know, I said, that the eyes, when a person directs them towards objects on which the light of day is no longer shining, but the moon and stars only, see dimly, and are nearly blind; they seem to have no clearness of vision in them?

Very true.

But when they are directed towards objects on which the sun shines, they see clearly and there is sight in them?

Certainly.

And the soul is like the eye: when resting upon that on which truth and being shine, the soul perceives and understands and is radiant with intelligence; but when turned towards the twilight of becoming and perishing, then she has opinion only, and goes blinking about, and is first of one opinion and then of another, and seems to have no intelligence?

Just so.

Now, that which imparts truth to the known and the power of knowing to the knower is what I would have you term the idea of good, and this you will deem to be the cause of science, and of truth in so far as the latter becomes the subject of knowledge; beautiful too, as are both truth and knowledge, you will be right in esteeming this other nature as more beautiful than either; and, as in the previous instance, light and sight may be truly said to be like the sun, and yet not to be the sun, so in this other sphere, science and truth may be deemed to be like the good, but not the good; the good has a place of honour yet higher.

What a wonder of beauty that must be, he said, which is the author of science and truth, and yet surpasses them in beauty; for you surely cannot mean to say that pleasure is the good?

God forbid, I replied; but may I ask you to consider the image in another point of view?

In what point of view?

You would say, would you not, that the sun is not only the author of visibility in all visible things, but of generation and nourishment and growth, though he himself is not generation?

Certainly.

In like manner the good may be said to be not only the author of knowledge to all things known, but of their being and essence, and yet the good is not essence, but far exceeds essence in dignity and power.

Glaucon said, with a ludicrous earnestness: By the light of heaven, how amazing!

Yes, I said, and the exaggeration may be set down to you; for you made me utter my fancies.

And pray continue to utter them; at any rate let us hear if there is anything more to be said about the similitude of the sun.

Yes, I said, there is a great deal more.

Then omit nothing, however slight.

I will do my best, I said; but I should think that a great deal will have to be omitted.

You have to imagine, then, that there are two ruling powers, and that one of them is set over the intellectual world, the other over the visible. . . . May I suppose that you have this distinction of the visible and intelligible fixed in your mind?

I have.

Now take a line which has been cut into two unequal parts, and divide each of them again in the same proportion, and suppose the two main divisions to answer, one to the visible and the other to the intelligible, and then compare the subdivisions in respect of their clearness and want of clearness, and you will find that the first section in the sphere of the visible consists of images. And by images I mean, in the first place, shadows, and in the second place, reflections in water and in solid, smooth and polished bodies and the like: Do you understand?

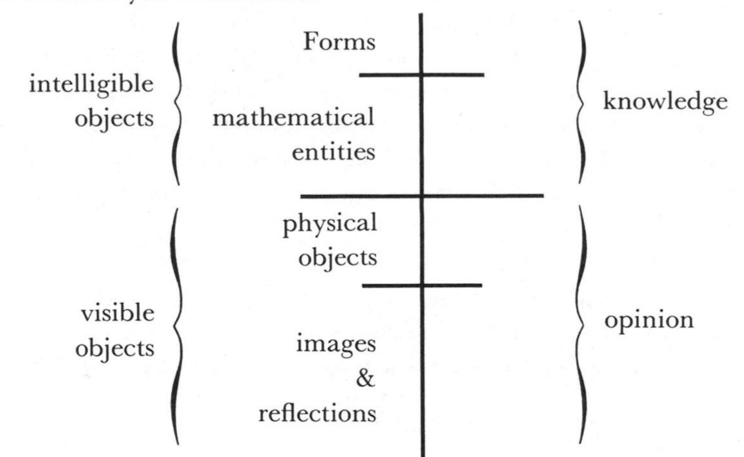

Yes, I understand.

Imagine, now, the other section, of which this is only the resemblance, to include the animals which we see, and everything that grows or is made.

Very good.

Would you not admit that both the sections of this division have different degrees of truth, and that the copy is to the original as the sphere of opinion is to the sphere of knowledge?

Most undoubtedly.

Next proceed to consider the manner in which the sphere of the intellectual is to be divided.

In what manner?

Thus:—There are two subdivisions, in the lower of which the soul uses the figures given by the former division as images; the enquiry can only be hypothetical, and instead of going upwards to a principle descends to the other end; in the higher of the two, the soul passes out of hypotheses, and goes up to a principle which is above hypotheses, making no use of images as in the former case, but proceeding only in and through the ideas themselves.

I do not quite understand your meaning, he said.

Then I will try again; you will understand me better when I have made some preliminary remarks. You are aware that students of geometry, arithmetic, and the kindred sciences assume the odd and the even and the figures and three kinds of angles and the like in their several branches of science; these are their hypotheses, which they and every body are supposed to know, and therefore they do not deign to give any account of them either to themselves or others; but they begin with them, and go on until they arrive at last, and in a consistent manner, at their conclusion?

Yes, he said, I know.

And do you not know also that although they make use of the visible forms and reason about them, they are thinking not of these, but of the ideals which they resemble; not of the figures which they draw, but of the absolute square and the absolute diameter, and so on—the forms which they draw or make, and which have shadows and reflections in water of their own, are converted by them into images, but they are really seeking to behold the things themselves, which can only be seen with the eye of the mind?

That is true.

And of this kind I spoke as the intelligible, although in the search after it the soul is compelled to use hypotheses; not ascending to a first principle, because she is unable to rise above the region of hypothesis, but employing the objects of which the shadows below are resemblances in their turn as images, they having in relation to the shadows and reflections of them a greater distinctness, and therefore a higher value.

I understand, he said, that you are speaking of the province of geometry and the sister arts.

And when I speak of the other division of the intelligible, you will understand me to speak of that other sort of knowledge which reason herself attains by the power of dialectic, using the hypotheses not as first principles, but only as hypotheses—that is to say, as steps and points of departure into a world which is above hypotheses, in order that she may soar beyond them to the first principle of the whole; and clinging to this and then to that which depends on this, by successive steps she descends again without the aid of any sensible object, from ideas, through ideas, and in ideas she ends.

I understand you, he replied; not perfectly, for you seem to me to be describing a task which is really tremendous; but, at any rate, I understand you to say that knowledge and being, which the science of dialectic contemplates, are clearer than the notions of the arts, as they are termed, which proceed from hypotheses only: these are also contemplated by the understanding, and not by the senses: yet, because they start from hypotheses and do not ascend to a principle, those who contemplate them appear to you not to exercise the higher reason upon them, although when a first principle is added to them they are cognizable by the higher reason. And the habit which is concerned with geometry and the cognate sciences I suppose that you would term understanding and not reason, as being intermediate between opinion and reason.

You have quite conceived my meaning, I said; and now, corresponding to these four divisions, let there be four faculties in the soul—reason answering to the highest, understanding to the second, faith (or conviction) to the third, and perception of shadows to the last—and let there be a scale of them, and let us suppose that the several faculties have clearness in the same degree that their objects have truth.

1.7 At any rate, we are satisfied, as before, to have four divisions; two for intellect and two for opinion, and to call the first division science, the second understanding, the third belief, and the fourth perception of shadows, opinion being concerned with becoming, and intellect with being; and so to make a proportion: *As being is to becoming, so is pure intellect to opinion. And as intellect is to opinion, so is science to belief, and understanding to the perception of shadows.* But let us defer the further correlation and subdivision of the subjects of opinion and of intellect, for it will be a long enquiry, many times longer than this has been.

As far as I understand, he said, I agree.

And do you also agree, I said, in describing the dialectician as one who attains a conception of the essence of each thing? And he who does not possess and is therefore unable to impart this conception, in whatever degree he fails, may in that degree also be said to fail in intelligence? Will you admit so much?

Yes, he said; how can I deny it?

And you would say the same of the conception of the good? Until the person is able to abstract and define rationally the idea of good, and unless he can run the gauntlet of all objections, and is ready to disprove them, not by appeals to opinion, but to absolute truth, never faltering at any step of the argument—unless he can do all this, you would say that he knows neither the idea of good nor any other good; he apprehends only a shadow, if anything at all, which is given by opinion and not by science—dreaming and slumbering in this life, before he is well awake here, he arrives at the world below, and has his final quietus.

In all that I should most certainly agree with you.

The metaphor of the cave.

1.8 And now, I said, let me show in a figure how far our nature is enlightened or unenlightened:—Behold! human beings living in an underground den, which has a mouth open towards the light and reaching all along the den; here they have been from their childhood, and have their legs and necks chained so that they cannot move, and can only see before them, being prevented by the chains from turning round their heads. Above and behind them a fire is blazing at a distance, and between the fire and the prisoners there is a raised way; and you will see, if you look, a low wall built along the way, like the screen which marionette players have in front of them, over which they show the puppets.

I see.

And do you see, I said, men passing along the wall carrying all sorts of vessels, and statues and figures of animals made of wood and stone and various materials, which appear over the wall? Some of them are talking, others silent.

You have shown me a strange image, and they are strange prisoners.

Like ourselves, I replied; and they see only their own shadows, or the shadows of one another, which the fire throws on the opposite wall of the cave?

True, he said; how could they see anything but the shadows if they were never allowed to move their heads?

And of the objects which are being carried in like manner they would only see the shadows?

Yes, he said.

And if they were able to converse with one another, would they not suppose that they were naming what was actually before them?

Very true.

And suppose further that the prison had an echo which came from the other side, would they not be sure to fancy when one of the passers-by spoke that the voice which they heard came from the passing shadow?

No question, he replied.

To them, I said, the truth would be literally nothing but the shadows of the images.

That is certain.

And now look again, and see what will naturally follow if the prisoners are released and disabused of their error. At first, when any of them is liberated and compelled suddenly to stand up and turn his neck round and walk and look towards the light, he will suffer sharp pains; the glare will distress him, and he will be unable to see the realities of which in his former state he had seen the shadows; and then conceive some one saying to him, that what he saw before was an illusion, but that now, when he is approaching nearer to being and his eye is turned towards more real existence, he has a clearer vision—what will be his reply? And you may further imagine that his instructor is pointing to the objects as they pass and requiring him to name them—will he not be perplexed? Will he not fancy that the shadows which he formerly saw are truer than the objects which are now shown to him?

Far truer.

And if he is compelled to look straight at the light, will he not have a pain in his eyes which will make him turn away to take refuge in the objects of vision which he can see, and which he will conceive to be in reality clearer than the things which are now being shown to him?

True, he said.

And suppose once more, that he is reluctantly dragged up a steep and rugged ascent, and held fast until he is forced into the presence of the sun himself, is he not likely to be pained and irritated? When he approaches the light his eyes will be dazzled, and he will not be able to see anything at all of what are now called realities.

Not all in a moment, he said.

He will require to grow accustomed to the sight of the upper world. And first he will see the shadows best, next the reflections of men and other objects in the water, and then the objects themselves; then he will gaze upon the light of the moon and the stars and the spangled heaven; and he will see the sky and the stars by night better than the sun or the light of the sun by day?

Certainly.

Last of all he will be able to see the sun, and not mere reflections of him in the water, but he will see him in his own proper place, and not in another; and he will contemplate him as he is.

Certainly.

He will then proceed to argue that this is he who gives the season and the years, and is the guardian of all that is in the visible world, and in a certain way the cause of all things which he and his fellows have been accustomed to behold?

Clearly, he said, he would first see the sun and then reason about him.

And when he remembered his old habitation, and the wisdom of the den and his fellow-prisoners, do you not suppose that he would felicitate himself on the change, and pity them?

Certainly, he would.

And if they were in the habit of conferring honours among themselves on those who were quickest to observe the passing shadows and to remark which of them went before, and which followed after, and which were together; and who were therefore best able to draw conclusions as to the future, do you think that he would care for such honours and glories, or envy the possessors of them? Would he not say with Homer,

Better to be the poor servant of a poor master,

and to endure anything, rather than think as they do and live after their manner?

Yes, he said, I think that he would rather suffer anything than entertain these false notions and live in this miserable manner.

Imagine once more, I said, such an one coming suddenly out of the sun to be replaced in his old situation; would he not be certain to have his eyes full of darkness?

To be sure, he said.

And if there were a contest, and he had to compete in measuring the shadows with the prisoners who had never moved out of the den, while his sight was still weak, and before his eyes had become steady (and the time which would be needed to acquire this new habit of sight might be very considerable), would he not be ridiculous? Men would say of him that up he went and down he came without his eyes; and that it was better not even to think of ascending; and if any one tried to loose another and lead him up to the light, let them only catch the offender, and they would put him to death.

No question, he said.

This entire allegory, I said, you may now append, dear Glaucon, to the previous argument; the prison-house is the world of sight, the light of the fire is the sun, and you will not misapprehend me if you interpret the journey upwards to be the ascent of the soul into the intellectual world according to my poor belief, which, at your desire, I have expressed—whether rightly or wrongly God knows. But, whether true or false, my opinion is that in the world of knowledge the idea of good appears last of all, and is seen only with an effort; and, when seen, is also inferred to be the universal author of all things beautiful and right, parent of light and of the lord of light in this visible world, and the immediate source of reason and truth in the intellectual; and that this is the power upon which he who would act rationally either in public or private life must have his eye fixed.

I agree, he said, as far as I am able to understand you.

Moreover, I said, you must not wonder that those who attain to this beatific vision are unwilling to descend to human affairs; for their souls are ever hastening into the upper world where they desire to dwell; which desire of theirs is very natural, if our allegory may be trusted.

The nature of mathematical knowledge.

1.9 [Socrates said:] . . . Arithmetic has a very great and elevating effect, compelling the soul to reason about abstract number, and rebelling against the introduction of visible or tangible objects into the argument. You know how steadily the masters of the art repel and ridicule any one who attempts to divide absolute unity when he is calculating, and if you divide, they multiply, taking care that one shall continue one and not become lost in fractions.

That is very true.

Now, suppose a person were to say to them: O my friends, what are these wonderful numbers about which you are reasoning, in which, as you say, there is a unity such as you demand, and each unit is equal, invariable, indivisible—what would they answer?

They would answer, as I should conceive, that they were speaking of those numbers which can only be realized in thought.

Then you see that this knowledge may be truly called necessary, necessitating as it clearly does the use of the pure intelligence in the attainment of pure truth?

Yes: that is a marked characteristic of it.

And have you further observed, that those who have a natural talent for calculation are generally quick at every other kind of knowledge; and even the dull, if they have had an arithmetical training, although they may derive no other advantage from it, always become much quicker than they would otherwise have been.

Very true, he said.

And indeed, you will not easily find a more difficult study, and not many as difficult.

You will not.

And, for all these reasons, arithmetic is a kind of knowledge in which the best natures should be trained, and which must not be given up.

I agree.

Let this then be made one of our subjects of education. And next, shall we enquire whether the kindred science also concerns us?

You mean geometry?

Exactly so.

Clearly, he said, we are concerned with that part of geometry which relates to war; for in pitching a camp, or taking up a position, or closing or extending the lines of an army, or any other military manoeuvre, whether in actual battle or on a march, it will make all the difference whether a general is or is not a geometrician.

Yes, I said, but for that purpose a very little of either geometry or calculation will be enough; the question relates rather to the greater and more advanced part of geometry—whether that tends in any degree to make more easy the vision of the idea of good; and thither, as I was saying, all things tend which compel the soul to turn her gaze towards that place, where is the full perfection of being, which she ought, by all means, to behold.

True, he said.

Then if geometry compels us to view being, it concerns us; if becoming only, it does not concern us?

Yes, that is what we assert.

Yet anybody who has the least acquaintance with geometry will not deny that such a conception of the science is in flat contradiction to the ordinary language of geometricians.

How so?

They have in view practice only, and are always speaking, in a narrow and ridiculous manner, of squaring and extending and applying and the like—they confuse the necessities of geometry with those of daily life; whereas knowledge is the real object of the whole science.

Certainly, he said.

Then must not a further admission be made?

What admission?

That the knowledge at which geometry aims is knowledge of the eternal, and not of aught perishing and transient.

That, he replied, may be readily allowed, and is true.

Then, my noble friend, geometry will draw the soul towards truth, and create the spirit of philosophy, and raise up that which is now unhappily allowed to fall down.

Nothing will be more likely to have such an effect.

How mathematical knowledge is acquired.

[A conversation between Socrates and Meno]

1.10 [Meno asked:] And how will you enquire, Socrates, into that which you do not know? What will you put forth as the subject of enquiry? And if you find what you want, how will you ever know that this is the thing which you did not know?

[Socrates answered:] I know, Meno, what you mean; but just see what a tiresome dispute you are introducing. You argue that a man cannot enquire either about that which he knows, or about that which he does not know; for if he knows, he has no need to enquire; and if not, he cannot; for he does not know the very subject about which he is to enquire.

Well, Socrates, and is not the argument sound?

I think not.

Why not?

I will tell you why: I have heard from certain wise men and women who spoke of things divine that—

What did they say?

They spoke of a glorious truth, as I conceive.

What was it? and who were they?

Some of them were priests and priestesses, who had studied how they might be able to give a reason of their profession: there have been poets also, who spoke of these things by inspiration, like Pindar, and many others who were inspired. And they say—mark, now, and see whether their words are true—they say that the soul of man is immortal, and at one time has an end, which is termed dying, and at another time is born again, but is never destroyed. And the moral is, that a man ought to live always in perfect holiness. *"For in the ninth year Persephone sends the souls of those from whom she has received the penalty of ancient crime back again from beneath into the light of the sun above, and these are they who become noble kings and mighty men and great in wisdom and are called saintly heroes in after ages."* The soul, then, as being immortal, and having been born again many times, and having seen all things that exist, whether in this world or in the world below, has knowledge of them all; and it is no wonder that she should be able to call to remembrance all that she ever knew about virtue, and about everything; for as all nature is akin, and the soul has learned all things, there is no difficulty in her eliciting or as men say learning, out of a single recollection all the rest, if a man is strenuous and does not faint; for all enquiry and all learning is but recollection. And therefore we ought not to listen to this sophistical argument about the impossibility of enquiry: for it will make us idle, and is sweet only to the sluggard; but the other saying will make us active and inquisitive. In that confiding, I will gladly enquire with you into the nature of virtue.

Yes, Socrates; but what do you mean by saying that we do not learn, and that what we call learning is only a process of recollection? Can you teach me how this is?

I told you, Meno, just now that you were a rogue, and now you ask whether I can teach you, when I am saying that there is no teaching, but only recollection; and thus you imagine that you will involve me in a contradiction.

Indeed, Socrates, I protest that I had no such intention. I only asked the question from habit; but if you can prove to me that what you say is true, I wish that you would.

It will be no easy matter, but I will try to please you to the utmost of my power. Suppose that you call one of your numerous attendants, that I may demonstrate on him.

Certainly. Come hither, boy.

He is Greek, and speaks Greek, does he not?

Yes, indeed; he was born in the house.

Attend now to the questions which I ask him, and observe whether he learns of me or only remembers.

I will.

Tell me, boy, do you know that a figure like this is a square?

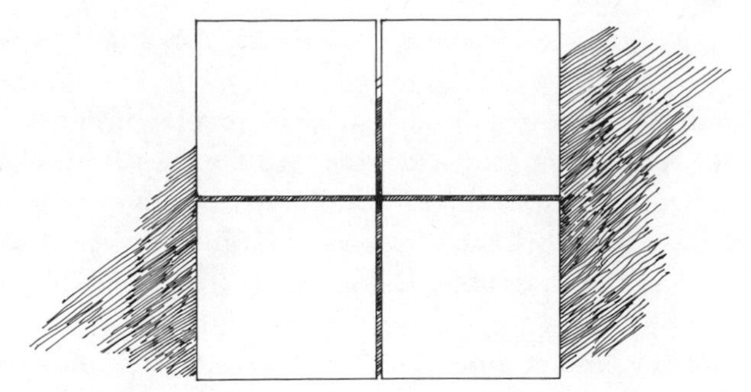

The boy replied, I do.

And you know that a square figure has these four lines equal?

Certainly.

And these lines which I have drawn through the middle of the square are also equal?

Yes.

A square may be of any size?

Certainly.

And if one side of the figure be of two feet, and the other side be of two feet, how much will the whole be? Let me explain: if in one direction the space was of two feet, and in the other direction of one foot, the whole would be of two feet taken once?

Yes.

But since this side is also of two feet, there are twice two feet?

There are.

Then the square is of twice two feet?

Yes.

And how many are twice two feet? count and tell me.

Four, Socrates.

And might there not be another square twice as large as this, and having like this the lines equal?

Yes.

And of how many feet will that be?

Of eight feet.

And now try and tell me the length of the line which forms the side of that double square: this is two feet—what will that be?

Clearly, Socrates, it will be double.

Do you observe, Meno, that I am not teaching the boy anything, but only asking him questions; and now he fancies that he knows how long a line is necessary in order to produce a figure of eight square feet; does he not?

Yes.

And does he really know?

Certainly not.

He only guesses that because the square is double, the line is double.

True.

Observe him while he recalls the steps in regular order. (To the Boy.) Tell me, boy, do you assert that a double space comes from a double line? Remember that I am not speaking of an oblong, but of a figure equal every way, and twice the size of this—that is to say of eight feet; and I want to know whether you still say that a double square comes from a double line?

Yes.

But does not this line become doubled if we add another such line here?

Certainly.

And four such lines will make a space containing eight feet?

Yes.

Let us describe such a figure: Would you not say that this is the figure of eight feet?

Yes.

And are there not these four divisions in the figure, each of which is equal to the figure of four feet?

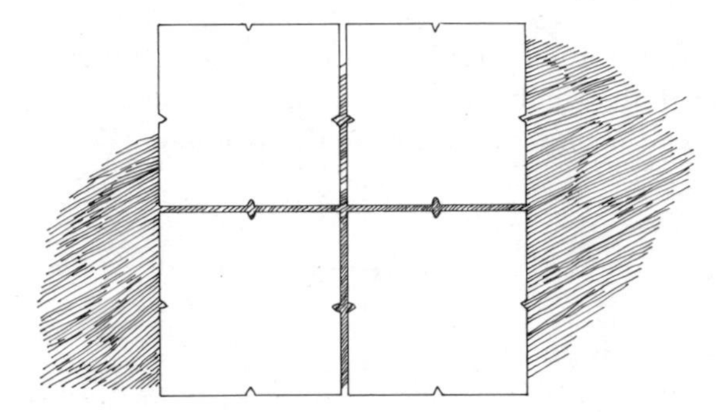

True.

And is not that four times four?

Certainly.

And four times is not double?

No, indeed.

But how much?

Four times as much.

Therefore the double line, boy, has given a space, not twice, but four times as much.

True.

Four times four are sixteen—are they not?

Yes.

What line would give you a space of eight feet, as this gives one of sixteen feet;—do you see?

Yes.

And the space of four feet is made from this half line?

Yes.

Good; and is not a space of eight feet twice the size of this, and half the size of the other?

Certainly.

Such a space, then, will be made out of a line greater than this one, and less than that one?

Yes; I think so.

Very good; I like to hear you say what you think. And now tell me, is not this a line of two feet and that of four?

Yes.

Then the line which forms the side of eight feet ought to be more than this line of two feet, and less than the other of four feet?

It ought.

Try and see if you can tell me how much it will be.

Three feet.

Then if we add a half to this line of two, that will be the line of three. Here are two and there is one; and on the other side, here are two also and there is one: and that makes the figure of which you speak?

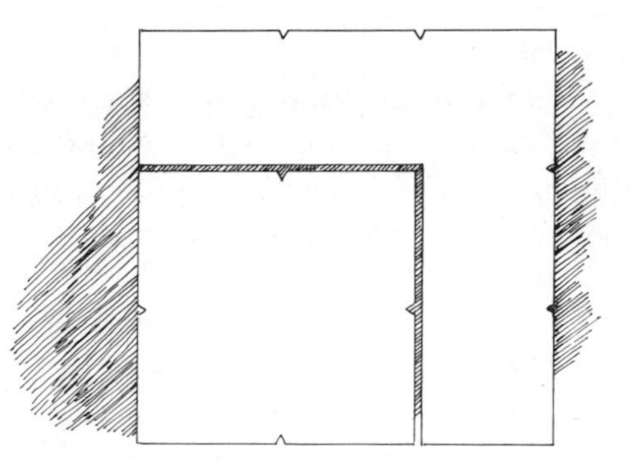

Yes.

But if there are three feet this way and three feet that way, the whole space will be three times three feet?

That is evident.

And how much are three times three feet?

Nine.

And how much is the double of four?

Eight.

Then the figure of eight is not made out of a line of three?

No.

But from what line?—tell me exactly; and if you would rather not reckon, try and show me the line.

Indeed, Socrates, I do not know.

Do you see, Meno, what advances he has made in his power of recollection? He did not know at first, and he does not know now, what is the side of a figure of eight feet: but then he thought that he knew, and answered confidently as if he knew, and had no difficulty; now he has a difficulty, and neither knows nor fancies that he knows.

True.

Is he not better off in knowing his ignorance?

I think that he is.

If we have made him doubt, and given him the "torpedo's shock," have we done him any harm?

I think not.

We have certainly, as would seem, assisted him in some degree to the discovery of the truth; and now he will wish to remedy his ignorance, but then he would have been ready to tell all the world again and again that the double space should have a double side.

True.

But do you suppose that he would ever have enquired into or learned what he fancied that he knew, though he was really ignorant of it, until he had fallen into perplexity under the idea that he did not know, and had desired to know?

I think not, Socrates.

. . . Mark now the farther development. I shall only ask him, and not teach him, and he shall share the enquiry with me: and do you watch and see if you find me telling or explaining anything to him, instead of eliciting his opinion. Tell me, boy, is not this a square of four feet which I have drawn?

Yes.

And now I add another square equal to the former one?

Yes.

And a third, which is equal to either of them?

Yes.

Suppose that we fill up the vacant corner?

Very good.

Here, then, there are four equal spaces?

Yes.

And how many times larger is this space than this other?

Four times.

But it ought to have been twice only, as you will remember.

True.

And does not this line, reaching from corner to corner, bisect each of these spaces?

Yes.

And are there not here four equal lines which contain this space?

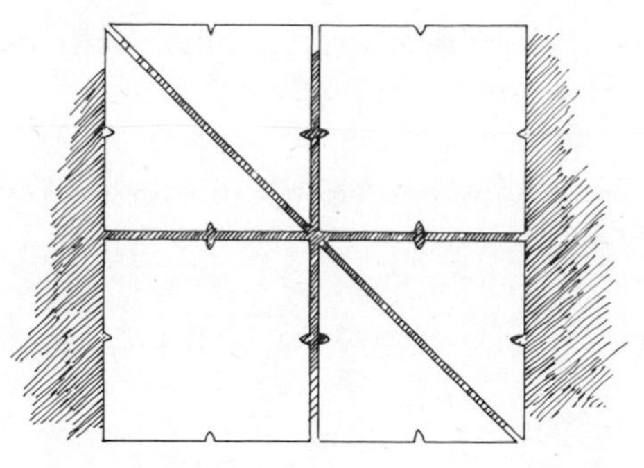

There are.

Look and see how much this space is.

I do not understand.

Has not each interior line cut off half of the four spaces?

Yes.

And how many spaces are there in this section?

Four.

And how many in this?

Two.

And four is how many times two?

Twice.

And this space is of how many feet?

Of eight feet.

And from what line do you get this figure?

From this.

That is, from the line which extends from corner to corner of the figure of four feet?

Yes.

And that is the line which the learned call the diagonal. And if this is the proper name, then you, Meno's slave, are prepared to affirm that the double space is the square of the diagonal?

Certainly, Socrates.

What do you say of him, Meno? Were not all these answers given out of his own head?

Yes, they were all his own.

And yet, as we were just now saying, he did not know?

True.

But still he had in him those notions of his—had he not?

Yes.

Then he who does not know may still have true notions of that which he does not know?

He has.

And at present these notions have just been stirred up in him, as in a dream; but if he were frequently asked the same questions, in different forms, he would know as well as any one at last?

I dare say.

Without any one teaching him he will recover his knowledge for himself, if he is only asked questions?

Yes.

And this spontaneous recovery of knowledge in him is recollection?

True.

And this knowledge which he now has must he not either have acquired or always possessed?

Yes.

But if he always possessed this knowledge he would always have known; or if he has acquired the knowledge he could not have acquired it in this life, unless he has been taught geometry; for he may be made to do the same with all geometry and every other branch of knowledge. Now, has any one ever taught him all this? You must know about him, if, as you say, he was born and bred in your house.

And I am certain that no one ever did teach him.

And yet he has the knowledge?

The fact, Socrates, is undeniable.

But if he did not acquire the knowledge in this life, then he must have had and learned it at some other time?

Clearly he must.

Which must have been the time when he was not a man?

Yes.

And if there have been always true thoughts in him, both at the time when he was and was not a man, which only need to be awakened into knowledge by putting questions to him, his soul must have always possessed this knowledge, for he always either was or was not a man?

Obviously.

And if the truth of all things always existed in the soul, then the soul is immortal. Wherefore be of good cheer, and try to recollect what you do not know, or rather what you do not remember.

I feel, somehow, that I like what you are saying.

And I, Meno, like what I am saying. Some things I have said of which I am not altogether confident. But that we shall be better and braver and less helpless if we think that we ought to enquire, than we should have been if we indulged in the idle fancy that there was no knowing and no use in seeking to know what we do not know;—that is a theme upon which I am ready to fight, in word and deed, to the utmost of my power.

There again, Socrates, your words seem to me excellent.

2

ARISTOTLE · Introduction

Aristotle (384–322 B.C.) was a student of Plato, and he ultimately became one of Plato's most severe critics. His writings are impressive both in the broad range of topics covered and in the depth of intellectual penetration into most of these topics. He devoted no single work specifically to the philosophy of mathematics, but like Plato he examined the nature of mathematical knowledge in the course of his studies of other topics. There is no evidence that Aristotle made any direct original contributions to the development of mathematics in the form of new methods, concepts, or proofs. However, his codification of the rules of deductive logic and his discussions of the 'scientific method' (see 2.28—2.34) paved the way for Euclid's *Elements*, which in turn served as the paradigm for most subsequent work in all areas of mathematics as well as for many treatises on philosophy and the natural sciences. Aristotle also presented detailed summaries, analyses, and criticisms of the philosophical theories of his predecessors. He assumed that there must be a certain measure of truth in each theory which had provided reason for its proponents to advocate it. In constructing his own system, Aristotle sought to extract what he considered to be the kernels of truth from each of his predecessor's theories, and then tied them together with his own original insights to give a comprehensive solution to almost all of the problems of philosophy.

Aristotle believed that the relation between ontology and epistemology, between the theory of what exists and the theory of what is known, is quite complex. This is made particularly evident in his discussions of the basic concept of substance (see 2.2—2.6). His treatment of this topic also brings out his interest in the interrelations between *language* and the *world.* Aristotle focused his attention on the subject-predicate form of declarative sentences which he believed reflects an important truth about the structure of the world; just as certain terms can be predicated of the subject-term of a sentence, so do certain properties belong to substances in the world. This interrelation is forcefully suggested by the ambivalence of the opening passage of the *Categories*: "Of things themselves some are predicable of a subject. . . ." It is no accident that one feels uncertain as to whether this statement refers to sentences or things in the world.

Aristotle drastically modified Plato's ontological and epistemological theories. He did not reject the concept of the Forms outright, but he denied that the Forms could exist independently of particular entities. Instead, he insisted that the 'real' world is the world perceived by the senses, and universals such as redness exist only as particular shades of red in particular physical objects such as balls, apples, roses, etc. In a similar manner, Aristotle denied that the human soul can exist independently of a particular human body. Instead, he asserted, the soul comes into existence at the birth of the individual; it is, in effect, the *form* of that person. (In several passages Aristotle possibly suggested that at least part of the soul might survive after the death of the body—but these passages have been interpreted in many different ways by scholars.) These features of Aristotle's ontology are directly reflected in his epistemology, particularly in his denial that certain knowledge (indeed, *any* knowledge) is gained by the soul's recollection of its perception of the Forms in a previous existence.

Aristotle agreed with Plato that the most important kind of knowledge after which men seek is that of general or universal truths, rather than of particular things. It is of much greater value to know that the sum of the squares of the sides of *any* right triangle is equal to the square of the hypotenuse than to know merely that this is the case for one particular triangle. For Plato, such universal knowledge could be acquired only by directly perceiving a Form, e.g., the Form of a right triangle. Man's knowledge of the abstract truths was acquired by the soul directly perceiving the non-physical and independently existing Forms during a previous existence; the soul was essentially *passive* in the reception of this knowledge. In contrast, Aristotle asserted that universal knowledge (or knowledge of universal truths) must be ultimately traceable to one or more of the five senses. Our senses may be passive in the reception of external stimuli, but since they give us knowledge only of particular things and their specific properties, something more is needed for the acquisition of knowledge of universal truths. If we have direct knowledge only of particular triangles, and if these particular triangles in some way 'contain' the form of triangularity, then the mind must perform some operation(s) on these particular sensations to enable it to know the universal form (see 2.22 and 2.34). Aristotle's account of this process of *abstraction* provided the paradigm for many theories which have been formulated during the last two thousand years.

In asserting that "all thought is either practical or productive or theoretical" (see 2.39), Aristotle was distinguishing what he considered to be the three basic kinds of knowledge. *Productive* knowledge is essentially what we might refer to today as possession of a skill or craft, such as the ability to build a house or paint a picture. *Practical* knowledge involves the ability to *act,* either individually or in groups, in such a way as to attain specific goals, particularly the state of happiness. *Theoretical* knowledge is knowledge about things, such as tables or rocks or men. Knowledge can be particular or general to varying degrees, depending on the extent to which one's powers of abstraction have been applied to the subject matter. Ethics and mathematics both involve knowledge of a quite abstract and general nature, but the former is practical knowledge whereas the latter is theoretical. Physics and mathematics, however, are both theoretical sciences but they differ in the degree of abstraction required, mathematics being more abstract (see 2.38). Both the physicist and the mathematician study the properties of things in general; neither is concerned with a particular table or planet. The physicist is concerned with motion or changes of things in general; the mathematician carries the process of abstraction even further and disregards the changes or motions of the particular things as well.

Aristotle also distinguished between arithmetic and geometry, maintaining that the former is more abstract than the latter. He defined arithmetic as the study of things *qua* non-extended and indivisible or discrete, and geometry as the study of things *qua* extended and infinitely divisible or continuous. Aristotle's formulation of this distinction was interpreted by some later philosophers and mathematicians as implying a sharp separation of these two disciplines, and this became a basic principle underlying the teaching and doing of mathematics for the next two thousand years. Some modern scholars doubt that Aristotle would have agreed with this interpretation of his original statements. This is just one illustration of the fact that it is important not only to know what an influential philosopher *actually* said, but also to consider what others *believed* he said.

From the eleventh century to the present, many philosophers have used concepts which can be traced back to Aristotle. For example, the philosophers known as the "British Empiricists" all utilized the concept of abstraction, although each of them gave a different account of the exact nature of the process itself, as is evidenced by the selections from Hobbes, Locke, Berkeley, Hume and Mill. However, very few philosophers and mathematicians since Kant (Mill was a notable exception) seriously considered trying to ground the concepts and laws of (pure) mathematics in sense experience. The debates among philosophers of mathematics at the beginning of the 20th century were essentially between Platonists and formalists (who claimed that pure mathematics is merely the manipulation of uninterpreted signs according to arbitrarily chosen rules). But in recent years there has been a renewed search for alternative approaches to the problems of the philosophy of mathematics, and several of the theories which have been suggested appear to be closer to the Aristotelean approach than to either Platonism or formalism. There has been a revival of interest in viewing mathematics as an empirical science like physics (although few modern philosophers view the physical sciences as Aristotle did); the essay in this volume by Sidney Axinn is one example of this approach. Other philosophers have approached the questions as being (in part, at least) questions about the nature and use of language in certain contexts, as Max Black does in his essay. (The present-day philosopher of course has a more sophisticated theory of language with which to work, but the basic approach is still similar to Aristotle's).

Some modern critics condemn Aristotle's discussions of the problems of philosophy as being hopelessly confused, while others praise them as being highly sophisticated as well as anticipatory of recent work by philosophers such as Carnap and Wittgenstein. We must always be careful about reading too much into the writings of philosophers of previous periods, since (hopefully) we are aware of things of which they could have had little or no knowledge. But at the same time we should not shortchange them and overlook the often surprising insights which they achieved with their limited resources and which have often been overlooked for centuries by their successors.

The following selections from the works of Aristotle are among the most difficult in this book. At first glance, many passages may appear to have little relevance to the problems of the philosophy of mathematics, and the reader may question the value of working his way through all of them. However, a careful examination of each passage in the general context of this section will reveal that it presents an essential element of Aristotle's overall philosophy of mathematics; if these selections are to be faulted at all it should be on the grounds that full justice still has not been done to Aristotle's theory. However, it would seem that the only way to avoid this latter charge would be to provide the reader with his collected works, rather than mere selections. More than in the case of any other philosopher, Aristotle's philosophy of mathematics is a capsule-statement of his general theory of knowledge and his ontology, and is closely tied to many other aspects of his philosophy. Thus, although difficult, a careful study of these selections will be particularly rewarding to anyone who is interested in the *philosophy* of mathematics in its fullest sense.

Sources

All selections are from the Oxford translation of the works of Aristotle, under the editorship of W. D. Ross.

2.1 Categories, 2 (1a17–1b9)
2.2 Prior Analytics, I, 27 (43a25–44)
2.3 Categories, 4 (1b25–2a10)
2.4 Metaphysics, V, 8 (1017b10–26)
2.5 Categories, 5 (2a11–18)
2.6 Categories, 5 (2a33–2b6)
2.7 Posterior Analytics, I, 4 (73a21–73b4)
2.8 Posterior Analytics, I, 4 (73b27–32)
2.9 Topics, I, 5 (101b39–102a17)

2.10 Metaphysics, VII, 10 (1034b20–1035a14)
2.11 Posterior Analytics, II, 3 (90b23–33)
2.12 On the Soul, II, 5 (416b32–417a21)
2.13 On the Soul, III, 2 (425b26–426a2)
2.14 On the Soul, II, 6 (418a4–19)
2.15 On the Soul, III, 1 (425a14–29)
2.16 Metaphysics, IV, 5 (1010b1–1011a2)
2.17 On the Soul, III, 4 (429a10–429b9)
2.18 On the Soul, III, 7 (431a14–19)
2.19 On the Soul, III, 7, 8 (431b13–432a9)
2.20 Physics, II, 3 (194b18–195a2, 195a15–27)
2.21 Metaphysics, I, 1 (980a22–980b12)
2.22 Physics, I, 1 (184a22–184b13)
2.23 On the Soul, II, 5 (417b17–28)
2.24 Posterior Analytics, I, 2 (71b8–72a8)
2.25 Posterior Analytics, I, 18 (81a38–81b9)
2.26 Posterior Analytics, I, 33 (88b30–89a10, 89a33–89b5)
2.27 Metaphysics, IX, 10 (1051a34–1051b17)
2.28 Posterior Analytics, I, 1 (71a1–71a9)
2.29 Posterior Analytics, I, 2 (72a15–25, 72a32–72b4)
2.30 Posterior Analytics, I, 10 (76a30–76b16)
2.31 Metaphysics, IV, 3 (1005b6–34)
2.32 Nicomachean Ethics, VI, 6 (1140b31–1141a9)
2.33 Nicomachean Ethics, VI, 11 (1143a31–1143b14)
2.34 Posterior Analytics, II, 17 (99b20–100b18)
2.35 Metaphysics, IV, 1 (1003a21–25)
2.36 Metaphysics, IV, 2 (1004a33–1004b17)
2.37 Metaphysics, XI, 4 (1061a17–33)
2.38 Physics, II, 2 (193b23–194a11)
2.39 Metaphysics, VI, 1 (1025b25–1026a33)
2.40 Metaphysics, XI, 3 (1061a29–1061b12)
2.41 Physics, II, 7 (198a16–31)
2.42 Metaphysics, VII, 11 (1036a26–1037a4)
2.43 Posterior Analytics, I, 27 (87a31–37)
2.44 Metaphysics, II, 3 (995a15–18)
2.45 Posterior Analytics, I, 12 (78a10–13)
2.46 Metaphysics, V, 13 (1020a8–33)
2.47 Categories, 6 (4b20–5a38)
2.48 Categories, 6 (6a19–35)
2.49 Metaphysics, V, 15 (1010b–1021a14)
2.50 Metaphysics, V, 6 (1016b11–1017a2)
2.51 Metaphysics, III, 4 (1001a3–9, 19–1001b7)

2.52 Metaphysics, III, 5 (1001b26–1002a14)
2.53 Metaphysics, XIII, 3 (1077b11–1078a14, 21–31)
2.54 Physics III, 4 (203b30–204a8)
2.55 Physics III, 5, 6 (205b24–206a24)
2.56 Physics III, 6 (206b13–19, 206b33–207a14)
2.57 Physics III, 7 (207b28–34)

ARISTOTLE · Selections

The objects of knowledge and their reflections in language.

2.1 Forms of speech are either simple or composite. Examples of the latter are such expressions as 'the man runs,' 'the man wins'; of the former 'man,' 'ox,' 'runs,' 'wins.'

Of things themselves some are predicable of a subject, and are never present in a subject. Thus 'man' is predicable of the individual man, and is never present in a subject.

By being 'present in a subject' I do not mean present as parts are present in a whole, but being incapable of existence apart from the said subject.

Some things, again, are present in a subject, but are never predicable of a subject. For instance, a certain point of grammatical knowledge is present in the mind, but is not predicable of any subject; or again, a certain whiteness may be present in the body (for colour requires a material basis), yet it is never predicable of anything.

Other things, again, are both predicable of a subject and present in a subject. Thus while knowledge is present in the human mind, it is predicable of grammar.

There is, lastly, a class of things which are neither present in a subject nor predicable of a subject, such as the individual man or the individual horse. But, to speak more generally, that which is individual and has the character of a unit is never predicable of a subject. Yet in some cases there is nothing to prevent such being present in a subject. Thus a certain point of grammatical knowledge is present in a subject.

2.2 Of all the things which exist some are such that they cannot be predicated of anything else truly and universally, e.g. Cleon and Callias, i.e. the individual and sensible, but other things may be predicated of them (for each of these is both man and animal); and some things are themselves predicated of others, but nothing prior is predicated of them; and some are predicated of others, and yet others of them, e.g. man of Callias and animal of man. It is clear then that some things are naturally not stated of anything: for as a rule each sensible thing is such that it cannot be predicated of anything, save incidentally: for we sometimes say that that white object is Socrates, or that that which approaches is Callias. We shall explain in another place that there is an upward limit also to the process of predicating: for the present we must assume this. Of these ultimate predicates it is not possible to demonstrate another predicate, save as a matter of opinion, but these may be predicated of other things. Neither can individuals be predicated of other things, though other things can be predicated of them. Whatever lies between these limits can be spoken of in both ways: they may be stated of others, and others stated of them. And as a rule arguments and inquiries are concerned with these things.

2.3 Expressions which are in no way composite signify substance, quantity, quality, relation, place, time, position, state, action, or affection. To sketch my meaning roughly, examples of substance are 'man' or 'the horse,' of quantity, such terms as 'two cubits long' or 'three cubits long,' of quality, such attributes as 'white,' 'grammatical.' 'Double,' 'half,' 'greater,' fall under the category of relation; 'in the market place,' 'in the Lyceum,' under that of place; 'yesterday,' 'last year,' under that of time; 'lying,' 'sitting,' are terms indicating position; 'shod,' 'armed,' state; 'to lance,' 'to cauterize,' action; 'to be lanced,' 'to be cauterized,' affection.

No one of these terms, in and by itself, involves an affirmation; it is by the combination of such terms that positive or negative statements arise. For every assertion must, as is admitted, be either true or false, whereas expressions which are not in any way composite, such as 'man,' 'white,' 'runs,' 'wins,' cannot be either true or false.

2.4 We call 'substance' (1) the simple bodies, i.e. earth and fire and water and everything of the sort, and in general bodies and the things composed of them, both animals and divine beings, and the parts of these. All these are called substance because they are not predicated of a subject but everything else is predicated of them.—(2) That which, being present in such things as are not predicated of a subject, is the cause of their being, as the soul is of the being of an animal.—(3) The parts which are present in such things, limiting them and marking them as individuals, and by whose destruction the whole is destroyed, as the body is by the destruction of the plane, as some say, and the plane by the destruction of the line; and in general number is thought by some to be of this nature; for if it is destroyed, they say, nothing exists, and it limits all things.—(4) The essence, the formula of which is a definition, is also called the substance of each thing.

It follows, then, that 'substance' has two senses, (A) the ultimate substratum, which is no longer predicated of anything else, and (B) that which, being a 'this,' is also separable—and of this nature is the shape or form of each thing.

2.5 Substance, in the truest and primary and most definite sense of the word, is that which is neither predicable of a subject nor present in a subject; for instance, the individual man or horse. But in a secondary sense those things are called substances within which, as species, the primary substances are included; also those which, as genera, include the species. For instance, the individual man is included in the species 'man,' and the genus to which the species belongs is 'animal'; these, therefore—that is to say, the species 'man' and the genus 'animal'—are termed secondary substances.

2.6 Everything except primary substances is either predicable of a primary substance or present in a primary substance. This becomes evident by reference to particular instances which occur. 'Animal' is predicated of the species 'man,' therefore of the individual man, for if there were no individual man of whom it could be predicated, it could not be predicated of the species 'man' at all. Again, colour is present in body, therefore in individual bodies, for if there were no individual body in which it was present, it could not be present in body at all. Thus everything except primary substances is either predicated of primary substances, or is present in them, and if these last did not exist, it would be impossible for anything else to exist.

Essential attributes and definitions.

2.7 Since the object of pure scientific knowledge cannot be other than it is, the truth obtained by demonstrative knowledge will be necessary. And since demonstrative knowledge is only present when we have a demonstration, it follows that demonstration is an inference from necessary premisses. So we must consider what are the premisses of demonstration—i.e. what is their character: and as a preliminary, let us define what we mean by an attribute 'true in every instance of its subject,' an 'essential' attribute, and a 'commensurate and universal' attribute. I call 'true in every instance' what is truly predicable of all instances—not of one to the exclusion of others—and at all times, not at this or that time only; e.g. if animal is truly predicable of every instance of man, then if it be true to say 'this is a man,' 'this is an animal' is also true, and if the one be true now the other is true now. A corresponding account holds if point is in every instance predicable as contained in line. There is evidence for this in the fact that the objection we raise against a proposition put to us as true in every instance is either an instance in which, or an occasion on which, it is not true. Essential attributes are (1) such as belong to their subject as elements in its essential nature (e.g. line thus belongs to triangle, point to line; for the very being or 'substance' of triangle and line is composed of these elements, which are contained in the formulae defining triangle and line): (2) such that, while they belong to certain subjects, the subjects to which they belong are contained in the attribute's own defining formula. Thus straight and curved belong to line, odd and even, prime and compound, square and oblong, to number; and also the formula defining any one of these attributes contains its subject—e.g. line or number as the case may be.

Extending this classification to all other attributes, I distinguish those that answer the above description as belonging essentially to their respective subjects; whereas attributes related in neither of these two ways to their subjects I call accidents or 'coincidents'; e.g. musical or white is a 'coincident' of animal.

2.8 I term 'commensurately universal' an attribute which belongs to every instance of its subject, and to every instance essentially and as such; from which it clearly follows that all commensurate universals inhere *necessarily* in their subjects. The essential attribute, and the attribute that belongs to its subject as such, are identical. E.g. point and straight belong to line essentially, for they belong to line as such; and triangle as such has two right angles, for it is *essentially* equal to two right angles.

2.9 A 'definition' is a phrase signifying a thing's essence. It is rendered in the form either of a phrase in lieu of a term, or of a phrase in lieu of another phrase; for it is sometimes possible to define the meaning of a phrase as well. People whose rendering consists of a term only, try it as they may, clearly do not render the definition of the thing in question, because a definition is always a phrase of a certain kind. One may, however, use the word 'definitory' also of such a remark as 'The "becoming" is "beautiful," ' and likewise also of the question, 'Are sensation and knowledge the same or different?,' for argument about definitions is mostly concerned with questions of sameness and difference. In a word we may call 'definitory' everything that falls under the same branch of inquiry as definitions; and that all the above-mentioned examples are of this character is clear on the face of them. For if we are able to argue that two things are the same or are different, we shall be well supplied by the same turn of argument with lines of attack upon their definitions as well: for when we have shown that they are not the same we shall have demolished the definition. Observe, please, that the converse of this last statement does not hold: for to show that they are the same is not enough to establish a definition. To show, however, that they are not the same is enough of itself to overthrow it.

2.10 Since a definition is a formula, and every formula has parts, and as the formula is to the thing, so is the part of the formula to the part of the thing, the question is already being asked whether the formula of the parts must be present in the formula of the whole or not. For in some cases the formulae of the parts are seen to be present, and in some not. The formula of the circle does not include that of the segments, but that of the syllable includes that of the letters; yet the circle is divided into segments as the syllable is into letters.—And further if the parts are prior to the whole, and the acute angle is a part of the right angle and the finger a part of the animal, the acute angle will be prior to the right angle and the finger to the man. But the latter are thought to be prior; for in the formula the parts are explained by reference to them, and in respect also of the power of existing apart from each other the wholes are prior to the parts.

Perhaps we should rather say that 'part' is used in several senses. One of these is 'that which measures another thing in respect of quantity.' But let this sense be set aside; let us inquire about the parts of which *substance* consists. If then matter is one thing, form another, the compound of these a third, and both the matter and the form and the compound are substance, even the matter is in a sense called part of a thing, while in a sense *it* is not, but only the elements of which the formula of the form consists. E.g. of concavity flesh (for this is the matter in which it is produced) is not a part, but of snubness it is a part; and the bronze is a part of the concrete statue, but not of the statue when this is spoken of in the sense of the form. (For the form, or the thing as having form, should be said to be the thing, but the material element by itself must never be said to be so.) And so the formula of the circle does not include that of the segments, but the formula of the syllable includes that of the letters; for the letters are pàrts of the formula of the form, and not matter, but the segments are parts in the sense of matter on which the form supervenes; yet they are nearer the form than the bronze is when roundness is produced in bronze.

2.11 Moreover, the basic premisses of demonstrations are definitions, and it has already been shown that these will be found indemonstrable; either the basic premisses will be demonstrable and will depend on prior premisses, and the regress will be endless; or the primary truths will be indemonstrable definitions.

But if the definable and the demonstrable are not wholly the same, may they yet be partially the same? Or is that impossible, because there can be no demonstration of the definable? There can be none, because definition is of the essential nature or being of something and all demonstrations evidently posit and assume the essential nature—mathematical demonstrations, for example, the nature of unity and the odd, and all the other sciences likewise.

The nature of perception.

2.12 . . . Let us now speak of sensation in the widest sense. Sensation depends as we have said, on a process of movement or affection from without, for it is held to be some sort of change of quality. Now some thinkers assert that like is affected only by like; in what sense this is possible and in what sense impossible, we have explained in our general discussion of acting and being acted upon.

Here arises a problem: why do we not perceive the senses themselves as well as the external objects of sense, or why without the stimulation of external objects do they not produce sensation, seeing that they contain in themselves fire, earth, and all the other elements, which are the direct or indirect objects of sense? It is clear that what is sensitive is so only potentially, not actually. The power of sense is parallel to what is combustible, for that never ignites itself spontaneously, but requires an agent which has the power of starting ignition; otherwise it could have set itself on fire, and would not have needed actual fire to set it ablaze.

In reply we must recall that we use the word 'perceive' in two ways, for we say (a) that what has the power to hear or see, 'sees' or 'hears,' even though it is at the moment asleep, and also (b) that what is actually seeing or hearing, 'sees' or 'hears.' Hence 'sense' too must have two meanings, sense potential, and sense actual. Similarly 'to be a sentient' means either (a) to have a certain power or (b) to manifest a certain activity. To begin with, for a time, let us speak as if there were no difference between (i) being moved or affected, and (ii) being active, for movement is a kind of activity—an imperfect kind, as has elsewhere been explained. Everything that is acted upon or moved is acted upon by an agent which is actually at work. Hence it is that in one sense, as has already been stated, what acts and what is acted upon are like, in another unlike, i.e. prior to and during the change the two factors are unlike, after it like.

2.13 The activity of the sensible object and that of the percipient sense is one and the same activity, and yet the distinction between their being remains. Take as illustration actual sound and actual hearing: a man may have hearing and yet not be hearing, and that which has a sound is not always sounding. But when that which can hear is actively hearing and that which can sound is sounding, then the actual hearing and the actual sound are merged in one (these one might call respectively hearkening and sounding).

2.14 In dealing with each of the senses we shall have first to speak of the objects which are perceptible by each. The term 'object of sense' covers three kinds of objects, two kinds of which are, in our language, directly perceptible, while the remaining one is only incidentally perceptible. Of the first two kinds one (a) consists of what is perceptible by a single sense, the other (b) of what is perceptible by any and all of the senses. I call by the name of special object of this or that sense that which cannot be perceived by any other sense than that one and in respect of which no error is possible; in this sense colour is the special object of sight, sound of hearing, flavour of taste. Touch, indeed, discriminates more than one set of different qualities. Each sense has one kind of object which it discerns, and never errs in reporting that what is before it is colour or sound (though it may err as to what it is that is coloured or where that is, or what it is that is sounding or where that is). Such objects are what we propose to call the special objects of this or that sense.

'Common sensibles' are movement, rest, number, figure, magnitude; these are not peculiar to any one sense, but are common to all. There are at any rate certain kinds of movement which are perceptible both by touch and by sight.

2.15 Further, there cannot be a special sense-organ for the common sensibles either, i.e. the objects which we perceive incidentally through this or that special sense, e.g. movement, rest, figure, magnitude, number, unity; for all these we perceive by movement, e.g. magnitude by movement, and therefore also figure (for figure is a species of magnitude), what is at rest by the absence of movement: number is perceived by the negation of continuity, and by the special sensibles; for each sense perceives one class of sensible objects. So that it is clearly impossible that there should be a special sense for any one of the common sensibles, e.g. movement; for, if that were so, our perception of it would be exactly parallel to our present perception of what is sweet by vision. *That* is so because we have a sense for each of the two qualities, in virtue of which when they happen to meet in one sensible object we are aware of both contemporaneously. If it were not like this our perception of the common qualities would always be incidental, i.e. as is the perception of Cleon's son, where we perceive him not as Cleon's son but as white, and the white thing which we really perceive happens to be Cleon's son.

But in the case of the common sensibles there is already in us a general sensibility which enables us to perceive them directly; there is therefore no special sense required for their perception: if there were, our perception of them would have been exactly like what has been above described.

2.16 Regarding the nature of truth, we must maintain that not everything which appears is true; firstly, because even if sensation—at least of the object peculiar to the sense in question—is not false, still appearance is not the same as sensation.—Again, it is fair to express surprise at our opponents' raising the question whether magnitudes are as great, and colours are of such a nature, as they appear to people at a distance, or as they appear to those close at hand, and whether they are such as they appear to the healthy or to the sick, and whether those things are heavy which appear so to the weak or those which appear so to the strong, and those things true which appear to the sleeping or to the waking. For obviously they do not think these to be open questions; no one, at least, if when he is in Libya he has fancied one night that he is in Athens, starts for the concert hall.—And again with regard to the future, as

Plato says, surely the opinion of the physician and that of the ignorant man are not equally weighty, for instance, on the question whether a man will get well or not.—And again, among sensations themselves the sensation of a foreign object and that of the appropriate object, or that of a kindred object and that of the object of the sense in question, are not equally authoritative, but in the case of colour sight, not taste, has the authority, and in the case of flavour taste, not sight; each of which senses never says at the same time of the same object that it simultaneously is 'so and not so.'—But not even at different times does one sense disagree about the quality, but only about that to which the quality belongs. I mean, for instance, that the same wine might seem, if either it or one's body changed, at one time sweet and at another time not sweet; but at least the sweet, such as it is when it exists, has never yet changed, but one is always right about it, and that which is to be sweet is of necessity of such and such a nature. Yet all these views destroy this necessity, leaving nothing to be of necessity, as they leave no essence of anything; for the necessary cannot be in this way and also in that, so that if anything is of necessity, it will not be 'both so and not so.'

And, in general, if only the sensible exists, there would be nothing if animate things were not; for there would be no faculty of sense. Now the view that neither the sensible qualities nor the sensations would exist is doubtless true (for they are affections of the perceiver), but that the substrata which cause the sensation should not exist even apart from sensation is impossible. For sensation is surely not the sensation of itself, but there is something beyond the sensation, which must be prior to the sensation; for that which moves is prior in nature to that which is moved, and if they are correlative terms, this is no less the case.

The nature of the knower.

2.17 Turning now to the part of the soul with which the soul knows and thinks (whether this is separable from the others in definition only, or spatially as well) we have to inquire (1) what differentiates this part, and (2) how thinking can take place.

If thinking is like perceiving, it must be either a process in which the soul is acted upon by what is capable of being thought, or a process different from but analogous to that. The thinking part of the soul must therefore be, while impassible, capable of receiving the form of an object; that is, must be potentially identical in character with its object without being the object. Mind must be related to what is thinkable, as sense is to what is sensible.

Therefore, since everything is a possible object of thought, mind in order, as Anaxagoras says, to dominate, that is, to know, must be pure from all admixture; for the co-presence of what is alien to its nature is a hindrance and a block: it follows that it too, like the sensitive part, can have no nature of its own, other than that of having a certain capacity. Thus that in the soul which is called mind (by mind I mean that whereby the soul thinks and judges) is, before it thinks, not actually any real thing. For this reason it cannot reasonably be regarded as blended with the body: if so, it would acquire some quality, e.g. warmth or cold, or even have an organ like the sensitive faculty: as it is, it has none. It was a good idea to call the soul 'the place of forms,' though (1) this description holds only of the intellective soul, and (2) even this is the forms only potentially, not actually.

Observation of the sense-organs and their employment reveals a distinction between the impassibility of the sensitive and that of the intellective faculty. After strong stimulation of a sense we are less able to exercise it than before, as e.g. in the case of a loud sound we cannot hear easily immediately after, or in the case of a bright colour or a powerful odour we cannot see or smell, but in the case of mind, thought about an object that is highly intelligible renders it more and not less able afterwards to think objects that are less intelligible: the reason is that while the faculty of sensation is dependent upon the body, mind is separable from it.

Once the mind has become each set of its possible objects, as a man of science has, when this phrase is used of one who is actually a man of science (this happens when he is now able to exercise the power on his own initiative), its condition is still one of potentiality, but in a different sense from the potentiality which preceded the acquisition of knowledge by learning or discovery: the mind too is then able to think *itself*.

2.18 To the thinking soul images serve as if they were contents of perception (and when it asserts or denies them to be good or bad it avoids or pursues them). That is why the soul never thinks without an image. The process is like that in which the air modifies the pupil in this or that way and the pupil transmits the modification to some third thing (and similarly in hearing), while the ultimate point of arrival is one, a single mean, with different manners of being.

2.19 The so-called abstract objects the mind thinks just as, if one had thought of the snub-nosed not as snub-nosed but as hollow, one would have thought of an actuality without the flesh in which it is embodied: it is thus that the mind when it is thinking the objects of Mathematics thinks as separate, elements which do not exist separate. In every case the mind which is actively thinking is the objects which it thinks.

Let us now summarize our results about soul, and repeat that the soul is in a way all existing things; for existing things are either sensible or thinkable, and knowledge is in a way what is knowable, and sensation is in a way what is sensible: in *what* way we must inquire.

Knowledge and sensation are divided to correspond with the realities, potential knowledge and sensation answering to potentialities, actual knowledge and sensation to actualities. Within the soul the faculties of knowledge and sensation are *potentially* these objects, the one what is knowable, the other what is sensible. They must be either the things themselves or their forms. The former alternative is of course impossible: it is not the stone which is present in the soul but its form.

It follows that the soul is analogous to the hand; for as the hand is a tool of tools, so the mind is the form of forms and sense the form of sensible things.

Since according to common agreement there is nothing outside and separate in existence from sensible spatial magnitudes, the objects of thought are in the sensible forms, viz. both the abstract objects and all the states and affections of sensible things. Hence (1) no one can learn or understand anything in the absence of sense, and (2) when the mind is actively aware of anything it is necessarily aware of it along with an image; for images are like sensuous contents except in that they contain no matter.

Scientific knowledge is of the universal abstracted from sense-experience

2.20 Knowledge is the object of our inquiry, and men do not think they know a thing till they have grasped the 'why' of it (which is to grasp its primary cause). So clearly we too must do this as regards both coming to be and passing away and every kind of physical change, in order that, knowing their principles, we may try to refer to these principles each of our problems.

In one sense, then (1) that out of which a thing comes to be and which persists, is called 'cause', e.g. the bronze of the statue, the silver of the bowl, and the genera of which the bronze and the silver are species.

In another sense (2) the form or the archetype, i.e. the statement of the essence, and its genera, are called 'causes' (e.g. of the octave the relation of 2:1, and generally number), and the parts in the definition.

Again (3) the primary source of the change or coming to rest; e.g. the man who gave advice is a cause, the father is cause of the child, and generally what makes of what is made and what causes change of what is changed.

Again (4) in the sense of end or 'that for the sake of which' a thing is done, e.g. health is the cause of walking about. ('Why is he walking about?' we say. 'To be healthy,' and, having said that, we think we have assigned the cause.) The same is true also of all the intermediate steps which are brought about through the action of something else as means towards the end, e.g. reduction of flesh, purging, drugs, or surgical instruments are means towards health. All these things are 'for the sake of' the end, though they differ from one another in that some are activities, others instruments.

All the causes now mentioned fall into four familiar divisions. The letters are the causes of syllables, the material of artificial products, fire, etc., of bodies, the parts of the whole, and the premisses of the conclusion, in the sense of 'that from which.' Of these pairs the one set are causes in the sense of substratum, e.g. the parts, the other set in the sense of essence—the whole and the combination and the form. But the seed and the doctor and the adviser, and generally the maker, are all sources whence the change or stationariness originates, while the others are causes in the sense of the end or the good of the rest; for 'that for the sake of which' means what is best and the end of the things that lead up to it. (Whether we say the 'good itself' or the 'apparent good' makes no difference.)

Such then is the number and nature of the kinds of cause.

2.21 All men by nature desire to know. An indication of this is the delight we take in our senses; for even apart from their usefulness they are loved for themselves; and above all others the sense of sight. For not only with a view to action, but even when we are not going to do anything, we prefer seeing (one might say) to everything else. The reason is that this, most of all the senses, makes us know and brings to light many differences between things.

By nature animals are born with the faculty of sensation, and from sensation memory is produced in some of them, though not in others. And therefore the former are more intelligent and apt at learning than those which cannot remember; those which are incapable of hearing sounds are intelligent though they cannot be taught, e.g. the bee, and any other race of animals that may be like it; and those which besides memory have the sense of hearing can be taught.

The animals other than man live by appearances and memories, and have but little of connected experience; but the human race lives also by art and reasonings. Now from memory experience is produced in men; for the several memories of the same thing produce finally the capacity for a single experience. And experience seems pretty much like science and art, but really science and art come to men *through* experience; for 'experience made art,' as Polus says, 'but inexperience luck.' Now art arises when from many notions gained by experience one universal judgment about a class of objects is produced. For to have a judgment that when Callias was ill of this disease this did him good, and similarly in the case of Socrates and in many individual cases, is a matter of experience; but to judge that it has done good to all persons of a certain constitution, marked off in one class, when they were ill of this disease, e.g. to phlegmatic or bilious people when burning with fever—this is a matter of art.

2.22 Now what is to us plain and obvious at first is rather confused masses, the elements and principles of which become known to us later by analysis. Thus we must advance from generalities to particulars; for it is a whole that is best known to sense-perception, and a generality is a kind of whole, comprehending many things within it, like parts. Much the same thing happens in the relation of the name to the formula. A name, e.g. 'round,' means vaguely a sort of whole: its definition analyses this into its particular senses. Similarly a child begins by calling all men 'father,' and all women 'mother,' but later on distinguishes each of them.

2.23 In the case of what is to possess sense, the first transition is due to the action of the male parent and takes place before birth so that at birth the living thing is, in respect of sensation, at the stage which corresponds to the *possession* of knowledge. Actual sensation corresponds to the stage of the exercise of knowledge. But between the two cases compared there is a difference; the objects that excite the sensory powers to activity, the seen, the heard, etc., are outside. The ground of this difference is that what actual sensation apprehends is individuals, while what knowledge apprehends is universals, and these are in a sense within the soul. That is why a man can exercise his knowledge when he wishes, but his sensation does not depend upon himself—a sensible object must be there. A similar statement must be made about our *knowledge* of what is sensible—on the same ground, viz. that the sensible objects are individual and external.

2.24 We suppose ourselves to possess unqualified scientific knowledge of a thing, as opposed to knowing it in the accidental way in which the sophist knows, when we think that we know the cause on which the fact depends, as the cause of that fact and of no other, and, further, that the fact could not be other than it is. Now that scientific knowing is something of this sort is evident—witness both those who falsely claim it and those who actually possess it, since the former merely imagine themselves to be, while the latter are also actually, in the condition described. Consequently the proper object of unqualified scientific knowledge is something which cannot be other than it is.

There may be another manner of knowing as well—that will be discussed later. What I now assert is that at all events we do know by demonstration. By demonstration I mean a syllogism productive of scientific knowledge, a syllogism, that is, the grasp of which is *eo ipso* such knowledge. Assuming then that my thesis as to the nature of scientific knowing is correct, the premisses of demonstrated knowledge must be true, primary, immediate, better known than and prior to the conclusion, which is further related to them as effect to cause. Unless these conditions are satisfied, the basic truths will not be 'appropriate' to the conclusion. Syllogism there may indeed be without these conditions, but such syllogism, not being productive of scientific knowledge, will not be demonstration. The premisses must be true: for that which is non-existent cannot be known—we cannot know, e.g., that the diagonal of a square is commensurate with its side. The premisses must be primary and indemonstrable; otherwise they will require demonstration in order to be known, since to have knowledge, if it be not accidental knowledge, of things which are demonstrable, means precisely to have a demonstration of them. The premisses must be the causes of the conclusion, better known than it, and prior to it; its causes, since we possess scientific knowledge of a thing only when we know its cause; prior, in order to be causes; antecedently known, this antecedent knowledge being not our mere understanding of the meaning, but knowledge of the fact as well. Now 'prior' and 'better known' are ambiguous terms, for there is a difference between what is prior and better known in the order of being and what is prior and better known to man. I mean that objects nearer to sense are prior and better known to man; objects without qualification prior and better known are those further from sense. Now the most universal causes are furthest from sense and particular causes are nearest to sense, and they are thus exactly opposed to one another. In saying that the premisses of demonstrated knowledge must be primary, I mean that they must be the 'appropriate' basic truths, for I identify primary premiss and basic truth. A 'basic truth' in a demonstration is an immediate proposition. An immediate proposition is one which has no other proposition prior to it.

2.25 It is also clear that the loss of any one of the senses entails the loss of a corresponding portion of knowledge, and that, since we learn either by induction or by demonstration, this knowledge cannot be acquired. Thus demonstration develops from universals, induction from particulars; but since it is possible to familiarize the pupil with even the so-called mathematical abstractions only through induction—i.e. only because each subject genus possesses, in virtue of a determinate mathematical character, certain properties which can be treated as separate even though they do not exist in isolation—it is consequently impossible to come to grasp universals except through induction. But induction is impossible for those who have not sense-perception. For it is sense-perception alone which is adequate for grasping the particulars: they cannot be objects of scientific knowledge, because neither can universals give us knowledge of them without induction, nor can we get it through induction without sense-perception.

2.26 Scientific knowledge and its object differ from opinion and the object of opinion in that scientific knowledge is commensurately universal and proceeds by necessary connexions, and that which is necessary cannot be otherwise. So though there are things which are true and real and yet can be otherwise, *scientific knowledge* clearly does not concern them: if it did, things which can be otherwise would be incapable of being otherwise. Nor are they any concern of *rational intuition*—by rational intuition I mean an originative source of scientific knowledge—nor of indemonstrable knowledge, which is the grasping of the immediate premiss. Since then rational intuition, science, and opinion, and what is revealed by these terms, are the only things that can be 'true,' it follows that it is *opinion* that is concerned with that which may be true or false, and can be otherwise: opinion in fact is the grasp of a premiss which is immediate but not necessary. This view also fits the observed facts, for opinion is unstable, and so is the kind of being we have described as its object. Besides, when a man thinks a truth incapable of being otherwise he always thinks that he knows it, never that he opines it. He thinks he opines when he thinks that a connexion, though actually so, may quite easily be otherwise; for he believes that such is the proper object of opinion, while the necessary is the object of knowledge.

The identity of the objects of knowledge and opinion is similar. Knowledge is the apprehension of, e.g. the attribute 'animal' as incapable of being otherwise, opinion the apprehension of 'animal' as capable of being otherwise—e.g. the apprehension that animal is an element in the essential nature of man is knowledge; the apprehension of animal as predicable of man but not as an element in man's essential nature is opinion: man is the subject in both judgments, but the mode of inherence differs.

This also shows that one cannot opine and know the same thing simultaneously; for then one would apprehend the same thing as both capable and incapable of being otherwise—an impossibility. Knowledge and opinion of the same thing can coexist in two different people in the sense we have explained, but not simultaneously in the same person. That would involve a man's simultaneously apprehending, e.g., (1) that man is essentially animal—i.e. cannot be other than animal—and (2) that man is not essentially animal, that is, we may assume, may be other than animal.

2.27 The terms 'being' and 'non-being' are employed firstly with reference to the categories, and secondly with reference to the potency or actuality of these or their non-potency or non-actuality, and thirdly in the sense of true and false. This depends, on the side of the objects, on their being combined or separated, so that he who thinks the separated to be separated and the combined to be combined has the truth, while he whose thought is in a state contrary to that of the objects is in error. This being so, when is what is called truth or falsity present, and when is it not? We must consider what we mean by these terms. It is not because we think truly that you are pale, that you *are* pale, but because you are pale we who say this have the truth. If, then, some things are always combined and cannot be separated, and others are always separated and cannot be combined, while others are capable either of combination or of separation, 'being' is being combined and one, and 'not being' is being not combined but more than one. Regarding contingent facts, then, the same opinion or the same statement comes to be false and true, and it is possible for it to be at one time correct and at another erroneous; but regarding things that cannot be otherwise opinions are not at one time true and at another false, but the same opinions are always true or always false.

The basic premises of every science—axioms, definitions, and hypotheses—are known by immediate intuition and are the most certain truths.

2.28 All instruction given or received by way of argument proceeds from pre-existent knowledge. This becomes evident upon a survey of all the species of such instruction. The mathematical sciences and all other speculative disciplines are acquired in this way, and so are the two forms of dialectical reasoning, syllogistic and inductive: for each of these latter makes use of old knowledge to impart new, the syllogism assuming an audience that accepts its premisses, induction exhibiting the universal as implicit in the clearly known particular. Again, the persuasion exerted by rhetorical arguments is in principle the same, since they use either example, a kind of induction, or enthymeme, a form of syllogism.

The pre-existent knowledge required is of two kinds. In some cases admission of the fact must be assumed, in others comprehension of the meaning of the term used, and sometimes both assumptions are essential. Thus, we assume that every predicate can be either truly affirmed or truly denied of any subject, and that 'triangle' means so and so; as regards 'unit' we have to make the double assumption of the meaning of the word and the existence of the thing. The reason is that these several objects are not equally obvious to us. Recognition of a truth may in some cases contain as factors both previous knowledge and also knowledge acquired simultaneously with that recognition—knowledge, this latter, of the particulars actually falling under the universal and therein already virtually known. For example, the student knew beforehand that the angles of every triangle are equal to two right angles; but it was only at the actual moment at which he was being led on to recognize this as true in the instance before him that he came to know 'this figure inscribed in the semicircle' to be a triangle. For some things (viz. the singulars finally reached which are not predicable of anything else as subject) are only learnt in this way, i.e. there is here no recognition through a middle of a minor term as subject to a major. Before he was led on to recognition or before he actually drew a conclusion, we should perhaps say that in a manner he knew, in a manner not.

If he did not in an unqualified sense of the term *know* the existence of this triangle, how could he *know* without qualification that its angles were equal to two right angles? No: clearly he *knows* not without qualification but only in the sense that he *knows* universally. If this distinction is not drawn, we are faced with the dilemma in the *Meno*: either a man will learn nothing or what he already knows; for we cannot accept the solution which some people offer. A man is asked, "Do you, or do you not, know that every pair is even?" He says he does know it. The questioner then produces a particular pair, of the existence, and so *a fortiori* of the evenness, of which he was unaware. The solution which some people offer is to assert that they do not know that every pair is even, but only that everything which they know to be a pair is even: yet what they know to be even is that of which they have demonstrated evenness, i.e. what they made the subject of their premiss, viz. not merely every triangle or number which they know to be such, but any and every number or triangle without reservation. For no premiss is ever couched in the form 'every number which you know to be such,' or 'every rectilinear figure which you know to be such': the predicate is always construed as applicable to any and every instance of the thing. On the other hand, I imagine there is nothing to prevent a man in one sense knowing what he is learning, in another not knowing it. The strange thing would be, not if in some sense he knew what he was learning, but if he were to know it in that precise sense and manner in which he was learning it.

2.29 I call an immediate basic truth of syllogism a 'thesis' when, though it is not susceptible of proof by the teacher, yet ignorance of it does not constitute a total bar to progress on the part of the pupil: one which the pupil must know if he is to learn anything whatever is an axiom. I call it an axiom because there are such truths and we give them the name of axioms *par excellence.* If a thesis assumes one part or the other of an enunciation, i.e. asserts either the existence or the non-existence of a subject, it is a hypothesis; if it does not so assert, it is a definition. Definition *is* a 'thesis' or a 'laying something down,' since the arithmetician lays it down that to be a unit is to be quantitatively indivisible; but it is not a hypothesis, for to define what a unit is is not the same as to affirm its existence.

Now a man cannot believe in anything more than in the things he knows, unless he has either actual knowledge of it or something better than actual knowledge. But we are faced with this paradox if a student whose belief rests on demonstration has not prior knowledge; a man must believe in some, if not in all, of the basic truths more than in the conclusion. Moreover, if a man sets out to acquire the scientific knowledge that comes through demonstration, he must not only have a better knowledge of the basic truths and a firmer conviction of them than of the connexion which is being demonstrated: more than this, nothing must be more certain or better known to him than these basic truths in their character as contradicting the fundamental premisses which lead to the opposed and erroneous conclusion. For indeed the conviction of pure science must be unshakable.

2.30 I call the basic truths of every genus those elements in it the existence of which cannot be proved. As regards both these primary truths and the attributes dependent on them the meaning of the name is assumed. The fact of their existence as regards the primary truths must be assumed; but it has to be proved of the remainder, the attributes. Thus we assume the meaning alike of unity, straight, and triangular; but while as regards unity and magnitude we assume also the fact of their existence, in the case of the remainder proof is required.

Of the basic truths used in the demonstrative sciences some are peculiar to each science, and some are common, but common only in the sense of analogous, being of use only in so far as they fall within the genus constituting the province of the science in question.

Peculiar truths are, e.g., the definitions of line and straight; common truths are such as 'take equals from equals and equals remain.' Only so much of these common truths is required as falls within the genus in question: for a truth of this kind will have the same force even if not used generally but applied by the geometer only to magnitudes, or by the arithmetician only to numbers. Also peculiar to a science are the subjects the existence as well as the meaning of which it assumes, and the essential attributes of which it investigates, e.g. in arithmetic units, in geometry points and lines. Both the existence and the meaning of the subjects are assumed by these sciences; but of their essential attributes only the meaning is assumed. For example arithmetic assumes the meaning of odd and even, square and cube, geometry that of incommensurable, or of deflection or verging of lines, whereas the existence of these attributes is demonstrated by means of the axioms and from previous conclusions as premisses. Astronomy too proceeds in the same way. For indeed every demonstrative science has three elements: (1) that which it posits, the subject genus whose essential attributes it examines; (2) the so-called axioms, which are primary premisses of its demonstration; (3) the attributes, the meaning of which it assumes.

2.31 Evidently then it belongs to the philosopher, i.e. to him who is studying the nature of all substance, to inquire also into the principles of syllogism. But he who knows best about each genus must be able to state the most certain principles of his subject, so that he whose subject is existing things *qua* existing must be able to state the most certain principles of all things. This is the philosopher, and the most certain principle of all is that regarding which it is impossible to be mistaken; for such a principle must be both the best known (for all men may be mistaken about things which they do not know), and non-hypothetical. For a principle which everyone must have who understands anything that is, is not a hypothesis; and that which every one must know who knows anything, he must already have when he comes to a special study. Evidently then such a principle is the most certain of all; which principle this is, let us proceed to say. It is, that the same attribute cannot at the same time belong and not belong to the same subject and in the same respect; we must presuppose to guard against dialectical objections, any further qualifications which might be added. This, then, is the most certain of all principles, since it answers to the definition given above. For it is impossible for any one to believe the same thing to be and not to be, as some think Heraclitus says. For what a man says, he does not necessarily believe; and if it is impossible that contrary attributes should belong at the same time to the same subject (the usual qualifications must be presupposed in this premiss too), and if an opinion which contradicts another is contrary to it, obviously it is impossible for the same man at the same time to believe the same thing to be and not to be; for if a man were mistaken on this point he would have contrary opinions at the same time. It is for this reason that all who are carrying out a demonstration reduce it to this as an ultimate belief; for this is naturally the starting-point even for all the other axioms.

2.32 Scientific knowledge is judgement about things that are universal and necessary, and the conclusions of demonstration, and all scientific knowledge, follow from first principles (for scientific knowledge involves apprehension of a rational ground). This being so, the first principle from which what is scientifically known follows cannot be an object of scientific knowledge, of art, or of practical wisdom; for that which can be scientifically known can be demonstrated, and art and practical wisdom deal with things that are variable. Nor are these first principles the objects of philosophic wisdom, for it is a mark of the philosopher to have *demonstration* about some things. If, then, the states of mind by which we have truth and are never deceived about things invariable or even variable are scientific knowledge, practical wisdom, philosophic wisdom, and intuitive reason, and it cannot be any of the three (i.e. practical wisdom, scientific knowledge, or philosophic wisdom), the remaining alternative is that it is *intuitive reason* that grasps the first principles.

2.33 Now all things which have to be done are included among particulars or ultimates; for not only must the man of practical wisdom know particular facts, but understanding and judgement are also concerned with things to be done, and these are ultimates. And intuitive reason is concerned with the ultimates in both directions; for both the first terms and the last are objects of intuitive reason and not of argument, and the intuitive reason which is presupposed by demonstrations grasps the unchangeable and first terms, while the intuitive reason involved in practical reasonings grasps the last and variable fact, i.e. the minor premiss. For these variable facts are the starting points for the apprehension of the end, since the universals are reached from the particulars; of these therefore we must have perception, and this perception is intuitive reason.

This is why these states are thought to be natural endowments—why, while no one is thought to be a philosopher by nature, people are thought to have by nature judgement, understanding, and intuitive reason. This is shown by the fact that we think our powers correspond to our time of life, and that a particular age brings with it intuitive reason and judgement; this implies that nature is the cause. [Hence intuitive reason is both beginning and end; for demonstrations are from these and about these.] Therefore we ought to attend to the undemonstrated sayings and opinions of experienced and older people or of people of practical wisdom not less than to demonstrations; for because experience has given them an eye they see aright.

2.34 We have already said that scientific knowledge through demonstration is impossible unless a man knows the primary immediate premisses. But there are questions which might be raised in respect of the apprehension of these immediate premisses: one might not only ask whether it is of the same kind as the apprehension of the conclusions, but also whether there is or is not scientific knowledge of both; or scientific knowledge of the latter, and of the former a different kind of knowledge; and, further, whether the developed states of knowledge are not innate but come to be in us, or are innate but at first unnoticed. Now it is strange if we possess them from birth; for it means that we possess apprehensions more accurate than demonstration and fail to notice them. If on the other hand we acquire them and do not previously possess them, how could we apprehend and learn without a basis of pre-existent knowledge? For that is impossible, as we used to find in the case of demonstration. So it emerges that neither can we possess them from birth, nor can they come to be in us if we are without knowledge of them to the extent of having no such developed state at all. Therefore we must possess a capacity of some sort, but not such as to rank higher in accuracy than these developed states. And this at least is an obvious characteristic of all animals, for they possess a congenital discriminative capacity which is called sense-perception. But though sense-perception is innate in all animals, in some the sense-impression comes to persist, in others it does not. So animals in which this persistence does not come to be have either no knowledge at all outside the act of perceiving, or no knowledge of objects of which no impression persists; animals in which it does come into being have perception and can continue to retain the sense-impression in the soul: and when such persistence is frequently repeated a further distinction at once arises between those which out of the persistence of such sense-impressions develop a power of systematizing them and those which do not. So out of sense-perception comes to be what we call memory, and out of frequently repeated memories of the same thing develops experience; for a number of memories constitute a single experience. From experience again—i.e., from the universal now stabilized in its entirety within the soul, the one beside the many which is a single identity within them all—originate the skill of the craftsman and the knowledge of the man of science, skill in the sphere of coming to be and science in the sphere of being.

We conclude that these states of knowledge are neither innate in a determinate form, nor developed from other higher states of knowledge, but from sense-perception. It is like a rout in battle stopped by first one man making a stand and then another, until the original formation has been restored. The soul is so constituted as to be capable of this process.

Let us now restate the account given already, though with insufficient clearness. When one of a number of logically indiscriminable particulars has made a stand, the earliest universal is present in the soul: for though the act of sense-perception is of the particular, its content is universal—is man, for example, not the man Callias. A fresh stand is made among these rudimentary universals, and the process does not cease until the indivisible concepts, the true universals, are established: e.g. such and such a species of animal is a step towards the genus animal, which by the same process is a step towards a further generalization.

Thus it is clear that we must get to know the primary premisses by induction; for the method by which even sense-perception implants the universal is inductive. Now of the thinking states by which we grasp truth, some are unfailingly true, others admit of error—opinion, for instance, and calculation, whereas scientific knowing and intuition are always true: further, no other kind of thought except intuition is more accurate than scientific knowledge, whereas primary premisses are more knowable than demonstrations, and all scientific knowledge is discursive. From these considerations it follows that there will be no scientific knowledge of the primary premisses and since except intuition nothing can be truer than scientific knowledge, it will be intuition that apprehends the primary premisses—a result which also follows from the fact that demonstration cannot be the originative source of demonstration, nor, consequently, scientific knowledge of scientific knowledge. If, therefore, it is the only other kind of true thinking except scientific knowing, intuition will be the originative source of scientific knowledge. And the originative source of science grasps the original basic premiss, while science as a whole is similarly related as originative source to the whole body of fact.

Mathematical knowledge is of the changeable qua *unchanging.*

2.35 There is a science which investigates being as being and the attributes which belong to this in virtue of its own nature. Now this is not the same as any of the so-called special sciences; for none of these others treats universally of being as being. They cut off a part of being and investigate the attribute of this part; this is what the mathematical sciences for instance do.

2.36 It is evident, then, that it belongs to one science to be able to give an account of these concepts as well as of substance (this was one of the questions in our book of problems), and that it is the function of the philosopher to be able to investigate all things. For if it is not the function of the philosopher, who is it who will inquire whether Socrates and Socrates seated are the same thing, or whether one thing has one contrary, or what contrariety is, or how many meanings it has? And similarly with all other such questions. Since, then, these are essential modifications of unity *qua* unity and of being *qua* being, not *qua* numbers or lines or fire, it is clear that it belongs to this science to investigate both the essence of these concepts and their properties. And those who study these properties err not by leaving the sphere of philosophy, but by forgetting that substance, of which they have no correct idea, is prior to these other things. For number *qua* number has peculiar attributes, such as oddness and evenness, commensurability and equality, excess and defect, and these belong to numbers either in themselves or in relation to one another. And similarly the solid and the motionless and that which is in motion and the weightless and that which has weight have other peculiar properties. So too there are certain properties peculiar to being as such, and it is about these that the philosopher has to investigate the truth.

2.37 Since even the mathematician uses the common axioms only in a special application, it must be the business of first philosophy to examine the principles of mathematics also. That when equals are taken from equals the remainders are equal, is common to all quantities, but mathematics studies a part of its proper matter which it has detached, e.g. lines or angles or numbers or some other kind of quantity—not, however, *qua* being but in so far as each of them is continuous in one or two or three dimensions; but philosophy does not inquire about particular subjects in so far as each of them has some attribute or other, but speculates about being, in so far as each particular thing *is*.—Physics is the same position as mathematics; for physics studies the attributes and the principles of the things that are, *qua* moving and not *qua* being (whereas the primary science, we have said, deals with these, only in so far as the underlying subjects are existent, and not in virtue of any other character); and so both physics and mathematics must be classed as *parts* of Wisdom.

2.38 The next point to consider is how the mathematician differs from the physicist. Obviously physical bodies contain surfaces and volumes, lines and points, and these are the subject-matter of mathematics.

Further, is astronomy different from physics or a department of it? It seems absurd that the physicist should be supposed to know the nature of sun or moon, but not to know any of their essential attributes, particularly as the writers on physics obviously do discuss their shape also and whether the earth and the world are spherical or not.

Now the mathematician, though he too treats of these things, nevertheless does not treat of them as the limits of a physical body; nor does he consider the attributes indicated as the attributes of such bodies. That is why he separates them; for in thought they are separable from motion, and it makes no difference, nor does any falsity result, if they are separated. The holders of the theory of Forms do the same, though they are not aware of it; for they separate the objects of physics, which are less separable than those of mathematics. This becomes plain if one tries to state in each of the two cases the definitions of the things and of their attributes. 'Odd' and 'even,' 'straight' and 'curved,' and likewise 'number,' 'line,' and 'figure,' do not involve motion; not so 'flesh' and 'bone' and 'man'—*these* are defined like 'snub nose,' not like 'curved.'

Similar evidence is supplied by the more physical of the branches of mathematics, such as optics, harmonics, and astronomy. These are in a way the converse of geometry. While geometry investigates physical lines but not *qua* physical, optics investigates mathematical lines, but *qua* physical, not *qua* mathematical.

2.39 . . . If all thought is either practical or productive or theoretical, physics must be a theoretical science, but it will theorize about such being as admits of being moved, and about substance-as-defined for the most part only as not separable from matter. Now, we must not fail to notice the mode of being of the essence and of its definition, for, without this, inquiry is but idle. Of things defined, i.e. of 'whats,' some are like 'snub,' and some like 'concave.' And these differ because 'snub' is bound up with matter (for what is snub is a concave *nose*), while concavity is independent of perceptible matter. If then all natural things are analogous to the snub in their nature—e.g. nose, eye, face, flesh, bone, and, in general, animal; leaf, root, bark, and, in general, plant (for none of these can be defined without reference to movement—they always have matter), it is clear how we must seek and define the 'what' in the case of natural objects, and also that it belongs to the student of nature to study even soul in a certain sense, i.e. so much of it as is not independent of matter.

That physics, then, is a theoretical science, is plain from these considerations. Mathematics also, however, is theoretical; but whether its objects are immovable and separable from matter, is not at present clear; still, it is clear that *some* mathematical theorems *consider* them *qua* immovable and *qua* separable from matter. But if there is some thing which is eternal and immovable and separable, clearly the knowledge of it belongs to a theoretical science—not, however, to physics (for physics deals with certain movable things) nor to mathematics, but to a science prior to both. For physics deals with things which exist separately but are not immovable, and some parts of mathematics deal with things which are immovable but presumably do not exist separately, but as embodied in matter; while the first science deals with things which both exist separately and are immovable. Now all causes must be eternal, but especially these; for they are the causes that operate on so much of the divine as appears to us. There must, then, be three theoretical philosophies, mathematics, physics, and what we may call theology, since it is obvious that if the divine is present anywhere, it is present in things of this sort. And the highest science must deal with the highest genus. Thus, while the theoretical sciences are more to be desired than the other sciences, this is more to be desired than the other theoretical sciences. For one might raise the question whether first philosophy is universal, or deals with one genus, i.e. some one kind of being; for not even the mathematical sciences are all alike in this respect—geometry and astronomy deal with a certain particular kind of thing, while universal mathematics applies alike to all. We answer that if there is no substance other than those which are formed by nature, natural science will be the first science; but if there is an immovable substance, the science of this must be prior and must be first philosophy, and universal in this way, because it is first. And it will belong to this to consider being *qua* being—both what it is and the attributes which belong to it *qua* being.

2.40 As the mathematician investigates abstractions (for before beginning his investigation he strips off all the sensible qualities, e.g. weight and lightness, hardness and its contrary, and also heat and cold and the other sensible contrarieties, and leaves only the quantitative and continuous, sometimes in one, sometimes in two, sometimes in three dimensions, and the attributes of these *qua* quantitative and continuous, and does not consider them in any other respect, and examines the relative positions of some and the attributes of these, and the commensurabilities and incommensurabilities of others, and the ratios of others; but yet we posit one and the same science of all these things—geometry)—the same is true with regard to being. For the attributes of this in so far as it is being, and the contrarieties in it *qua* being, it is the business of no other science than philosophy to investigate; for to physics one would assign the study of things not *qua* being, but rather *qua* sharing in movement; while dialectic and sophistic deal with the attributes of things that are, but not of things *qua* being, and not with being itself in so far as it is being; therefore it remains that it is the philosopher who studies the things we have named, in so far as they are being.

2.41 The 'why' is referred ultimately either (1) in things which do not involve motion, e.g. in mathematics, to the 'what' (to the definition of 'straight line' or 'commensurable,' etc.) or (2) to what initiated a motion, e.g. 'why did they go to war?—because there had been a raid'; or (3) we are inquiring 'for the sake of what?'—'that they may rule'; or (4), in the case of things that come into being, we are looking for the matter. The causes, therefore, are these and so many in number.

Now, the causes being four, it is the business of the physicist to know about them all, and if he refers his problems back to all of them, he will assign the 'why' in the way proper to his science—the matter, the form, the mover, 'that for the sake of which.' The last three often coincide; for the 'what' and 'that for the sake of which' are one, while the primary source of motion is the same in species as these (for man generates man), and so too, in general, are all things which cause movement by being themselves moved; and such as are not of this kind are no longer inside the province of physics, for they cause motion not by possessing motion or a source of motion in themselves, but being themselves incapable of motion. Hence there are three branches of study, one of things which are incapable of motion, the second of things in motion, but indestructible, the third of destructible things.

2.42 Another question is naturally raised, viz. what sort of parts belong to the form and what sort not to the form, but to the concrete thing. Yet if this is not plain it is not possible to define any thing; for definition is of the universal and of the form. If then it is not evident what sort of parts are of the nature of matter and what sort are not, neither will the formula of the thing be evident. In the case of things which are found to occur in specifically different materials, as a circle may exist in bronze or stone or wood, it seems plain that these, the bronze or the stone, are no part of the essence of the circle, since it is found apart from them. Of things which are *not* seen to exist apart, there is no reason why the same may not be true, just as if all circles that had ever been seen were of bronze; for none the less the bronze would be no part of the form; but it is hard to eliminate it in thought. E.g. the form of man is always found in flesh and bones and parts of this kind; are these then also parts of the form and the formula? No, they are matter; but because man is not found also in other matters we are unable to perform the abstraction.

Since this is thought to be possible, but it is not clear *when* it is the case, some people, already raise the question even in the case of the circle and the triangle, thinking that it is not right to define these by reference to lines and to the continuous, but that all these are to the circle or the triangle as flesh and bones are to man, and bronze or stone to the statue; and they reduce all things to numbers, and they say the formula of 'line' is that of 'two.' And of those who assert the Ideas some make 'two' the line-itself, and others make it the Form of the line; for in some cases they say the Form and that of which it is the Form are the same, e.g. 'two' and the Form of two; but in the case of 'line' they say this is no longer so.

It follows then that there is one Form for many things whose form is evidently different (a conclusion which confronted the Pythagoreans also); and it is possible to make one thing the Form-itself of all, and to hold that the others are not Forms; but thus all things will be one.

We have pointed out, then, that the question of definitions contains some difficulty, and why this is so. And so to reduce all things thus to Forms and to eliminate the matter is useless labour; for some things surely are a particular form in a particular matter, or particular things in a particular state. And the comparison which Socrates the younger used to make in the case of 'animal' is not sound; for it leads away from the truth, and makes one suppose that man can possibly exist without his parts, as the circle can without the bronze. But the case is not similar; for an animal is something perceptible, and it is not possible to define it without reference to movement—nor, therefore, without reference to the parts' being in a certain state. For it is not a hand in any and every state that is a part of man, but only when it can fulfill its work, and therefore only when it is alive; if it is not alive it is not a part.

Regarding the objects of mathematics, why are the formulae of the parts not parts of the formulae of the wholes; e.g. why are not the semicircles included in the formula of the circle? It cannot be said, 'because these parts are perceptible things'; for they are not. But perhaps this makes no difference; for even some things which are not perceptible must have matter; indeed there is some matter in everything which is not an essence and a bare form but a 'this.' The semicircles, then, will not be parts of the universal circle, but will be parts of the individual circles, as has been said before; for while one kind of matter is perceptible, there is another which is intelligible.

2.43 A science such as arithmetic, which is not a science of properties *qua* inhering in a substratum, is more exact than and prior to a science like harmonics, which is a science of properties inhering in a substratum; and similarly a science like arithmetic, which is constituted of fewer basic elements, is more exact than and prior to geometry, which requires additional elements. What I mean by 'additional elements' is this: a unit is substance without position, while a point is substance with position; the latter contains an additional element.

2.44 The minute accuracy of mathematics is not to be demanded in all cases, but only in the case of things which have no matter. Hence its method is not that of natural science; for presumably the whole of nature has matter.

2.45 Reciprocation of premisses and conclusion is more frequent in mathematics, because mathematics takes definitions, but never an accident, for its premisses—a second characteristic distinguishing mathematical reasoning from dialectical disputations.

The objects of mathematical knowledge–unity, quantity, and magnitude in the abstract.

2.46 'Quantum' means that which is divisible into two or more constituent parts of which each is by nature a 'one' and a 'this.' A quantum is a plurality if it is numerable, a magnitude if it is measurable. 'Plurality' means that which is divisible potentially into non-continuous parts, 'magnitude' that which is divisible into continuous parts; of magnitude, that which is continuous in one dimension is length, in two breadth, in three depth. Of these, limited plurality is number, limited length is a line, breadth a surface, depth a solid.

Again, some things are called quanta in virtue of their own nature, others incidentally; e.g. the line is a quantum by its own nature, the musical is one incidentally. Of the things that are quanta by their own nature some are so as substances, e.g. the line is a quantum (for 'a certain kind of quantum' is present in the definition which states what it is), and others are modifications and states of this kind of substance, e.g. much and little, long and short, broad and narrow, deep and shallow, heavy and light, and all other such attributes. And also great and small, and greater and smaller, both in themselves and when taken relatively to each other, are by their own nature attributes of what is quantitative; but these names are transferred to other things also. Of things that are quanta incidentally, some are so called in the sense in which it was said that the musical and the white were quanta, viz. because that to which musicalness and whiteness belong is a quantum, and some are quanta in the way in which movement and time are so; for these also are called quanta of a sort and continuous because the things of which these are attributes are divisible. I mean not that which is moved, but the space through which it is moved; for because that is a quantum movement also is a quantum, and because this is a quantum time is one.

2.47 Quantity is either discrete or continuous. Moreover, some quantities are such that each part of the whole has a relative position to the other parts: others have within them no such relation of part to part.

Instances of discrete quantities are number and speech; of continuous, lines, surfaces, solids, and, besides these, time and place.

In the case of the parts of a number, there is no common boundary at which they join. For example: two fives make ten, but the two fives have no common boundary, but are separate; the parts three and seven also do not join at any boundary. Nor, to generalize, would it ever be possible in the case of number that there should be a common boundary among the parts; they are always separate. Number, therefore, is a discrete quantity.

The same is true of speech. That speech is a quantity is evident: for it is measured in long and short syllables. I mean here that speech which is vocal. Moreover, it is a discrete quantity, for its parts have no common boundary. There is no common boundary at which the syllables join, but each is separate and distinct from the rest.

A line, on the other hand, is a continuous quantity, for it is possible to find a common boundary at which its parts join. In the case of the line, this common boundary is the point; in the case of the plane, it is the line: for the parts of the plane have also a common boundary. Similarly you can find a common boundary in the case of the parts of a solid, namely either a line or a plane.

Space and time also belong to this class of quantities. Time, past, present, and future, forms a continuous whole. Space, likewise, is a continuous quantity: for the parts of a solid occupy a certain space, and these have a common boundary; it follows that the parts of space also, which are occupied by the parts of the solid, have the same common boundary as the parts of the solid. Thus, not only time, but space also, is a continuous quantity, for its parts have a common boundary.

Quantities consist either of parts which bear a relative position each to each, or of parts which do not. The parts of a line bear a relative position to each other, for each lies somewhere, and it would be possible to distinguish each, and to state the position of each on the plane and to explain to what sort of part among the rest each was contiguous. Similarly the parts of a plane have position, for it could similarly be stated what was the position of each and what sort of parts were contiguous. The same is true with regard to the solid and to space. But it would be impossible to show that the parts of a number had a relative position each to each, or a particular position, or to state what parts were contiguous. Nor could this be done in the case of time, for none of the parts of time has an abiding existence, and that which does not abide can hardly have position. It would be better to say that such parts had a relative order, in virtue of one being prior to another. Similarly with number: in counting, 'one' is prior to 'two,' and 'two' to 'three,' and thus the parts of number may be said to possess a relative order, though it would be impossible to discover any distinct position for each. This holds good also in the case of speech. None of its parts has an abiding existence: when once a syllable is pronounced, it is not possible to retain it, so that, naturally, as the parts do not abide, they cannot have position. Thus, some quantities consist of parts which have position, and some of those which have not.

Strictly speaking, only the things which I have mentioned belong to the category of quantity: everything else that is called quantitative is a quantity in a secondary sense.

2.48 Quantity does not, it appears, admit of variation of degree. One thing cannot be two cubits long in a greater degree than another. Similarly with regard to number: what is 'three' is not more truly three than what is 'five' is five; nor is one set of three more truly three than another set. Again, one period of time is not said to be more truly time than another. Nor is there any other kind of quantity, of all that have been mentioned, with regard to which variation of degree can be predicated. The category of quantity, therefore, does not admit of variation of degree.

The most distinctive mark of quantity is that equality and inequality are predicated of it. Each of the aforesaid quantities is said to be equal or unequal. For instance, one solid is said to be equal or unequal to another; number, too, and time can have these terms applied to them, as indeed can all those kinds of quantity that have been mentioned.

That which is not a quantity can by no means, it would seem, be termed equal or unequal to anything else. One particular disposition or one particular quality, such as whiteness, is by no means compared with another in terms of equality and inequality but rather in terms of similarity. Thus it is the distinctive mark of quantity that it can be called equal and unequal.

2.49 Things are 'relative' (1) as double to half, and treble to a third, and in general that which contains something else many times to that which is contained many times in something else, and that which exceeds to that which is exceeded; (2) as that which can heat to that which can be heated, and that which can cut to that which can be cut, and in general the active to the passive; (3) as the measurable to the measure, and the knowable to knowledge, and the perceptible to perception.

(1) Relative terms of the first kind are numerically related either indefinitely or definitely, to numbers themselves or to 1. E.g. the double is in a definite numerical relation to 1, and that which is 'many times as great' is in a numerical, but not a definite, relation to 1, i.e. not in this or in that numerical relation to it; the relation of that which is half as big again as something else to that something is a definite numerical relation to a number; that which is $\frac{n+1}{n}$ times something else is in an indefinite relation to that something, as that which is 'many times as great' is in an indefinite relation to 1; the relation of that which exceeds to that which is exceeded is numerically quite indefinite; for number is always commensurate, and 'number' is not predicated of that which is not commensurate, but that which exceeds is, in relation to that which is exceeded, so much and something more; and this something is indefinite; for it can, indifferently be either equal or not equal to that which is exceeded.—All these relations, then, are numerically expressed and are determinations of number, and so in another way are the equal and the like and the same. For all refer to unity. Those things are the same whose substance is one; those are like whose quality is one; those are equal whose quantity is one; and 1 is the beginning and measure of number, so that all these relations imply number, though not in the same way.

2.50 While in a sense we call anything one if it is a quantity and continuous, in a sense we do not unless it is a whole, i.e. unless it has unity of form; e.g. if we saw the parts of a shoe put together anyhow we should not call them one all the same (unless because of their continuity); we do this only if they are put together so as to be a shoe and to have already a certain single form. This is why the circle is of all lines most truly one, because it is whole and complete.

The *essence* of what is one is to be some kind of beginning of number; for the first measure is the beginning, since that by which we first know each class is the first measure of the class; the one, then, is the beginning of the knowable regarding each class. But the one is not the same in all classes. For here it is a quarter-tone, and there it is the vowel or the consonant; and there is another unit of weight and another of movement. But everywhere the one is indivisible either in quantity or in kind. Now that which is indivisible in quantity is called a unit if it is not divisible in any dimension and is without position, a point if it is not divisible in any dimension, and has position, a line if it is divisible in one dimension, a plane if in two, a body if divisible in quantity in all—i.e. in three—dimensions. And, reversing the order, that which is divisible in two dimensions is a plane, that which is divisible in one a line, that which is in no way divisible in quantity is a point or a unit—that which has not position a unit, that which has position a point.

Again, some things are one in number, others in species, others in genus, others by analogy; in number those whose matter is one, in species those whose definition is one, in genus those to which the same figure of predication applies, by analogy those which are related as a third thing is to a fourth. The latter kinds of unity are always found when the former are; e.g. things that are one in number are also one in species, while things that are one in species are not all one in number; but things that are one in species are all one in genus, while things that are so in genus are not all one in species but are all one by analogy; while things that are one by analogy are not all one in genus.

2.51 The inquiry that is both the hardest of all and the most necessary for knowledge of the truth is whether being and unity are the substances of things, and whether each of them, without being anything else, is being or unity respectively, *or* we must inquire what being and unity are, with the implication that they have some other underlying nature. For some people think they are of the former, others think they are of the latter character.

(A) If we do not suppose unity and being to be substances, it follows that none of the other universals is a substance; for these are most universal of all, and if there is no unity-itself or being-itself, there will scarcely be in any *other* case anything apart from what are called the individuals. Further, if unity is not a substance, evidently number also will not exist as an entity separate from the individual things; for number is units, and the unit is precisely a certain kind of one.

But (B) if there is a unity-itself and a being-itself, unity and being must be their substance; fot it is not something else that is predicated universally of the things that are and are one, but just unity and being. But if there *is* to be a being-itself and a unity-itself, there is much difficulty in seeing how there will be anything else besides these—I mean, how things will be more than one in number. For what is different from being does not exist, so that it necessarily follows, according to the argument of Parmenides, that all things that are are one and this is being.

There are objections to both views. For whether unity is not a substance or there *is* a unity-itself, number cannot be a substance. We have already said why this result follows if unity is not a substance; and if it is, the same difficulty arises as arose with regard to being. For whence is there to be another one besides unity-itself? It must be not-one; but all things are either one or many, and of the many each is one.

2.52 A question connected with these is whether numbers and bodies and planes and points are substances of a kind, or not. If they are not, it baffles us to say what being is and what the substances of things are. For modifications and movements and relations and dispositions and ratios do not seem to indicate the substance of anything; for all are predicated of a subject, and none is a 'this.' And as to the things which might seem most of all to indicate substance, water and earth and fire and air, of which composite bodies consist, heat and cold and the like are modifications of these, not substances, and the body which is thus modified alone persists as something real and as a substance. But, on the other hand, the body is surely less of a substance than the surface, and the surface than the line, and the line than the unit and the point. For the body is bounded by these; and they are thought to be capable of existing without body, but body incapable of existing without these. This is why, while most of the philosophers and the earlier among them thought that substance and being were identical with *body,* and that all other things were modifications of this, so that the first principles of bodies were the first principles of being, the more recent and those who were held to be wiser thought *numbers* were the first principles. As we said, then, if these are not substance, there is no substance and no being at all; for the *accidents* of these it cannot be right to call beings.

2.53 It has, then, been sufficiently pointed out that the objects of mathematics are not substances in a higher degree than bodies are, and that they are not prior to sensibles in being, but only in definition, and that they cannot exist somewhere apart. But since it was not possible for them to exist *in* sensibles either, it is plain that they either do not exist at all or exist in a special sense and therefore do not 'exist' without qualification. For 'exist' has many senses. For just as the universal propositions of mathematics deal not with objects which exist separately, apart from extended magnitudes and from numbers, but with magnitudes and numbers, not however *qua* such as to have magnitude or to be divisible, clearly it is possible that there should also be both propositions and demonstrations about sensible magnitudes, not however *qua* sensible but *qua* possessed of certain definite qualities. For as there are many propositions about things merely considered as in motion, apart from what each such thing is and from their accidents, and as it is not therefore necessary that there should be either a mobile separate from sensibles, or a distinct mobile entity in the sensibles, so too in the case of mobiles there will be propositions and sciences, which treat them however not *qua* mobile but only *qua* bodies, or again only *qua* planes, or only *qua* lines, or *qua* divisibles, or *qua* indivisibles having position, or only *qua* indivisibles. Thus since it is true to say without qualification that not only things which are separable but also things which are inseparable exist (for instance, that mobiles exist), it is true also to say without qualification that the objects of mathematics exist, and with the character ascribed to them by mathematicians. And as it is true to say of the other sciences too, without qualification, that they deal with such and such a subject—not with what is accidental to it (e.g. not with the pale, if the healthy thing is pale, and the science has the healthy as its subject), but with that which is the subject of each science—with the healthy if it treats its object *qua* healthy, with man if *qua* man:—so too is it with geometry; if its subjects happen to be sensible, though it does not treat them *qua* sensible, the mathematical sciences will not for that reason be sciences of sensibles—nor, on the other hand, of other things separate from sensibles. Many properties attach to things in virtue of their own nature as possessed of each such character; e.g. there are attributes peculiar to the animal *qua* female or *qua* male (yet there is no 'female' nor 'male' separate from animals); so that there are also attributes which belong to things merely as lengths or as planes. And in proportion as we are dealing with things which are prior in definition and simpler, our knowledge has more accuracy, i.e. simplicity. Therefore a science which abstracts from spatial magnitude is more precise than one which takes it into account; and a science is most precise if it abstracts from movement, but if it takes account of movement, it is most precise if it deals with the primary movement, for this is the simplest; and of this again uniform movement is the simplest form.

Each question will be best investigated in this way—by setting up by an act of separation what is not separate, as the arithmetician and the geometer do. For a man *qua* man is one indivisible thing; and the arithmetician supposed one indivisible thing, and then considered whether any attribute belongs to a man *qua* indivisible. But the geometer treats him neither *qua* man nor *qua* indivisible, but as a solid. For evidently the properties which would have belonged to him even if perchance he had not been indivisible, can belong to him even apart from these attributes. Thus, then, geometers speak correctly; they talk about existing things, and their subjects do exist; for being has two forms—it exists not only in complete reality but also materially.

Potential, but not actual, infinities exist.

2.54 But the problem of the infinite is difficult; many contradictions result whether we suppose it to exist or not to exist. If it exists, we have still to ask *how* it exists; as a substance or as the essential attribute of some entity? Or in neither way, yet none the less is there something which is infinite or some things which are infinitely many?

The problem, however, which specially belongs to the physicist is to investigate whether there is a sensible magnitude which is infinite.

We must begin by distinguishing the various senses in which the term 'infinite' is used.

(1) What is incapable of being gone through, because it is not its nature to be gone through (the sense in which the voice is 'invisible').
(2) What admits of being gone through, the process however having no termination, or (3) what scarcely admits of being gone through.
(4) What naturally admits of being gone through but is not actually gone through or does not actually reach an end.

Further, everything that is infinite may be so in respect of addition or division or both.

2.55 In general, the view that there is an infinite body is plainly incompatible with the doctrine that there is necessarily a proper place for each kind of body, if every sensible body has either weight or lightness, and if a body has a natural locomotion towards the centre if it is heavy, and upwards if it is light. This would need to be true of the infinite also. But neither character can belong to it: it cannot be either as a whole, nor can it be half the one and half the other. For how should you divide it? or how can the infinite have the one part up and the other down, or an extremity and a centre?

Further, every sensible body is in place, and the kinds or differences of place are up-down, before-behind, right-left; and these distinctions hold not only in relation to us and by arbitrary agreement, but also in the whole itself. But in the infinite body they cannot exist. In general, if it is impossible that there should be an infinite place, and if every body is in place, there cannot be an infinite body.

Surely what is in a special place is in place, and what is in place is in a special place. Just, then, as the infinite cannot be quantity—that would imply that it has a particular quantity, e.g. two or three cubits; quantity just means these—so a thing's being in place means that it is *some*where, and that is either up or down or in some other of the six differences of position: but each of these is a limit.

It is plain from these arguments that there is no body which is *actually* infinite.

But on the other hand to suppose that the infinite does not exist in any way leads obviously to many impossible consequences: there will be a beginning and an end of time, a magnitude will not be divisible into magnitudes, number will not be infinite. If, then, in view of the above considerations, neither alternative seems possible, an arbiter must be called in; and clearly there is a sense in which the infinite exists and another in which it does not.

We must keep in mind that the word 'is' means either what *potentially* is or what *fully* is.

Further, a thing is infinite either by addition or by division.

Now, as we have seen, magnitude is not actually infinite. But by division it is infinite. (There is no difficulty in refuting the theory of indivisible lines.) The alternative then remains that the infinite has a potential existence.

But the phrase 'potential existence' is ambiguous. When we speak of the potential existence of a statue we mean that there will be an actual statue. It is not so with the infinite. There will not be an actual infinite. The word 'is' has many senses, and we say that the infinite 'is' in the sense in which we say 'it is day' or 'it is the games,' because one thing after another is always coming into existence. For of these things too the distinction between potential and actual existence holds. We say that there are Olympic games, both in the sense that they may occur and that they are actually occurring.

2.56 The infinite, then, exists in no other way, but in this way it does exist, potentially and by reduction. It exists fully in the sense in which we say 'it is day' or 'it is the games'; and potentially as matter exists, not independently as what is finite does.

By addition then, also, there is potentially an infinite, namely, what we have described as being in a sense the same as the infinite in respect of division. For it will always be possible to take something *ab extra.* Yet the sum of the parts taken will not exceed every determinate magnitude, just as in the direction of division every determinate magnitude is surpassed in smallness and there will be a smaller part.

The infinite turns out to be the contrary of what it is said to be. It is not what has nothing outside it that is infinite but what always has something outside it. This is indicated by the fact that rings also that have no bezel are described as 'endless,' because it is always possible to take a part which is outside a given part. The description depends on a certain similarity, but it is not true in the full sense of the word. This condition alone is not sufficient: it is necessary also that the next part which is taken should never be the same. In the circle, the latter condition is not satisfied: it is only the adjacent part from which the new part is different.

Our definition then is as follows:

A quantity is infinite if it is such that we can always take a part outside what has been already taken. On the other hand, what has nothing outside it is complete and whole. For thus we define the whole—that from which nothing is wanting, as a whole man or a whole box. What is true of each particular is true of the whole as such—the whole is that of which nothing is outside. On the other hand that from which something is absent and outside, however small that may be, is not 'all.' 'Whole' and 'complete' are either quite identical or closely akin. Nothing is complete (teleion) which has no end (telos); and the end is a limit.

2.57 Our account does not rob the mathematicians of their science, by disproving the actual existence of the infinite in the direction of increase, in the sense of the untraversable. In point of fact they do not need the infinite and do not use it. They postulate only that the finite straight line may be produced as far as they wish. It is possible to have divided in the same ratio as the largest quantity another magnitude of any size you like. Hence, for the purposes of proof, it will make no difference to them to have such an infinite instead, while its existence will be in the sphere of real magnitudes.

3

DESCARTES · Introduction

The absence of any selections in this volume from the two thousand year period between Aristotle and René Descartes (1596–1650) must not be interpreted as reflecting a total lack of work or progress in either mathematics or philosophy. An examination of any comprehensive history of philosophy or of mathematics will reveal that in fact much significant work was done in both areas during this period. Most of the work in philosophy was grounded on the Platonic and Aristotelean systems as they were interpreted by Moslem, Jewish, and Christian intellectuals. New developments in the areas of logic and philosophy of language by thinkers such as Duns Scotus and William of Occam were of particular relevance to later work in the philosophy of mathematics.

The end of the 16th and the beginning of the 17th centuries were a period of unprecedented rebellion in the intellectual world. The Aristotelean philosophy, along with the whole Scholastic theological-philosophical system that had been built around it, was subjected to penetrating criticisms and even ridicule. Copernicus, Bruno, Kepler, and Galileo (to name only some of the most prominent figures) successfully attacked the two thousand year old geocentric cosmology of Aristotle and Ptolemy. Luther, Calvin, and others were challenging the doctrines and traditions of the Roman Church. And a revival of the classical Greek school of Scepticism, led by the eloquent Montaigne, challenged *everything*.

This general atmosphere of intellectual rebelliousness was accompanied by a resurgence of intellectual creativity. Not only were the old theories attacked, but exciting and ingenious new ones were created to replace them. In addition, totally new theories and techniques were formulated in areas which had not even been considered by earlier generations. One such radical innovation was René Descartes' invention of what is today referred to as "analytic geometry," which involved the application of arithmetical techniques to geometrical problems. This was not only an act of creative genius, but it also reflected the general reaction of the times against the doctrines of Aristotle. For almost 2000 years it had been assumed (correctly or not) that Aristotle had taught that the analytic techniques of arithmetic can *not* be used in geometry. In doing just this, Descartes was reacting against Aristoteleanism, if not against Aristotle.

Descartes did not reject all of the work of all of his predecessors. Many of his ideas can be traced back to Plato, Augustine, and others; he was also influenced by certain of his contemporaries. His most significant *original* contributions to philosophy include his "method of doubt" (see 3.1—3.3) and his formulation of a sharp dichotomy between mental and physical substance. He never seriously questioned the validity of the mathematical method (as epitomized by Euclid's *Geometry*), and he attempted to use this method for constructing a philosophical system which would provide an unshakable foundation for all kinds of knowledge.

Using his own method of systematic, universal, philosophical doubt, Descartes discovered what he considered to be a set of absolutely certain axioms from which all other knowledge could be derived as theorems. Unfortunately, his use of the term "certainty" (as well as several other crucial terms) was both vague and ambiguous. In some contexts he used the term to refer to the subjective feeling of certainty, while in others he referred to the criteria of clarity and distinctness (see 3.8–3.11) as an objective ground of certainty. However, his formulations of these criteria were also quite vague, and on most interpretations they appear to be essentially subjective as well. In attempting to strengthen these criteria by arguing that *God* would not permit one to be deceived by one's feeling of certainty or by the clarity and distinctness of one's perceptions, Descartes introduced what some critics consider to be a vicious circularity into his argument in so far as his proofs of the existence of God are themselves dependent on these very criteria.

Descartes believed that the only immediate objects of perception and knowledge are *ideas,* among which he distinguished three kinds—adventitious, factitious, and innate (see 3.19—3.20). He also held that the only objects of certain knowledge are innate ideas. He explicitly asserted that innate ideas are *not* received through the five external senses; thus even though ideas of shapes and numbers are perceived (as properties of and caused by physical substances) through the senses, these are of no relevance to one's understanding of mathematics. Descartes also indicated that innate ideas are *not* created by the mind; they are placed in each mind at the moment God creates that mind, although one obviously does not become conscious of these ideas until later in life. A highly controversial corollary to this theory of innate knowledge of mathematical ideas is that the laws of mathematics would be different if God had chosen (as he could have done, according to Descartes) to put different ideas in our minds (see 3.34—3.36).

The attacks on Descartes' criteria of certainty and on the circularity of his proofs of the existence of God are only two of the many kinds of criticism which other philosophers have directed against him. It might seem strange that anyone whose theories were subjected to so much criticism could be considered a significant figure, let alone be honored with the title "the father of modern philosophy." However, Descartes' most valuable contributions were his challenging statements of the problems, rather than his ultimate solutions. More than anything, it was his provocative rendering of the question "What can man know with certainty, and how can he know it?" that motivated much of the philosophical activity of the next three centuries. It would be difficult, if not impossible, to name any major post-Cartesian Western philosopher who was not influenced in some significant way by Descartes' formulations of the basic problems of philosophy.

Any attempt to dissect Descartes' philosophy in order to more closely examine individual elements of it necessarily does a great injustice to it. Many of the criticisms of it (including even the charges of circular reasoning) result in part at least from the critics' failure to appreciate the importance of viewing his work as a whole rather than in bits and pieces. It is hoped that the following selections do justice to Descartes' treatment of the problems of the philosophy of mathematics, which involves such a complete integration of the concepts of epistemology and ontology that his discussions of this special topic are virtually inseparable from the rest of his philosophy.

Sources

All selections are from *The Philosophical Works of Descartes,* two volumes, translated and edited by Elizabeth S. Haldane and G. R. T. Ross, Cambridge University Press, Cambridge, England, 1967 (abbreviated to H & R), or *Descartes—Philosophical Letters,* translated and edited by Anthony Kenny, Clarendon Press, Oxford, 1970 (abbreviated to K).

3.1 Discourse IV (H & R, I, pp. 100–101)
3.2 Meditation I (H & R, I, p. 147)
3.3 Meditation I (H & R, I, pp. 148–149)
3.4 Meditation II (H & R, I, pp. 149–150
3.5 Meditation II (H & R, I, p. 153)
3.6 Meditation II (H & R, I, pp. 153–155
3.7 Meditation II (H & R, I, pp. 157)
3.8 Meditation III (H & R, I, pp. 157–158)
3.9 Principles 45 & 46 (H & R, I, p. 237)
3.10 Meditation III (H & R, I, pp. 161–162)
3.11 Meditation III (H & R, I, pp. 158–159)
3.12 Meditation III (H & R, I, pp. 165–167)
3.13 Meditation III (H & R, I, pp. 170–171)
3.14 Letter to Mersenne, July, 1641 (K, p. 105)
3.15 Letter to Elizabeth, May, 1643 (K, p. 138)
3.16 Rules XII, (H & R, I, pp. 40–41
3.17 Principles 49 (H & R, I, pp. 238–239)
3.18 Rules XII (H & R, I, p. 43)

3.19 Meditation III (H & R, I, p. 161)
3.20 Letter to Mersenne, June, 1641 (K, p. 104)
3.21 Rules IV (H & R, I, p. 12)
3.22 Letter to Hyperaspistes, 1641 (K, p. 111)
3.23 Notes, 1647 (H & R, I, p. 442)
3.24 Reply to Objections V, (H & R, II, pp. 227–228)
3.25 Meditation V, (H & R, I, pp. 179–180)
3.26 Rules XIV (H & R, I, pp. 59–60)
3.27 Meditation IV, (H & R, I, pp. 175–176)
3.28 Meditation IV (H & R, I, pp. 176–177)
3.29 Principles 30 (H & R, I, p. 231)
3.30 Reply to Objections II (H & R, II, p. 39)
3.31 Rules III (H & R, I, p. 7)
3.32 Rules III (H & R, I, pp. 7–8)
3.33 Rules II (H & R, I, pp. 4–5)
3.34 Letter to Mersenne, May 1630 (K, pp. 14–15)
3.35 Letter to Mersenne, April, 1630 (K, p. 11)
3.36 Letter to Mesland, May, 1644 (K, p. 151)

DESCARTES · Selections

The Method—systematic, universal, intellectual doubt.

3.1 I do not know that I ought to tell you of the first meditations there made by me, for they are so metaphysical and so unusual that they may perhaps not be acceptable to everyone. And yet at the same time, in order that one may judge whether the foundations which I have laid are sufficiently secure, I find myself constrained in some measure to refer to them. For a long time I had remarked that it is sometimes requisite in common life to follow opinions which one knows to be most uncertain, exactly as though they were indisputable, as has been said above. But because in this case I wished to give myself entirely to the search after Truth, I thought that it was necessary for me to take an apparently opposite course, and to reject as absolutely false everything as to which I could imagine the least ground of doubt, in order to see if afterwards there remained anything in my belief that was entirely certain. Thus, because our senses sometimes deceive us, I wished to suppose that nothing is just as they cause us to imagine it to be; and because there are men who deceive themselves in their reasoning and fall into paralogisms, even concerning the simplest matters of geometry, and judging that I was as subject to error as was any other, I rejected as false all the reasons formerly accepted by me as demonstrations. And since all the same thoughts and conceptions which we have while awake may also come to us in sleep, without any of them being at that time true, I resolved to assume that everything that ever entered into my mind was no more true than the illusions of my dreams.

3.2 That is possibly why our reasoning is not unjust when we conclude from this that Physics, Astronomy, Medicine and all other sciences which have as their end the consideration of composite things, are very dubious and uncertain; but that Arithmetic, Geometry and other sciences of that kind which only treat of things that are very simple and very general, without taking great trouble to ascertain whether they are actually existent or not, contain some measure of certainty and an element of the indubitable. For whether I am awake or asleep, two and three together always form five, and the square can never have more than four sides, and it does not seem possible that truths so clear and apparent can be suspected of any falsity or uncertainty.

Nevertheless I have long had fixed in my mind the belief that an all-powerful God existed by whom I have been created such as I am. But how do I know that He has not brought it to pass that there is no earth, no heaven, no extended body, no magnitude, no place, and that nevertheless I possess the perceptions of all these things and that they seem to me to exist just exactly as I now see them? And, besides, as I sometimes imagine that others deceive themselves in the things which they think they know best, how do I know that I am not deceived every time that I add two and three, or count the sides of a square, or judge of things yet simpler, if anything simpler can be imagined? But possibly God has not desired that I should be thus deceived, for He is said to be supremely good. If, however, it is contrary to His goodness to have made me such that I constantly deceive myself, it would also appear to be contrary to His goodness to permit me to be sometimes deceived, and nevertheless I cannot doubt that He does permit this.

3.3 I shall then suppose, not that God who is supremely good and the fountain of truth, but some evil genius not less powerful than deceitful, has employed his whole energies in deceiving me; I shall consider that the heavens, the earth, colours, figures, sound, and all other external things are naught but the illusions and dreams of which this genius has availed himself in order to lay traps for my credulity; I shall consider myself as having no hands, no eyes, no flesh, no blood, nor any senses, yet falsely believing myself to possess all these things; I shall remain obstinately attached to this idea, and if by this means it is not in my power to arrive at the knowledge of any truth, I may at least do what is in my power, i.e. suspend my judgment, and with firm purpose avoid giving credence to any false thing, or being imposed upon by this arch deceiver, however powerful and deceptive he may be.

The fundamental principle—Cogito, ergo sum.

3.4 I suppose, then, that all the things that I see are false; I persuade myself that nothing has ever existed of all that my fallacious memory represents to me. I consider that I possess no senses; I imagine that body, figure, extension, movement and place are but the fictions of my mind. What, then, can be esteemed as true? Perhaps nothing at all, unless that there is nothing in the world that is certain.

But how can I know there is not something different from those things that I have just considered, of which one cannot have the slightest doubt? Is there not some God, or some other being by whatever name we call it, who puts these reflections into my mind? That is not necessary, for is it not possible that I am capable of producing them myself? I myself, am I not at least something? But I have already denied that I had senses and body. Yet I hesitate, for what follows from that? Am I so dependent on body and senses that I cannot exist without these? But I was persuaded that there was nothing in all the world, that there was no heaven, no earth, that there were no minds, nor any bodies: was I not then likewise persuaded that I did not exist? Not at all; of a surety I myself did exist since I persuaded myself of something or merely because I thought of something. But there is some deceiver or other, very powerful and very cunning, who ever employs his ingenuity in deceiving me. Then without doubt I exist also if he deceives me, and let him deceive me as much as he will, he can never cause me to be nothing so long as I think that I am something. So that after having reflected well and carefully examined all things, we must come to the definite conclusion that this proposition: I am, I exist, is necessarily true each time that I pronounce it, or that I mentally conceive it.

3.5 But what then am I? A thing which thinks. What is a thing which thinks? It is a thing which doubts, understands, conceives, affirms, denies, wills, refuses, which also imagines and feels.

True knowledge is not gained through the senses.

3.6 From this time I begin to know what I am with a little more clearness and distinctness than before; but nevertheless it still seems to me, and I cannot prevent myself from thinking, that corporeal things, whose images are framed by thought, which are tested by the senses, are much more distinctly known than that obscure part of me which does not come under the imagination. Although really it is very strange to say that I know and understand more distinctly these things whose existence seems to me dubious, which are unknown to me, and which do not belong to me, than others of the truth of which I am convinced, which are known to me and which pertain to my real nature, in a word, than myself. But I see clearly how the case stands: my mind loves to wander, and cannot yet suffer itself to be retained within the just limits of truth. Very good, let us once more give it the freest rein, so that, when afterwards we seize the proper occasion for pulling up, it may the more easily be regulated and controlled.

Let us begin by considering the commonest matters, those which we believe to be the most distinctly comprehended, to wit, the bodies which we touch and see; not indeed bodies in general, for these general ideas are usually a little more confused, but let us consider one body in particular. Let us take, for example, this piece of wax: it has been taken quite freshly from the hive, and it has not yet lost the sweetness of the honey which it contains; it still retains somewhat of the odour of the flowers from which it has been culled; its colour, its figure, its size are apparent; it is hard, cold, easily handled, and if you strike it with the finger, it will emit a sound. Finally all the things which are requisite to cause us distinctly to recognise a body, are met within it. But notice that while I speak and approach the fire what remained of the taste is exhaled, the smell evaporates, the colour alters, the figure is destroyed, the size increases, it becomes liquid, it heats, scarcely can one handle it, and when one strikes it, no sound is emitted. Does the same wax remain after this change? We must confess that it remains; none would judge otherwise. What then did I know so distinctly in this piece of wax? It could certainly be nothing of all that the senses brought to my notice, since all these things which fall under taste, smell, sight, touch, and hearing, are found to be changed, and yet the same wax remains.

Perhaps it was what I now think, viz. that this wax was not that sweetness of honey, nor that agreeable scent of flowers, nor that particular whiteness, nor that figure, nor that sound, but simply a body which a little while before appeared to me as perceptible under these forms, and which is now perceptible under others. But what, precisely, is it that I imagine when I form such conceptions? Let us attentively consider this, and, abstracting from all that does not belong to the wax, let us see what remains. Certainly nothing remains excepting a certain extended thing which is flexible and movable. But what is the meaning of flexible and movable? Is it not that I imagine that this piece of wax being round is capable of becoming square and of passing from a square to a triangular figure? No, certainly it is not that, since I imagine it admits of an infinitude of similar changes, and I nevertheless do not know how to compass the infinitude by my imagination, and consequently this conception which I have of the wax is not brought about by the faculty of imagination. What now is this extension? Is it not also unknown? For it becomes greater when the wax is melted, greater when it is boiled, and greater still when the heat increases; and I should not conceive clearly according to truth what wax is, if I did not think that even this piece that we are considering is capable of receiving more variations in extension than I have ever imagined. We must then grant that I could not even understand through the imagination what this piece of wax is, and that it is my mind alone which perceives it. I say this piece of wax in particular, for as to wax in general it is yet clearer. But what is this piece of wax which cannot be understood excepting by the understanding or mind? It is certainly the same that I see, touch, imagine, and finally it is the same which I have always believed it to be from the beginning. But what must particularly be observed is that its perception is neither an act of vision, nor of touch, nor of imagination, and has never been such although it may have appeared formerly to be so, but only an intuition of the mind, which may be imperfect and confused as it was formerly, or clear and distinct as it is at present, according as my attention is more or less directed to the elements which are found in it, and of which it is composed.

3.7 But finally here I am, having insensibly reverted to the point I desired, for, since it is now manifest to me that even bodies are not properly speaking known by the senses or by the faculty of imagination, but by the understanding only, and since they are not known from the fact that they are seen or touched, but only because they are understood, I see clearly that there is nothing which is easier for me to know than my mind.

The criteria of certainty—clarity and distinctness.

3.8 And in the little that I have just said, I think I have summed up all that I really know, or at least all that hitherto I was aware that I knew. In order to try to extend my knowledge further, I shall now look around more carefully and see whether I cannot still discover in myself some other things which I have not hitherto perceived. I am certain that I am a thing which thinks; but do I not then likewise know what is requisite to render me certain of a truth? Certainly in this first knowledge there is nothing that assures me of its truth, excepting the clear and distinct perception of that which I state, which would not indeed suffice to assure me that what I say is true, if it could ever happen that a thing which I conceived so clearly and distinctly could be false; and accordingly it seems to me that already I can establish as a general rule that all things which I perceive very clearly and very distinctly are true.

3.9 There are even a number of people who throughout all their lives perceive nothing so correctly as to be capable of judging of it properly. For the knowledge upon which a certain and incontrovertible judgment can be formed, should not alone be clear but also distinct. I term that clear which is present and apparent to an attentive mind, in the same way as we assert that we see objects clearly when, being present to the regarding eye, they operate upon it with sufficient strength. But the distinct is that which is so precise and different from all other objects that it contains within itself nothing but what is clear.

When, for instance, a severe pain is felt, the perception of this pain may be very clear, and yet for all that not distinct, because it is usually confused by the sufferers with the obscure judgment that they form upon its nature, assuming as they do that something exists in the part affected, similar to the sensation of pain of which they are alone clearly conscious. In this way perception may be clear without being distinct, and cannot be distinct without being also clear.

3.10 . . . I cannot doubt that which the natural light causes me to believe to be true, as, for example, it has shown me that I am from the fact that I doubt, or other facts of the same kind. And I possess no other faculty whereby to distinguish truth from falsehood, which can teach me that what this light shows me to be true is not really true, and no other faculty that is equally trustworthy. But as far as apparently natural impulses are concerned, I have frequently remarked, when I had to make active choice between virtue and vice, that they often enough led me to the part that was worse; and this is why I do not see any reason for following them in what regards truth and error.

3.11 But when I took anything very simple and easy in the sphere of arithmetic or geometry into consideration, e.g. that two and three together made five, and other things of the sort, were not these present to my mind so clearly as to enable me to affirm that they were true? Certainly if I judged that since such matters could be doubted, this would not have been so for any other reason than that it came into my mind that perhaps a God might have endowed me with such a nature that I may have been deceived even concerning things which seemed to me most manifest. But every time that this preconceived opinion of the sovereign power of a God presents itself to my thought, I am constrained to confess that it is easy to Him, if He wishes it, to cause me to err, even in matters in which I believe myself to have the best evidence. And, on the other hand, always when I direct my attention to things which I believe myself to perceive very clearly, I am so persuaded of their truth that I let myself break out into words such as these: Let who will deceive me, He can never cause me to be nothing while I think that I am, or some day cause it to be true to say that I have never been, it being true now to say that I am, or that two and three make more or less than five, or any such thing in which I see a manifest contradiction. And, certainly, since I have no reason to believe that there is a God who is a deceiver, and as I have not yet satisfied myself that there is a God at all, the reason for doubt which depends on this opinion alone is very slight, and so to speak metaphysical. But in order to be able altogether to remove it, I must inquire whether there is a God as soon as the occasion presents itself; and if I find that there is a God, I must also inquire whether He may be a deceiver; for without a knowledge of these two truths I do not see that I can ever be certain of anything.

The guarantee of all other certainty—certain knowledge of the existence of God.

3.12 By the name God I understand a substance that is infinite, eternal, immutable, independent, all-knowing, all-powerful, and by which I myself and everything else, if anything else does exist, have been created. Now all these characteristics are such that the more diligently I attend to them, the less do they appear capable of proceeding from me alone; hence, . . . we must conclude that God necessarily exists.

For although the idea of substance is within me owing to the fact that I am substance, nevertheless I should not have the idea of an infinite substance—since I am finite—if it had not proceeded from some substance which was veritably infinite.

Nor should I imagine that I do not perceive the infinite by a true idea, but only by the negation of the finite, just as I perceive repose and darkness by the negation of movement and of light; for, on the contrary, I see that there is manifestly more reality in infinite substance than in finite, and therefore that in some way I have in me the notion of the infinite earlier than the finite—to wit, the notion of God before that of myself. For how would it be possible that I should know that I doubt and desire, that is to say, that something is lacking to me, and that I am not quite perfect, unless I had within me some idea of a Being more perfect than myself, in comparison with which I should recognise the deficiencies of my nature?

And we cannot say that this idea of God is perhaps materially false and that consequently I can derive it from naught, i.e. that possibly it exists in me because I am imperfect, as . . . is the case with ideas of heat, cold and other such things; for, on the contrary, as this idea is very clear and distinct and contains within it more objective reality than any other, there can be none which is of itself more true, nor any in which there can be less suspicion of falsehood. The idea, I say, of this Being who is absolutely perfect and infinite, is entirely true; for although, perhaps, we can imagine that such a Being does not exist, we cannot nevertheless imagine that His idea represents nothing real to me, as I have said of the idea of cold. This idea is also very clear and distinct; since all that I conceive clearly and distinctly of the real and the true, and of what conveys some perfection, is in its entirety contained in this idea. And this does not cease to be true although I do not comprehend the infinite, or though in God there is an infinitude of things which I cannot comprehend, nor possibly even reach in any way by thought; for it is of the nature of the infinite that my nature, which is finite and limited, should not comprehend it; and it is sufficient that I should understand this, and that I should judge that all things which I clearly perceive and in which I know that there is some perfection, and possibly likewise an infinitude of properties of which I am ignorant, are in God formally or eminently, so that the idea which I have of Him may become the most true, most clear, and most distinct of all the ideas that are in my mind.

But possibly I am something more than I suppose myself to be, and perhaps all those perfections which I attribute to God are in some way potentially in me, although they do not yet disclose themselves, or issue an action. As a matter of fact I am already sensible that my knowledge increases and perfects itself little by little, and I see nothing which can prevent it from increasing more and more into infinitude; nor do I see, after it has thus been increased or perfected, anything to prevent my being able to acquire by its means all the other perfections of the Divine nature; nor finally why the power I have of acquiring these perfections, if it really exists in me, shall not suffice to produce the ideas of them.

At the same time I recognise that this cannot be. For, in the first place, although it were true that every day my knowledge acquired new degrees of perfection, and that there were in my nature many things potentially which are not yet there actually, nevertheless these excellences do not pertain to or make the smallest approach to the idea which I have of God in whom there is nothing merely potential but in whom all is present really and actually; for it is an infallible token of imperfection in my knowledge that it increases little by little. And further, although my knowledge grows more and more, nevertheless I do not for that reason believe that it can ever be actually infinite, since it can never reach a point so high that it will be unable to attain to any greater increase. But I understand God to be actually infinite, so that He can add nothing to His supreme perfection. And finally I perceive that the objective being of an idea cannot be produced by a being that exists potentially only, which properly speaking is nothing, but only by a being which is formal or actual.

3.13 It only remains to me to examine into the manner in which I have acquired this idea from God; for I have not received it through the senses, and it is never presented to me unexpectedly, as is usual with the ideas of sensible things when these things present themselves, or seem to present themselves, to the external organs of my senses; nor is it likewise a fiction of my mind, for it is not in my power to take from or to add anything to it; and consequently the only alternative is that it is innate in me, just as the idea of myself is innate in me.

And one certainly ought not to find it strange that God, in creating me, placed this idea within me to be like the mark of the workman imprinted on his work; and it is likewise not essential that the mark shall be something different from the work itself. For from the sole fact that God created me it is most probable that in some way He has placed His image and similitude upon me, and that I perceive this similitude (in which the idea of God is contained) by means of the same faculty by which I perceive myself—that is to say, when I reflect on myself I not only know that I am something imperfect, incomplete and dependent on another, which incessantly aspires after something which is better and greater than myself, but I also know that He on whom I depend possesses in Himself all the great things towards which I aspire and the ideas of which I find within myself, and that not indefinitely or potentially alone, but really, actually and infinitely; and that thus He is God. And the whole strength of the argument which I have here made use of to prove the existence of God consists in this, that I recognize that it is not possible that my nature should be what it is, and indeed that I should have in myself the idea of a God, if God did not veritably exist—a God, I say, whose idea is in me, i.e. who possesses all those supreme perfections of which our mind may indeed have some idea but without understanding them all, who is liable to no errors or defect and who has none of all those marks which denote imperfection. From this it is manifest that He cannot be a deceiver, since the light of nature teaches us that fraud and deception necessarily proceed from some defect.

The objects of human knowledge—ideas or 'simple natures.'

3.14 . . . By 'idea' I do not just mean the images depicted in the imagination, indeed, in so far as these images are in the corporeal fancy, I do not use that term for them at all. Instead, by the term 'idea' I mean in general everything which is in our mind when we conceive something, no matter how we conceive it.

. . . God cannot be conceived by the imagination. But if He is not conceived by the imagination, then either one conceives nothing when one speaks of God (which would be a sign of terrible blindness) or one conceives Him in another manner; but whatever way we conceive Him, we have the idea of Him. For we cannot express anything by our words, when we understand what we are saying, without its being certain *eo ipso* that we have in us the idea of the thing which is signified by our words.

3.15 First I observe that there are in us certain primitive notions which are as it were models on which all our other knowledge is patterned. There are very few such notions. First, there are the most general ones, such as being, number, and duration, which apply to everything we can conceive. Then, as regards body in particular, we have only the notion of extension which entails the notions of shape and motion; and as regards soul in particular we have only the notion of thought, which includes the conceptions of the intellect and the inclinations of the will. Finally, as regards soul and body together, we have only the notion of their union, on which depends our notion of the soul's power to move the body, and the body's power to act on the soul and cause sensations and passions.

I observe next that all human scientific knowledge consists solely in clearly distinguishing these notions and attaching each of them only to the things to which it applies. For if we try to solve a problem by means of a notion that does not apply, we cannot help going wrong. Similarly we go wrong if we try to explain one of these notions by another, for since they are primitive notions each of them can only be understood by itself.

3.16 Finally, then, we assert that relatively to our knowledge single things should be taken in an order different from that in which we should regard them when considered in their more real nature. Thus, for example, if we consider a body as having extension and figure, we shall indeed admit that from the point of view of the thing itself it is one and simple. For we cannot from that point of view regard it as compounded of corporeal nature, extension and figure, since these elements have never existed in isolation from each other. But relatively to our understanding we call it a compound constructed out of these three natures, because we have thought of them separately before we were able to judge that all three were found in one and the same subject. Hence here we shall treat of things only in relation to our understanding's awareness of them, and shall call those only simple, the cognition of which is so clear and so distinct that they cannot be analysed by the mind into others more distinctly known. Such are figure, extension, motion, etc.; all others we conceive to be in some way compounded out of these. This principle must be taken so universally as not even to leave out those objects which we sometimes obtain by abstraction from the simple natures themselves. . . .

. . . Those things which relatively to our understanding are called simple, are either purely intellectual or purely material, or else common both to intellect and to matter. Those are purely intellectual which our understanding apprehends by means of a certain inborn light, and without the aid of any corporeal image. That a number of such things exist is certain; and it is impossible to construct any corporeal idea which shall represent to us what the act of knowing is, what doubt is, what ignorance, and likewise what the action of the will is which it is possible to term volition, and so with other things. Yet we have a genuine knowledge of all these things, and know them so easily that in order to recognize them it is enough to be endowed with reason. Those things are purely material which we discern only in bodies; e.g. figure, extension, motion, etc. Finally those must be styled common which are ascribed now to corporeal things, now to spirits, without distinction. Such are existence, unity, duration and the like. To this group also we must ascribe those common notions which are, as it were, bonds for connecting together the other simple natures, and on whose evidence all the inferences which we obtain by reasoning depend. The following are examples:—things that are the same as a third thing are the same as one another. So too:—things which do not bear the same relation to a third thing, have some diversity from each other, etc. As a matter of fact these common notions can be discerned by the understanding either unaided or when it is aware of the images of material things.

3.17 We must now talk of what we know as eternal truths.

When we apprehend that it is impossible that anything can be formed of nothing, the proposition *ex nihilo nihil fit* is not to be considered as an existing thing, or the mode of a thing, but as a certain eternal truth which has its seat in our mind, and is a common notion or axiom. Of the same nature are the following: 'It is impossible that the same thing can be and not be at the same time,' and that 'what has been done cannot be undone,' 'that he who thinks must exist while he thinks,' and very many other propositions the whole of which it would not be easy to enumerate. But this is not necessary since we cannot fail to recognise them when the occasion presents itself for us to do so, and if we have no prejudices to blind us.

3.18 . . . No knowledge is at any time possible of anything beyond those simple natures and what may be called their intermixture or combination with each other. Indeed it is often easier to be aware of several of them in union with each other, than to separate one of them from the others. For, to illustrate, I am able to know what a triangle is, though I have never thought that in that knowledge was contained the knowledge of an angle, a line, the number three, figure, extension, etc. But that does not prevent me from saying that the nature of the triangle is composed of all these natures, and that they are better known than the triangle since they are the elements which we comprehend in it. It is possible also that in the triangle many other features are involved which escape our notice, such as the magnitude of the angles, which are equal to two right angles, and the innumerable relations which exist between the sides and the angles, or the size of the area, etc.

The sources and kinds of ideas.

3.19 . . . I have noticed that in many cases there was a great difference between the object and its idea. I find, for example, two completely diverse ideas of the sun in my mind; the one derives its origin from the senses, and should be placed in the category of adventitious ideas; according to this idea the sun seems to be extremely small; but the other is derived from astronomical reasonings, i.e. is elicited from certain notions that are innate in me, or else it is formed by me in some other manner; in accordance with it the sun appears to be several times greater than the earth. These two ideas cannot, indeed, both resemble the same sun, and reason makes me believe that the one which seems to have originated directly from the sun itself, is the one which is most dissimilar to it.

3.20 I use the word 'idea' to mean everything which can be in our thought, and I distinguish three kinds. Some are adventitious, such as the idea we commonly have of the sun; others are constructed or factitious, in which class we can put the idea which the astronomers construct of the sun by their reasoning; and others are innate, such as the idea of God, mind, body, triangle, and in general all those which represent true immutable and eternal essences. Now if from a constructed idea I were to conclude to what I explicitly put into it when I was constructing it, I would obviously be begging the question; but it is not the same if I draw out from an innate idea something which was implicitly contained in it but which I did not at first notice in it. Thus I can draw out from the idea of triangle that its three angles equal two right angles, and from the idea of God that He exists. So far from being a begging of the question, this method of argument, in which the true definition of a thing occurs as the middle term, is even according to Aristotle the most perfect of all.

3.21 . . . I am convinced that certain primary germs of truth implanted by nature in human minds—though in our case the daily reading and hearing of innumerable diverse errors stifle them—had a very great vitality in that rude and unsophisticated age of the ancient world. Thus the same mental illumination which let them see that virtue was to be preferred to pleasure, and honour to utility, although they knew not why this was so, made them recognize true notions in Philosophy and Mathematics, although they were not yet able thoroughly to grasp these sciences.

3.22 . . . The nature or essence of soul consists in the fact that it is thinking, just as the essence of body consists in the fact that it is extended. Now nothing can ever be deprived of its own essence; so it seems to me that a man who denies that his soul was thinking at times when he does not remember noticing it thinking, deserves no more attention than a man who denied that his body was extended while he did not notice that it had extension. This does not mean that I believe that the mind of an infant meditates on metaphysics in its mother's womb; not at all. We know by experience that our minds are so closely joined to our bodies as to be almost always acted upon by them; and though in an adult and healthy body the mind enjoys some liberty to think of other things than those presented by the senses, we know there is not the same liberty in those who are sick or asleep or very young; and the younger they are the less liberty they have. So if one may conjecture on such an unexplored topic, it seems most reasonable to think that a mind newly united to an infant's body is wholly occupied in perceiving or feeling the ideas of pain, pleasure, heat, cold and other similar ideas which arise from its union and intermingling with the body. Nevertheless, it has in itself the ideas of God, itself, and all such truths as are called self-evident, in the same way as adult humans have when they are not attending to them; it does not acquire these ideas later on, as it grows older. I have no doubt that if it were taken out of the prison of the body it would find them within itself.

3.23 . . . I never wrote or concluded that the mind required innate ideas which were in some sort different from its faculty of thinking; but when I observed the existence in me of certain thoughts which proceeded, not from extraneous objects nor from the determination of my will, but solely from the faculty of thinking which is within me, then, that I might distinguish the ideas or notions (which are the forms of these thoughts) from other thoughts *adventitious* or *factitious,* I termed the former *'innate.'* In the same sense we say that in some families generosity is innate, in others certain diseases like gout or gravel, not that on this account the babes of these families suffer from these diseases in their mother's womb, but because they are born with a certain disposition or propensity for contracting them.

All mathematical ideas and truths are innate.

3.24 . . . As to the essences which are clearly and distinctly conceived, such as that of the triangle or of any other geometrical figure, I shall easily compel you to acknowledge that the ideas existing in us of those things, are not derived from particulars; for here you say that they are false. . . .

That is forsooth in your opinion, because you suppose the nature of things to be such that these essences cannot be conformable to it. But, unless you also maintain that the whole of geometry is a fiction, you cannot deny that many truths are demonstrated of them, which, being always the same, are rightly styled immutable and eternal. But though they happen not to be conformable to the nature of things as it exists in your conception, as they likewise fail to agree with the atomic theory constructed by Democritus and Epicurus, this is merely an external attribute relatively to them and makes no difference to them; they are, nevertheless, conformable certainly with the real nature of things which has been established by the true God. But this does not imply that there are substances in existence which possess length without breadth, or breadth without depth, but merely that the figures of geometry are considered not as substances but as the boundaries within which substance is contained.

Meanwhile, moreover, I do not admit *that the ideas of these figures have at any time entered our minds through the senses,* as is the common persuasion. For though, doubtless, figures such as the Geometers consider can exist in reality, I deny that any can be presented to us except such minute ones that they fail altogether to affect our senses. For, let us suppose that these figures consist as far as possible of straight lines; yet it will be quite impossible for any really straight part of the line to affect our senses, because when we examine with a magnifying glass those lines that appear to us to be most straight, we find them to be irregular and bending everywhere in an undulating manner. Hence when first in infancy we see a triangular figure depicted on paper, this figure cannot show us how a real triangle ought to be conceived, in the way in which geometricians consider it, because the triangle is contained in this figure, just as the statue of Mercury is contained in a rough block of wood. But because we already possess within us the idea of a true triangle, and it can be more easily conceived by our mind than the more complex figure of the triangle drawn on paper, we, therefore, when we see that composite figure, apprehend not it itself, but rather the authentic triangle. This is exactly the same as when we look at a piece of paper on which little strokes have been drawn with ink to represent a man's face; for the idea produced in us in this way is not so much that of the lines of the sketch as of the man. But this could not have happened unless the human face had been known to us by other means, and we had been more accustomed to think of it than of those minute lines, which indeed we often fail to distinguish from each other when they are moved to a slightly greater distance away from us. So certainly we should not be able to recognize the Geometrical triangle by looking at that which is drawn on paper, unless our mind possessed an idea of it derived from other sources.

3.25 And not only do I know these things with distinctness when I consider them in general, but, likewise however little I apply my attention to the matter, I discover an infinitude of particulars respecting numbers, figures, movements, and other such things, whose truth is so manifest, and so well accords with my nature, that when I begin to discover them, it seems to me that I learn nothing new, or recollect what I formerly knew—that is to say, that I for the first time perceive things which were already present to my mind, although I had not as yet applied my mind to them.

And what I here find to be most important is that I discover in myself an infinitude of ideas of certain things which cannot be esteemed as pure negations, although they may possibly have no existence outside of my thought, and which are not framed by me, although it is within my power either to think or not to think them, but which possess natures which are true and immutable. For example, when I imagine a triangle, although there may nowhere in the world be such a figure outside my thought, or ever have been, there is nevertheless in this figure a certain determinate nature, form, or essence, which is immutable and eternal, which I have not invented, and which in no wise depends on my mind, as appears from the fact that diverse properties of that triangle can be demonstrated, viz. that its three angles are equal to two right angles, that the greatest side is subtended by the greatest angle, and the like, which now, whether I wish it or do not wish it, I recognise very clearly as pertaining to it, although I never thought of the matter at all when I imagined a triangle for the first time, and which therefore cannot be said to have been invented by me.

Nor does the objection hold good that possibly this idea of a triangle has reached my mind through the medium of my senses, since I have sometimes seen bodies triangular in shape; because I can form in my mind an infinitude of other figures regarding which we cannot have the least conception of their ever having been objects of sense, and I can nevertheless demonstrate various properties pertaining to their nature as well as to that of the triangle, and these must certainly all be true since I conceive them clearly. Hence they are something, and not pure negation; for it is perfectly clear that all that is true is something, and I have already fully demonstrated that all that I know clearly is true. And even although I had not demonstrated this, the nature of my mind is such that I could not prevent myself from holding them to be true so long as I conceive them clearly; and I recollect that even when I was still strongly attached to the objects of sense, I counted as the most certain those truths which I conceived clearly as regards figures, numbers, and the other matters which pertain to arithmetic and geometry, and, in general, to pure and abstract mathematics.

3.26 But we should carefully note that in all other propositions in which these terms, though retaining the same signification and similarly employed in abstraction from their subject matter, do not exclude or deny anything from which they are not really distinct, it is both possible and necessary to use the imagination as an aid. The reason is that even though the understanding in the strict sense attends merely to what is signified by the name, the imagination nevertheless ought to fashion a real idea of the object, in order that the very understanding itself may be able to fix upon other features belonging to it that are not expressed by the name in question, whenever there is occasion to do so, and may never imprudently believe that they have been excluded. Thus, if number be the question, we imagine an object which we can measure by summing a plurality of units. Now though it is allowable for the understanding to confine its attention for the present solely to the multiplicity displayed by the object, we must be on our guard nevertheless not on that account afterwards to come to any conclusion which implies that the object which we have described numerically has been excluded from our concept. But this is what those people do who ascribe mysterious properties to number, empty inanities in which they certainly would not believe so strongly, unless they conceived that number was something distinct from the things we number. In the same way, if we are dealing with figure, let us remember that we are concerned with an extended subject though we restrict ourselves to conceiving it merely as possessing figure.

The source of error.

3.27 Whence then come my errors? They come from the sole fact that since the will is much wider in its range and compass than the understanding, I do not restrain it within the same bounds, but extend it also to things which I do not understand: and as the will is of itself indifferent to these, it easily falls into error and sin, and chooses the evil for the good, or the false for the true.

3.28 But if I abstain from giving my judgment on any thing when I do not perceive it with sufficient clearness and distinctness, it is plain that I act rightly and am not deceived. But if I determine to deny or affirm, I no longer make use as I should of my free will, and if I affirm what is not true, it is evident that I deceive myself; even though I judge according to truth, this comes about only by chance, and I do not escape the blame of misusing my freedom; for the light of nature teaches us that the knowledge of the understanding should always precede the determination of the will. And it is in the misuse of the free will that the privation which constitutes the characteristic nature of error is met with. Privation, I say, is found in the act, in so far as it proceeds from me, but it is not found in the faculty which I have received from God, nor even in the act in so far as it depends on Him.

3.29 Whence it follows that the light of nature, or the faculty of knowledge which God has given us, can never disclose to us any object which is not true, inasmuch as it comprehends it, that is, inasmuch as it apprehends it clearly and distinctly. Because we should have had reason to think God a deceiver if He had given us this faculty perverted, or such that we should take the false for the true when using the faculty aright. And this should deliver us from the supreme doubt which encompassed us when we did not know whether our nature had been such that we had been deceived in things that seemed most clear. It should also protect us against all the other reasons already mentioned which we had for doubting. The truths of mathematics should now be above suspicion, for they are of the clearest. And if we perceive anything by our senses, either waking or sleeping, if it is clear and distinct, and if we separate it from what is obscure and confused, we shall easily assure ourselves of what is the truth.

3.30 That *an atheist can know clearly that the three angles of a triangle are equal to two right angles,* I do not deny, I merely affirm that, on the other hand, such knowledge on his part cannot constitute true science, because no knowledge that can be rendered doubtful should be called science. Since he is, as supposed, an Atheist, he cannot be sure that he is not deceived in the things that seem most evident to him, as has been sufficiently shown; and though perchance the doubt does not occur to him, nevertheless it may come up, if he examine the matter, or if another suggests it; he can never be safe from it unless he first recognises the existence of a God.

And it does not matter though he think he has demonstrations proving that there is no God. Since they are by no means true, the errors in them can always be pointed out to him, and when this takes place he will be driven from his opinion.

Direct versus indirect knowledge.

3.31 But lest we in turn should slip into the same error, we shall here take note of all those mental operations by which we are able, wholly without fear of illusion, to arrive at the knowledge of things. Now I admit only two, viz. intuition and deduction.

By *intuition* I understand, not the fluctuating testimony of the senses, nor the misleading judgment that proceeds from the blundering constructions of imagination, but the conception which an unclouded and attentive mind gives us so readily and distinctly that we are wholly freed from doubt about that which we understand. Or, what comes to the same thing, *intuition* is the undoubting conception of an unclouded and attentive mind, and springs from the light of reason alone; it is more certain than deduction itself, in that it is simpler, though deduction as we have noted above, cannot by us be erroneously conducted. Thus each individual can mentally have intuition of the fact that he exists, and that he thinks; that the triangle is bounded by three lines only, the sphere by a single superficies, and so on. Facts of such a kind are far more numerous than many people think, disdaining as they do to direct their attention upon such simple matters.

3.32 This evidence and certitude, however, which belongs to intuition, is required not only in the enunciation of propositions, but also in discursive reasoning of whatever sort. For example consider this consequence: 2 and 2 amount to the same as 3 and 1. Now we need to see intuitively not only that 2 and 2 make 4, and that likewise 3 and 1 make 4, but further that the third of the above statements is a necessary conclusion from these two.

Hence now we are in a position to raise the question as to why we have, besides intuition, given this supplementary method of knowing, viz. knowing by *deduction,* by which we understand all necessary inference from other facts that are known with certainty. This, however, we could not avoid, because many things are known with certainty, though not by themselves evident, but only deduced from true and known principles by the continuous and uninterrupted action of a mind that has a clear vision of each step in the process. It is in a similar way that we know that the last link in a long chain is connected with the first, even though we do not take in by means of one and the same act of vision all the intermediate links on which that connection depends, but only remember that we have taken them successively under review and that each single one is united to its neighbour, from the first even to the last. Hence we distinguish this mental intuition from deduction by the fact that into the conception of the latter there enters a certain movement or succession, into that of the former there does not. Further deduction does not require an immediately presented evidence such as intuition possesses; its certitude is rather conferred upon it in some way by memory. The upshot of the matter is that it is possible to say that those propositions indeed which are immediately deduced from first principles are known now by intuition, now by deduction, i.e. in a way that differs according to our point of view. But the first principles themselves are given by intuition alone, while, on the contrary, the remote conclusions are furnished only by deduction.

These two methods are the most certain routes to knowledge, and the mind should admit no others. All the rest should be rejected as suspect of error and dangerous.

3.33 . . . None of the mistakes which men can make (men, I say, not beasts) are due to faulty inference; they are caused merely by the fact that we found upon a basis of poorly comprehended experiences, or that propositions are posited which are hasty and groundless.

This furnishes us with an evident explanation of the great superiority in certitude of Arithmetic and Geometry to other sciences. The former alone deal with an object so pure and uncomplicated, that they need make no assumptions at all which experience renders uncertain, but wholly consist in the rational deduction of consequences. They are on that account much the easiest and clearest of all, and possess an object such as we require, for in them it is scarce humanly possible for anyone to err except by inadvertence. And yet we should not be surprised to find that plenty of people of their own accord prefer to apply their intelligence to other studies, or to Philosophy. The reason for this is that every person permits himself the liberty of making guesses in the matter of an obscure subject with more confidence than in one which is clear, and that it is much easier to have some vague notion about any subject, no matter what, than to arrive at the real truth about a single question however simple that may be.

But one conclusion now emerges out of these considerations, viz. not, indeed, that Arithmetic and Geometry are the sole sciences to be studied, but only that in our search for the direct road towards truth we should busy ourselves with no object about which we cannot attain a certitude equal to that of the demonstrations of Arithmetic and Geometry.

Mathematical truths are dependent on the will of God.

3.34 You ask me by what kind of causality God established the eternal truths. I reply: by the same kind of causality as He created all things, that is to say, as their efficient and total cause. For it is certain that He is no less the author of creatures' essence than He is of their existence; and this essence is nothing other than the eternal truths. I do not conceive them as emanating from God like rays from the sun; but I know that God is the author of everything and that these truths are something and consequently that he is their author. I say that I know this, not that I can conceive it or comprehend it; because it is possible to know that God is infinite and all-powerful although our soul, being finite, cannot comprehend or conceive Him. In the same way we can touch a mountain with our hands but we cannot put our arms around it as we could put them around a tree or something else not too large for them. To comprehend something is to embrace it in one's thought; to know something it is sufficient to touch it with one's thought.

You ask also what necessitated God to create these truths; and I reply that just as He was free not to create the world, so He was no less free to make it untrue that all the lines drawn from the centre of a circle to its circumference are equal. And it is certain that these truths are no more necessarily attached to his essence than other creatures are. You ask what God did in order to produce them. I reply that from all eternity he willed and understood them to be, and by that very fact he created them. Or, if you reserve the word *created* for the existence of things, then he established them and made them. In God, willing, understanding, and creating are all the same thing without one being prior to the other even conceptually.

3.35 The mathematical truths which you call eternal have been laid down by God and depend on Him entirely no less than the rest of his creatures. Indeed to say that these truths are independent of God is to talk of Him as if He were Jupiter or Saturn and to subject Him to the Styx and the Fates. Please do not hesitate to assert and proclaim everywhere that it is God who has laid down these laws in nature just as a king lays down laws in his kingdom. There is no single one that we cannot understand if our mind turns to consider it. They are all *inborn in our minds* just as a king would imprint his laws on the hearts of all his subjects if he had enough power to do so. The greatness of God, on the other hand, is something which we cannot comprehend even though we know it. But the very fact that we judge it incomprehensible makes us esteem it more greatly; just as a king has more majesty when he is less familiarly known by his subjects, provided of course that they do not get the idea that they have no king—they must know him enough to be in no doubt about that.

3.36 I turn to the difficulty of conceiving how it was free and indifferent for God to make it not be true that the three angles of a triangle were equal to two right angles, or in general that contradictories could not be true together. It is easy to dispel this difficulty by considering that the power of God cannot have any limits, and that our mind is finite and so created as to be able to conceive as possible things which God has wished to be in fact possible, but not to be able to conceive as possible things which God could have made possible, but which he has in fact wished to make impossible. The first consideration shows us that God cannot have been determined to make it true that contradictories cannot be true together, and therefore that he could have done the opposite. The second consideration shows us that even if this be true, we should not try to comprehend it since our nature is incapable of doing so. And even if God has willed that some truths should be necessary, this does not mean that he willed them necessarily; for it is one thing to will that they be necessary, and quite another to will them necessarily, or to be necessitated to will them. I agree that there are contradictions which are so evident that we cannot put them before our minds without judging them entirely impossible, like the one which you suggest: *that God might have made creatures independent of him.* But if we would know the immensity of his power we should not put these thoughts before our minds, nor should we conceive any precedence or priority between his understanding and his will; for the idea which we have of God teaches us that there is in him only a single activity, entirely simple and entirely pure. This is well expressed by the words of St. Augustine: *They are so because you see them to be so;* because in God *seeing and willing* are one and the same thing.

4

HOBBES · Introduction

Thomas Hobbes (1588–1679), the first major English philosopher to respond to Descartes, did not begin to seriously study the problems of philosophy until after the age of 40. It is significant that his interest in *philosophy* arose out of his discovery of Euclid's *Elements of Geometry* during a visit to Paris around 1630. His close friend and biographer John Aubrey related that Hobbes came upon an open copy of the *Elements* by chance in a library, and read a random theorem.

> " 'By God,' said he, 'this is impossible!' So he reads the demonstration of it, which referred him back to such a proposition; which proposition he read. That referred him back to another, which he also read. *Et sic deinceps* that at last he was demonstratively convinced of that truth. This made him in love with geometry."

Like most philosophers of the time, Hobbes was a man of many interests, and during his long life he managed to involve himself in heated public disputes on almost every conceivable topic. He is most frequently remembered today as the author of the famous political treatise, the *Leviathan.* Hobbes also addressed himself to the basic questions of epistemology and ontology; he presented comprehensive theories of physics, psychology, and ethics; and he formulated a relatively sophisticated philosophy of language. There are numerous common threads running through all of Hobbes' theories and one's understanding and appreciation of any one aspect of his philosophy is inevitably enhanced by a careful study of all of his works.

One of the most significant of the threads which run through Hobbes' work is his philosophy of language, which is grounded on the assumption that a language is a system of signs. He asserted that the meaning of a statement is essentially equal to the sum of the meanings of the individual words comprising the statement, and that (with a few exceptions such as "nothing") the meanings of the individual words are the *ideas* signified or denoted by the words. Hobbes also argued that the sign-relations between words and ideas are *arbitrarily created by human beings* to serve specific ends, such as communication. Although this may seem to be a quite innocent statement to the modern reader, it was quite shocking and revolutionary in the 17th century, when it was generally believed (even by intellectuals) that all languages had been created by God and given to man as described in the Biblical story of the Tower of Babel. Moreover, it was also widely assumed that each child is born with an innate knowledge of his "native" language, since one must study to learn only "foreign" languages. In rejecting these doctrines, Hobbes was laying the groundwork for new work not only in linguistics, but also in philosophy of mathematics, in so far as mathematics was at this time just beginning to be considered the "*language* of nature."

Hobbes accepted Descartes' principle that ideas are the only immediate objects of knowledge, but he explicitly asserted that all ideas are images received from our senses or else memories of such images. This not only rules out the possibility of innate ideas but also eliminates certain kinds of processes of abstraction by means of which the mind would be able to acquire or create universal or general ideas such as that of a triangle-in-general. Further, it implies that to be meaningful, general words must signify or denote collections or groups of particular ideas, i.e. sense-images or after-images. Thus the term "triangle" does not signify a single idea of an 'ideal' triangle; rather it denotes the whole set of particular ideas of particular triangles which some individual has acquired via his senses. And the term "geometrical point" when defined as "dimensionless" is completely meaningless according to Hobbes, since it is theoretically impossible to perceive such a point with our senses.

In addition to denying the possibility of any universal or general idea, Hobbes also denied that men have certain *particular* ideas. His denial of the existence of a positive idea (particular or general) of infinity is of special importance for his philosophy of mathematics, but it also is relevant to his general theory of knowledge and his ontology. The reader will remember that Descartes grounded one of his proofs of the existence of God on the *fact* that he possessed a "true idea" of infinity which was something *more* than "the negation of the finite," and that he asserted that "in some way I have in me the notion of the infinite earlier than the finite" (see 3.12). Hobbes denied that any man could have such an idea, and maintained that "infinite" is meaningful only in so far as it is defined as "not finite." Further debate is certainly difficult when one party asserts that he has a certain idea, and the other denies that he has it. There is no way that either can 'see' the other's ideas. But the ramifications of this disagreement are many, particularly with regard to problems such as those of innate knowledge and the existence of God.

Hobbes agreed with the Cartesian principle that many of our ideas are caused by substances external to and independent of the knower, but he denied that there are any kinds of substance other than material or corporeal substance. Human minds are nothing more than certain arrangements of material substances. And mathematics is nothing more than the study of ideas of certain properties of physical objects. If his philosophy of mathematics had been accepted by mathematicians as the basis for their work, it would have had a truly revolutionary effect. Hobbes' solution to the problem of squaring the circle (that is, of using a straight-edge and compass to construct a square which has exactly the same area as a given circle) provides a good illustration of this. Since for Hobbes all mathematical problems are concerned with ideas of sensible physical objects, it follows that the equality or difference in area between two figures must be ascertainable by the senses. Thus, the problem of squaring the circle reduces simply to the problem of constructing a square whose area is not *perceptibly* different from the area of some circle, and it was exactly this which he proceeded to do. This suggested solution led to a heated debate between Hobbes and the mathematicians with each side charging the other with simple mathematical incompetence. Both sides failed to recognize that the disagreement was really over the philosophical questions of the nature of the objects of geometry. Few historians have done Hobbes full justice in their accounts of this debate; he is most often described as having been either incompetent or naive, rather than as having held a radical, but essentially coherent, philolsophical position on the nature of mathematics.

Hobbes influenced many later philosophers including Locke, Newton, Berkeley, Leibniz, and Mill, and some of his theories are still being discussed today. Much of the recent interest in materialist ontologies has resulted from developments in the areas of computer science and artificial intelligence, which again reflects the overlap between mathematics and philosophy. Some philosophers have also credited Hobbes with the first formulation of what is today known as the "conventionalist" theory of truth, but it is doubtful that such claims are supportable. Hobbes only recognized that the connection between a word and the idea it signifies is established arbitrarily and by convention. He believed that once the meanings of the words in a sentence have been established, then the truth or falsity of the sentence is necessarily fixed. Although there are several different theories which have been formulated in recent years all of which are called "conventionalistic," none of them places any significant emphasis on the relation between words and ideas; rather, they assume that this relation is arbitrary, and then go much further to discuss the arbitrariness of the truth and falsity of entire sentences. This is merely one example of the difficulty of relating theories from previous centuries to present-day discussions on the basis of mere terminological similarities.

Sources

All selections are from *The English Works of Thomas Hobbes*, edited by W. Molesworth, London, 1839–45. The numerals in parentheses refer to the volume and page, respectively.

4.1 Leviathan I, 1 (III, 1)
4.2 Leviathan, I, 1 (III, 1–2)
4.3 Leviathan, I, 2 (III, 4–5)
4.4 Leviathan, I, 2 (III, 5–6)
4.5 Leviathan, I, 3 (III, 17)
4.6 Elements of Philosophy (hereafter Elements), I, i, 3 (I, 4)
4.7 Elements I, i, 4, 5 (I, 5–6)
4.8 Six Lessons to the Professors of the Mathematics (hereafter 6 Lessons), (VII, 192–194)
4.9 Elements II, xii, 2, 5 (I, 139, 141)
4.10 Elements II, vii, 4 (I, 105)
4.11 6 Lessons (VII, 211)
4.12 6 Lessons (VII, 202)
4.13 Elements I, iii, 4 (I, 33–34)
4.14 Human Nature (IV, 27)
4.15 Elements I, v, 1 (I, 56–57)
4.16 Leviathan I, iii (III, 16)
4.17 Leviathan I, iv (III, 21–22)
4.18 Elements I, ii, 4 (I, 16)
4.19 Elements I, ii, 2 (I, 14–15)
4.20 Elements I, ii, 9 (I, 19–20)
4.21 Leviathan I, iv (III, 25–27)
4.22 Elements, I, ii, 6 (I, 17)
4.23 Leviathan I, iv (III, 27–28)
4.24 Leviathan I, iv (III, 23)
4.25 Elements I, iii, 7, 8 (I, 35–36)
4.26 Elements I, iii, 8, 9 (I, 36–37)
4.27 Elements I, iii, 10 (I, 37–38)
4.28 Philosophical Rudiments (II, 303)

HOBBES · Selections

The source of all knowledge—sense experience caused by material bodies.

4.1 Concerning the thoughts of man, I will consider them first singly, and afterwards in train, or dependence upon one another. Singly, they are every one a *representation* or *appearance,* or some quality, or other accident of a body without us, which is commonly called an *object.* Which object works on the eyes, ears, and other parts of a man's body; and by diversity of working, produces diversity of appearances.

The original of them all, is that which we call *sense,* for there is no conception in a man's mind, which has not at first, totally, or by parts, been begotten upon the organs of sense. The rest are derived from that original.

4.2 The cause of sense, is the external body, or object, which presses the organ proper to each sense, either immediately, as in the taste and touch; or mediately, as in seeing, hearing, and smelling; which pressure, by the mediation of nerves, and other strings and membranes of the body, continued inwards to the brain and heart, causes there a resistance, or counterpressure, or endeavor of the heart to deliver itself, which endeavor, because *outward,* seems to be some matter without. And this *seeming,* or *fancy,* is that which men call *sense.* . . .

4.3 When a body is once in motion, it moves, unless something else hinder it, eternally; and whatsoever hinders it, cannot in an instant, but in time, and by degrees, quite extinguish it; and as we see in the water, though the wind cease, the waves give not over rolling for a long time after: so also it happens in that motion, which is made in the internal parts of a man, then, when he sees, dreams, etc. For after the object is removed, or the eye shut, we still retain an image of the thing seen, though more obscure than when we see it. And this is it, the Latins call *imagination,* from the image made in seeing; and apply the same, though improperly, to all the other senses. But the Greeks call it *fancy;* which signifies *appearance,* and is as proper to one sense, as to another. *Imagination* therefore is nothing but *decaying sense;* and is found in men, and many other living creatures, as well sleeping, as waking.

4.4 This *decaying sense,* when we would express the thing itself, I mean *fancy* itself, we call *imagination,* as I said before: but when we would express the decay, and signify that the sense is fading, old, and past, it is called *memory.* So that imagination and memory, are but one thing, which for diverse considerations has diverse names.

Much memory, or memory of many things, is called *experience.* Again, imagination being only of those things which have been formerly perceived by sense, either all at once, or by parts at several times; the former, which is the imagining the whole object, as it was presented to the sense, is *simple* imagination, as when one imagines a man, or horse, which he has seen before. The other is *compounded;* as when from the sight of a man at one time, and of a horse at another, we conceive in our mind a Centaur.

The objects of mathematical knowledge—certain properties of material objects.

4.5 . . . Because whatsoever, as I said before, we conceive, has been perceived first by sense, either all at once, or by parts; a man can have no thought, representing any thing, not subject to sense. No man therefore can conceive any thing, but he must conceive it in some place; and indued with some determinate magnitude; and which may be divided into parts; nor that anything is all in this place, and all in another place at the same time; nor that two, or more things can be in one, and the same place at once: for none of these things ever have, nor can be incident to sense; but are absurd speeches, taken upon credit, without any signification at all, from deceived philosophers, and deceived, or deceiving schoolmen.

4.6 If therefore a man see something afar off and obscurely, although no appellation had yet been given to anything, he will, notwithstanding, have the same idea of that thing for which now, by imposing a name on it, we call it *body*. Again, when, by coming nearer, he sees the same thing thus and thus, now in one place and now in another, he will have a new idea thereof, namely, that for which we now call such a thing *animated*. Thirdly, when standing nearer, he perceives the figure, hears the voice, and sees other things which are signs of a rational mind, he has a third idea, though it have yet no appellation, namely, that for which we now call anything *rational*. Lastly, when, by looking fully and distinctly upon it, he conceives all that he has seen as one thing, the idea he has now is compounded of his former ideas, which are put together in the mind in the same order in which these three single names, *body, animated, rational,* are in speech compounded into this one name, *body-animated-rational,* or *man*. In like manner, of the several conceptions of *four sides, equality of sides, and right angles,* is compounded the conception of a *square*. For the mind may conceive a figure of four sides without any conception of their equality, and of that equality without conceiving a right angle; and may join together all these single conceptions into one conception or one idea of a square. And thus we see how the conceptions of the mind are compounded.

4.7 But *effects* and the *appearances* of things to sense, are faculties or powers of bodies, which make us distinguish them from one another; that is to say, conceived one body to be equal or unequal, like or unlike to another body; as in the example above, when by coming near enough to any body, we perceive the motion and going of the same, we distinguish it thereby from a tree, a column, and other fixed bodies; and so that motion or going is the *property* thereof, as being proper to living creatures, and a faculty by which they make us distinguish them from other bodies.

How the knowledge of any effect may be gotten from the knowledge of the generation thereof, may easily be understood by the example of a circle: for if there be set before us a plain figure, having, as near as may be, the figure of a circle, we cannot possibly perceive by sense whether it be a true circle or no; than which, nevertheless, nothing is more easy to be known to him that knows first the generation of the propounded figure. For let it be known that the figure was made by the circumduction of a body whereof one end remained unmoved, and we may reason thus; a body carried about, retaining always the same length, applies itself first to one radius, then to another, to a third, a fourth, and successively to all; and, therefore, the same length, from the same point, toucheth the circumference in every part thereof, which is as much as to say, as all the radii are equal. We know, therefore, that from such generation proceeds a figure, from whose one middle point all the extreme points are reached unto by equal radii. And in like manner, by knowing first what figure is set before us, we may come by ratiocination to some generation of the same, though perhaps not that by which it was made, yet that by which it might have been made; for he that knows that a circle has the property above declared, will easily know whether a body carried about, as is said, will generate a circle or no.

4.8 Quantity is that which is signified by what we answer to him that asketh, *how much* anything is? and thereby determines the magnitude thereof.

And because for the computing of the magnitudes of bodies, it is not necessary that the bodies themselves should be present, the ideas and memory of them supplying their presence, we reckon upon those imaginary bodies, which are the quantities themselves. . . .

So also is number quantity; but in no other sense than as a line is quantity divided into equal parts.

4.9 And *quantity* is determined two ways; one, by the sense, when some sensible object is set before it; as when a line, a superfices or solid, of a foot or cubit, marked out in some matter is objected to the eyes; which way of determining, is called *exposition,* and the quantity so known is called *exposed quantity;* the other by memory, that is, by comparison with some exposed quantity.

Number is exposed, either by the exposition of points, or of the names of number, *one, two, three, etc.;* and those points must not be contiguous, so as that they cannot be distinguished by notes, but they must be so placed that they may be *discerned* one from another; for, from this it is, that number is called *discrete quantity,* whereas all quantity, which is designed by motion, is called *continual quantity.*

4.10 The *extension* of a body is the same thing with the *magnitude* of it, or that which some call *real space.* But this magnitude does not depend upon our cogitation, as imaginary space doth; for this is an effect of our imagination, but magnitude is the cause of it; [imaginary space] is an accident of the mind, [magnitude is a property] of a body existing out of the mind.

4.11 Euclid in the definitions of a point, a line, and a superfices, did not intend that a point should be nothing, or a line without latitude, or a superfices without thickness; for if he did, his petitions are not only unreasonable to be granted, but also impossible to be performed. For lines are not drawn but by motion, and motion is of body only. And therefore his meaning was, that the quantity of a point, the breadth of a line, and the thickness of a superfices were not to be *considered,* that is to say, not to be reckoned in the demonstration of any theorems concerning the quantity of bodies . . . .

4.12 The second definition [in Euclidean geometry] is of a line . . . *"a line is a length which hath no breadth";* and if candidly interpreted, sound enough, though rigorously not so. . . . One path may be broader than another path, but not one mile longer than another mile; and it is not the path but the mile which is the way's length. If therefore a man have any ingenuity he will understand it thus, that a line is a body whose length is considered without its breadth, else we must say . . . untruly, that there be bodies which have length and yet no breadth. . . .

4.13 But the abuse proceeds from this, that some men seeing they can consider, that is bring into account the increasings and decreasings of quantity, heat, and other accidents, without considering their bodies or subjects (which they call *abstracting,* or making to exist apart by themselves) they speak of accidents, as if they might be separated from all bodies. And from hence proceed the gross errors of writers of metaphysics; for because they can consider thought without considerations of body, they infer there is no need of a thinking body; and because quantity may be considered without considering body, they think also that quantity may be without body, and body without quantity; and that a body has quantity by the addition of quantity to it.

The importance of language.

4.14 There be *two* kinds of knowledge, whereof the one is nothing else but *sense,* or knowledge *original,* and *remembrance* of the same; the other is called *science* or knowledge of the *truth of propositions,* and how things are called, and is derived from *understanding.* Both of these sorts are but *experience;* the former being the experience of the effects of things that work on us from *without,* and the latter experience men have from the proper use of *names* in language: and all experience being, as I have said, but remembrance, all knowledge is remembrance.

4.15 . . . Neither things, nor imaginations of things, can be said to be false, seeing they are truly what they are; nor do they, as signs, promise any thing which they do not perform; for they indeed do not promise at all, but we from them; nor do the clouds, but we, from seeing the clouds, say it shall rain.

4.16 There is no other act of man's mind, that I can remember, naturally planted in him, so as to need no other thing to the exercise of it, but to be born a man, and live with the use of his five senses. Those other faculties, of which I shall speak by and by, and which seem proper to man only, are acquired and increased by study and industry; and of most men learned by instruction, and discipline; and proceed all from the invention of words, and speech. For besides sense, and thoughts, and the train of thoughts, the mind of man has no other motion; though by the help of speech, and method, the same faculties may be improved to such a height, as to distinguish men from all other living creatures.

4.17 By this imposition of names, some of larger, some of stricter signification, we turn the reckoning of the consequences of things imagined in the mind, into a reckoning of the consequences of appellations. For example: a man that has no use of speech at all, such as is born and remains perfectly deaf and dumb, if he set before his eyes a triangle, and by it two right angles, such as are the corners of a square figure, he may, by meditation, compare and find, that the three angles of that triangle, are equal to those two right angles that stand by it. But if another triangle be shown him, different in shape from the former, he cannot know, without a new labor, whether the three angles of that also be equal to the same. But he that has the use of words, when he observes, that such equality was consequent, not to the length of the sides, nor to any other particular thing in his triangle; but only to this, that the sides were straight, and the angles three; and that that was all, for which he named it a triangle; will boldly conclude universally, that such equality of angles is in all triangles whatsoever; and register his invention in these general terms, *every triangle has its three angles equal to two right angles.* And thus the consequences found in one particular, comes to be registered and remembered, as a universal rule, and discharges our mental reckoning, of time and place, and delivers us from all labor of the mind, saving the first, and makes that which was found true *here,* and *now,* to be true in *all times* and *places.*

But the use of words in registering our thoughts is in nothing so evident as in numbering. A natural fool that could never learn by heart the order of numeral words, as *one, two,* and *three,* may observe every stroke of the clock, and nod to it, or say *one, one, one,* but can never know what hour it strikes. And it seems, there was a time when those names of numbers were not in use; and men were fain to apply their fingers of one or both hands, to those things they desired to keep account of; and that thence it proceeded, that now our numeral words are but ten, in any nation, and in some but five; and then they begin again. And he that can tell ten, if he recite them out of order, will lose himself, and not know when he has done. Much less will he be able to add, and subtract, and perform all other operations of arithmetic. So that without words there is no possibility of reckoning of numbers; much less of magnitudes, of swiftness, of force, and other things, the reckonings whereof are necessary to the being, or well-being of mankind.

The origin and nature of language—names arbitrarily chosen by man.

4.18 *A name is a word taken at pleasure to serve for a mark, which may raise in our mind a thought like to some thought we had before, and which being pronounced to others, may be to them a sign of what thought the speaker had, or had not before in his mind.* And it is for brevity's sake that I suppose the original of names to be arbitrary, judging it a thing that may be assumed as unquestionable. For considering that new names are daily made, and old ones laid aside; that diverse nations use different names, and how impossible it is either to observe similitude, or make any comparison betwixt a name and a thing, how can any man imagine that the names of things were imposed from their natures? For though some names of living creatures and other things, which our first parents used, were taught by God himself; yet they were by him arbitrarily imposed, and afterwards, both at the Tower of Babel, and since, in process of time, growing everywhere out of use, are quite forgotten, and in their room have succeeded others, invented and received by men at pleasure. Moreover, whatsoever the common use of words be, yet philosophers, who were to teach their knowledge to others, had always the liberty, and sometimes they both had and will have a necessity, of taking to themselves such names as they please for the signifying of their meaning, if they would have it understood. Nor had mathematicians need to ask leave of any but themselves to name the figures they invented, *parabolas, hyperboles, cissoeides, quadratices,* etc. or to call one magnitude A, another B.

4.19 . . . Those things we call signs are the antecedents of their consequents, and the consequents of their antecedents, as often as we observe them to go before or follow after in the same manner. For example, a thick cloud is a sign of rain to follow, and rain a sign that a cloud has gone before, for this reason only, that we seldom see clouds without the consequence of rain, nor rain at any time but when a cloud has gone before. And of signs, some are *natural,* whereof I have already given an example, others are *arbitrary,* namely, those we make choice of at our own pleasure, as a bush hung up, signifies that wine is to be sold there; a stone set in the ground signifies the bound of a field, and words so and so connected, signify the cogitations and motions of our minds.

The meanings of words are dependent on the kinds of things they name.

4.20 . . . Of names, some are *common* to many things, as a *man*, a *tree;* others *proper* to one thing, as *he that writ the Iliad, Homer, this man, that man.* And a common name, being the name of many things severally taken, but not collectively of all together (as man is not the name of all mankind, but of every one, as of Peter, John, and the rest severally) is therefore called an *universal name*; and therefore this word *universal* is never the name of any thing existent in nature, nor of any idea or phantasm formed in the mind, but always the name of some word or name; so that when *a living creature, a stone, a spirit,* or any other thing, is said to be *universal,* it is not to be understood, that any man, stone, etc. ever was or can be universal, but only that these words, *living creature, stone, etc.* are *universal names,* that is, names common to many things; and the conceptions answering them in our mind, are the images and phantasms of several living creatures, or other things. And therefore, for the understanding of the extent of an universal name, we need no other faculty but that of our imagination, by which we remember that such names bring sometimes one thing, sometimes another, into our mind. Also of common names, some are more, some less common. *More common,* is that which is the name of more things; *less common,* the name of fewer things; as *living creature* is more common than *man,* or *horse,* or *lion,* because it comprehends them all: and therefore a more common name, in respect of a less common, is called the *genus,* or a *general name;* and this in respect of that, the *species,* or a *special name.*

4.21 The Greeks have but one word for both *speech* and *reason;* not that they thought there was no speech without reason, but no reasoning without speech: and the act of reasoning they called *syllogism,* which signifies summing up of the consequences of one saying to another. And because the same things may enter into account for diverse accidents, their names are, to show that diversity, diversely wrested and diversified. This diversity of names may be reduced to four general heads.

First, a thing may enter into account for *matter* or *body;* as *living, sensible, rational, hot, cold, moved, quiet;* with all which names the word *matter,* or *body* is understood; all such being names of matter.

Secondly, it may enter into account, or be considered, for some accident or quality which we conceive to be in it; as for *being moved,* for *being so long,* for *being hot,* etc.; and then, of the name of the thing itself, by a little change or wresting, we make a name for that accident, which we consider: and for *living* put into the account *life;* for *moved, motion;* for *hot, heat;* for *long, length,* and the like: and all such names are the names of the accidents and properties by which one matter and body is distinguished from another. These are called *names abstract,* because severed, not from matter, but from the account of matter.

Thirdly, we bring into account the properties of our own bodies, whereby we make such distinction; as when anything is seen by us, we reckon not the thing itself, but the sight, the color, the idea of it in the fancy: and when anything is heard, we reckon it not, but the hearing, or sound only, which is our fancy or conception of it by the ear; and such are names of fancies.

Fourthly, we bring into account, consider, and give names, to *names* themselves, and to *speeches:* for, *general, universal, special, equivocal,* are names of names. And *affirmation, interrogation, commandment, narration, syllogism, sermon, oration,* and many other such, are names of speeches. And this is all the variety of names *positive*; which are put to mark somewhat which is in nature, or may be feigned by the mind of man, as bodies that are, or may be conceived to be; or of bodies, the properties that are, or may be feigned to be; or words and speech.

4.22 Nor, indeed, is it at all necessary that every name should be the name of something. For as these, a *man,* a *tree,* a *stone,* are the names of the things themselves, so the images of a man, of a tree, and of a stone, which are represented to men sleeping, have their names also, though they be not things, but only fictions and phantasms of things. For we can remember these; and, therefore, it is no less necessary that they have names to mark and signify them, than the things themselves. Also this word *future* is a name, but no future thing has yet any being, nor do we know whether that which we call future, shall ever have a being or no. Nevertheless, seeing we use in our mind to knit together things past with those that are present, the name *future* serves to signify such knitting together. Moreover, that which neither is, nor has been, nor ever shall, or ever can be, has a name, namely, *that which neither is nor has been,* etc.; or more briefly this, *impossible.* To conclude; this word *nothing* is a name, which yet cannot be the name of any thing: for when, for example, we substract 2 and 3 from 5, and so nothing remaining, we would call that subtraction to mind, this speech *nothing remains,* and in it the word *nothing* is not unuseful. And for the same reason we say truly, *less than nothing* remains, when we subtract more from less; for the mind feigns such remains as these for doctrine's sake, and desires, as often as is necessary, to call the same to memory. But seeing every name has some relation to that which is named, though that which we name be not always a thing that has a being in nature, yet it is lawful for doctrine's sake to apply the word *thing* to whatsoever we name: as if it were all one whether that thing be truly existent, or be only feigned.

4.23 All other names, are but insignificant sounds; and those of two sorts. One when they are new, and yet their meaning not explained by definition; whereof there have been abundance coined by schoolmen, and puzzled philosophers.

Another, when men make a name of two names, whose significations are contradictory and inconsistent; as this name, an *incorporeal body,* or, which is all one, an *incorporeal substance,* and a great number more. For whensoever any affirmation is false, the two names of which it is composed, put together and made one, signify nothing at all. For example, if it be a false affirmation to say *a quadrangle is round,* the word *round quadrangle* signifies nothing; but is a mere sound. . . .

When a man, upon the hearing of any speech, has those thoughts which the words of that speech and their connection were ordained and constituted to signify, then he is said to understand it; *understanding* being nothing else but conception caused by speech.

Language and knowledge—truth and falsity are attributable only to statements.

4.24 When two names are joined together into a consequence, or affirmation, as thus, *A man is a living creature;* or thus, *if he be a man, he is a living creature;* if the latter name, *living creature,* signify all that the former name *man* signifies, then the affirmation, or consequence, is *true;* otherwise *false.* For *true* and *false* are attributes of speech, not of things. And where speech is not, there is neither *truth* nor *falsehood; error* there may be, as when we expect that which shall not be, or suspect what has not been; but in neither case can a man be charged with untruth.

4.25 A *true* proposition is that, whose predicate contains, or comprehends its subject, or whose predicate is the name of every thing, of which the subject is the name; as *man is a living creature* is therefore a true proposition, because whatsoever is called *man,* the same is also called *living creature;* and *some man is sick,* is true, because *sick* is the name of *some man.* That which is not true, or that whose predicate does not contain its subject, is called a *false* proposition, as *man is a stone.*

Now these words *true, truth,* and *true proposition,* are equivalent to one another; for truth consists in speech, and not in the things spoken of; and though *true* be sometimes opposed to *apparent* or *feigned,* yet it is always to be referred to the truth of proposition; for the image of a man in a glass, or a ghost, is therefore denied to be a very man, because this proposition, *a ghost is a man,* is not true; for it cannot be denied but that a ghost is a very ghost. And therefore truth or verity is not any affection of the thing, but of the proposition concerning it. As for that which the writers of metaphysics say, that *a thing, one thing,* and *a very thing,* are equivalent to one another, it is but trifling and childish; for who does not know, that *a man, one man,* and *a very man,* signify the same.

And from hence it is evident, that truth and falsity have no place but amongst such living creatures as use speech. For though some brute creatures, looking upon the image of a man in a glass, may be affected with it, as if it were the man himself, and for this reason fear it or fawn upon it in vain; yet they do not apprehend it as true or false, but only as like; and in this they are not deceived. Wherefore, as men owe all their true ratiocination to the right understanding of speech; so also they owe their errors to the misunderstanding of the same; and as all the ornaments of philosophy proceed only from man, so from man also is derived the ugly absurdity of false opinions.

Primary truths (axioms) are 'arbitrary' and 'necessary' because they are definitions.

4.26 The first truths were arbitrarily made by those that first of all imposed names upon things, or received them from the imposition of others. For it is true (for example) that *man is a living creature,* but it is for this reason, that it pleased men to impose both those names on the same thing.

Propositions are distinguished into *primary* and *not primary. Primary* is that wherein the subject is explicated by a predicate of many names, as *man is a body, animated, rational;* for that which is comprehended in the name *man,* is more largely expressed in the names *body, animated,* and *rational,* joined together; and it is called *primary,* because it is first in ratiocination; for nothing can be proved, without understanding first the name of the thing in question. Now *primary* propositions are nothing but definitions, or parts of definitions, and these only are the principles of demonstration, being truths constituted arbitrarily by the inventors of speech, and therefore not to be demonstrated.

4.27 . . . Propositions are distinguished into *necessary,* that is, necessarily true; and true, but not necessarily, which they call *contingent.* A *necessary* proposition is when nothing can at any time be conceived or feigned, whereof the subject is the name, but the predicate also is the name of the same thing; as *man is a living creature* is a necessary proposition, because at what time soever we suppose the name *man* agrees with any thing, at that time the name *living-creature* also agrees with the same. But a *contingent* proposition is that, which at one time may be true, at another time false; as *every crow is black;* which may perhaps be true now, but false hereafter. Again, in every *necessary* proposition, the predicate is either equivalent to the subject, as in this, *man is a rational living creature;* or part of an equivalent name, as in this, *man is a living creature,* for the name *rational-living-creature,* or *man,* is compounded of these two, *rational* and *living-creature.* But in a *contingent* proposition this cannot be; for though this were true, *every man is a liar,* yet because the word *liar* is no part of a compounded name equivalent to the name *man,* that proposition is not to be called *necessary,* but *contingent,* though it should happen to be true always. And therefore those propositions only are *necessary,* which are of sempiternal truth, that is, true at all times. From hence also it is manifest, that truth adheres not to things, but to speech only, for some truths are eternal; for it will be eternally true, *if man, then living-creature*; but that any *man,* or *living-creature,* should exist eternally, is not necessary.

4.28 If it be propounded that *two and three make five;* and by calling to mind, that the order of numeral words is so appointed by the common consent of them who are of the same language with us, (as it were, by a certain contract necessary for human society), that *five* shall be the name of so many unities as are contained in two and three taken together, a man assent that this is therefore true, because two and three together are the same with five: this assent shall be called knowledge. And to know this truth is nothing else, but to acknowledge that it is made by ourselves. For by whose will and rules of speaking the number || is called *two,* ||| is called *three,* and ||||| is called *five;* by their will also it comes to pass that this proposition is true, *two and three taken together make five.*

5

LOCKE · Introduction

John Locke (1632–1704) was educated at Oxford, where he developed a strong antipathy toward the Scholastic Aristotelianism which flavored most of the teaching there at that time. His intellectual interests were primarily in philosophy (his inspiration being Descartes rather than the Scholastics), chemistry and physics (he was a friend of Robert Boyle and other "natural philosophers" of the time), medicine (he was trained as a physician but never entered general practice), and political theory. There is little reason to believe that Locke was particularly interested in mathematics or even that he was very well trained in it, and apparently he made no contributions to mathematics in the way of new concepts, proofs, or methods. Nonetheless, he devoted a significant amount of attention to the nature of mathematical knowledge in his major philosophical work, the *Essay Concerning Human Understanding*. Despite his own lack of experience in the field of mathematics, his discussions of the philosophy of mathematics are as sophisticated and intelligent as those of outstanding mathematicians such as Descartes and Leibniz.

Although Locke is most frequently described as an empiricist, the major influence on his thinking would appear to have been Descartes. Over all, it is probably easier to list the points on which Descartes and Locke *differ* than those on which they agree. The most fundamental disagreement between the two was over innate ideas. Locke emphatically denied the existence of any such ideas, and he devoted the entirety of one of the four parts of his *Essay* to examining every aspect of this issue, and to refuting numerous arguments given by the proponents of the theory of innate knowledge.

Despite this disagreement with Descartes over the origins of ideas, Locke tended to follow his broader interpretation of the basic meaning of the term "idea," and he rejected Hobbes' assertion that ideas are only images or after-images from *sense* experience. In particular, Locke recognized a second kind of experience, namely *introspection,* which provides what he referred to as "ideas of reflection." He generally accepted the fundamental Cartesian distinction between mental and physical substance, grounding it in the two sources of ideas of reflection and sensation. He differed with Descartes over certain secondary issues such as the relation between physical substance and space.

But his adherence to the Cartesian concepts of substance was generally so close that this became one of the central issues on which he was criticized by later empiricists.

Like Hobbes, Locke also accepted the Cartesian assumptions that the meaning of any word is the *idea* that it denotes or signifies, and that the meaning of a sentence is essentially the sum of the meanings of the individual terms of which it is composed. In contrast to Hobbes, he asserted that general words such as "man" signify a single idea—an abstract general idea. Locke asserted that these ideas are neither innate nor objects of immediate sense perception; rather, they must be produced by a process of abstracting from the particular concrete ideas derived from sensation or reflection. His accounts of the nature of abstract general ideas (see 5.17) became an attractive target for later critics, mainly on the grounds of vagueness and ambiguity.

Although much of his philosophy was an elaboration on or variation of previously suggested themes, Locke's philosophy of mathematics was strikingly original. Whereas his predecessors had generally maintained that mathematical knowledge was either innate (e.g., Plato and Descartes) or else derived by induction and/or abstraction from experience (e.g., Aristotle and Hobbes), Locke argued that mathematics is concerned with ideas *constructed* by the mind (see 5.23—5.24). He adhered to his fundamental assumption that all ideas must be ultimately traceable to experience, but he also asserted that it is irrelevant whether or not there are 'real' entities existing in the world which correspond to abstract general mathematical ideas. He reconciled the potential inconsistency between these views with his theory of complex modes (see 5.15). Only the simple ideas of unity and of the geometrical point are derived (by abstraction) from experience, and the complex mathematical ideas are constructed out of them by repeating, combining, and arranging them in various ways. Thus it is possible to construct an idea of a *perfect* circle, geometrically defined as the locus of points equidistant from a fixed point, out of the simple idea of a geometrical point abstracted from sense experience. Since the objects of mathematical knowledge are abstract general ideas created by the mind itself, there is no need to look for anything in experience corresponding to these ideas (except the sensations from which we originally abstracted the simple ideas of unity and geometrical point).

Locke believed that only his theory could account for what he considered to be the basic features of mathematical knowledge—universality, absolute certainty, and significance—without appealing to innate ideas or Platonic Forms. If the objects of mathematical knowledge are abstract *general* ideas, then this knowledge must be universally valid. It is absolutely certain because it is concerned only with the ideas which the mind directly perceives—ideas which do not necessarily refer to or represent anything beyond themselves, as ideas of physical entities do. According to Locke's distinction between real and nominal essences (see 5.30), it is only in mathematics and ethics that the nominal essence of a thing is identical to its real essence. And mathematical statements are more than mere definitions or identity-statements; the concept of the predicate term is not "contained in" the subject-concept (see 5.28).

It was almost a century later that Kant classified such universal, necessary, and significant knowledge as "synthetic a priori." Either Kant was not aware of Locke's work on this topic, or else he did not understand it, for he claimed credit for being the first to have had this insight. And it was in fact Kant, rather than Locke, who influenced later work on "constructivist" philosophies of mathematics. Even today most discussions of synthetic a priori knowledge fail to give any recognition to Locke's anticipation of Kant's work. Locke's greatest direct influence was on his contemporaries and immediate successors, including Leibniz and Berkeley; but none of this influence was in the area of the philosophy of mathematics. Despite his failure to influence later work on philosophy of mathematics, or perhaps because of it, Locke deserves recognition for having been well ahead of his time, rather than to be passed over, as he has been by many historians, as a relatively unimaginative moderate or conservative thinker.

Sources

All selections are from John Locke's *Essay Concerning Human Understanding.* The numerals refer to the book, chapter, and section, respectively.

5.1 IV, i, 1
5.2 IV, iv, 3
5.3 I, i, 8
5.4 I, ii, 1
5.5 I, ii, 15–16
5.6 I, ii, 18
5.7 II, i, 2–4
5.8 II, xi, 17
5.9 II, i, 23
5.10 II, viii, 7–10
5.11 II, viii, 15
5.12 II, ii, 1–2
5.13 II, iii, 1
5.14 II, xii, 1
5.15 II, xii, 3–5
5.16 II, xi, 9
5.17 IV, vii, 9
5.18 III, iii, 6
5.19 II, xxix, 4
5.20 III, ii, 2
5.21 IV, v, 5
5.22 IV, iii, 30
5.23 IV, xvi, 1–8
5.24 IV, iv, 6
5.25 IV, i, 2–5
5.26 IV, vii, 2–3
5.27 IV, vii, 6
5.28 IV, viii, 7–8
5.29 III, iii, 15
5.30 III, iii, 18
5.31 IV, ii, 1–2
5.32 IV, ii, 4
5.33 IV, xvii, 4

LOCKE · Selections

The object of all knowledge—ideas.

5.1 Since the mind in all its thoughts and reasonings hath no other immediate object but its own ideas, which it alone does or can contemplate, it is evident that our knowledge is only conversant about them.

5.2 It is evident the mind knows not things immediately, but only by the intervention of the ideas it has of them. Our knowledge, therefore, is real only so far as there is a conformity between our ideas and the reality of things.

5.3 . . . Before I proceed on to what I have thought on this subject, I must here in the entrance beg pardon of my reader for the frequent use of the word *'idea'* which he will find in the following treatise. It being that term which, I think, serves best to stand for whatsoever is the *object* of the understanding when a man thinks, I have used it to express whatever is meant by *phantasm, notion, species,* or *whatever it is which the mind can be employed about in thinking*; and I could not avoid frequently using it.

I presume it will be easily granted me that there are such *ideas* in men's minds: everyone is conscious of them in himself; and men's words and actions will satisfy him that they are in others.

Our first inquiry then shall be how they come into the mind.

The source of all ideas—experience.

5.4 It is an established opinion amongst some men that there are in the understanding certain *innate principles,* some primary notions, characters, as it were stamped upon the mind of man, which the soul receives in its very first being, and brings into the world with it. It would be sufficient to convince unprejudiced readers of the falseness of this supposition, if I should only show (as I hope I shall in the following parts of this Discourse) how men, barely by the use of their natural faculties, may attain to all the knowledge they have, without the help of any innate impressions, and may arrive at certainty, without any such original notions or principles.

5.5 The senses at first let in *particular* ideas, and furnish the yet empty cabinet, and, the mind by degrees growing familiar with some of them, they are lodged in the memory, and names got to them. Afterwards, the mind proceeding further abstracts them, and by degrees learns the use of general names. In this manner the mind comes to be furnished with ideas and language, the materials about which to exercise its discursive faculty. And the use of reason becomes daily more visible, as these materials that give it employment increase. But, though the having of general ideas and the use of general words and reason usually grow together, yet I see not how this any way proves them innate. The knowledge of some truths, I confess, is very early in the mind; but in a way that shows them not to be innate. For, if we will observe, we shall find it still to be about ideas, not innate, but acquired; it being about those first which are imprinted by external things, with which infants have earliest to do, which make the most frequent impressions on their senses. In ideas thus got the mind discovers that some agree and others differ, probably as soon as it has any use of memory, as soon as it is able to retain and receive distinct ideas. But whether it be then or no, this is certain, it does so long before it has the use of words, or comes to that which we commonly call "the use of reason." For a child knows as certainly before it can speak the difference between the ideas of sweet and bitter (i.e. that sweet is not bitter), as it knows afterwards (when it comes to speak) that wormwood and sugarplums are not the same thing.

A child knows not that three and four are equal to seven, till he comes to be able to count to seven, and has got the name and idea of equality; and then, upon explaining those words, he presently assents to, or rather perceives the truth of, that proposition. But neither does he then readily assent because it is an innate truth, nor was his assent wanting till then because he wanted the use of reason; but the truth of it appears to him as soon as he has settled in his mind the clear and distinct ideas that these names stand for. And then he knows the truth of that proposition upon the same grounds and by the same means, that he knew before that a rod and cherry are not the same thing; and upon the same grounds also that he may come to know afterwards "That it is impossible for the same thing to be and not to be." So that the later it is before anyone comes to have those general ideas about which those maxims are, or to know the signification of those general terms that stand for them, or to put together in his mind the ideas they stand for, the later also will it be before he comes to assent to those maxims;—whose terms, with the ideas they stand for, being no more innate than those of a cat or a weasel, he must stay till time and observation have acquainted him with them; and then he will be in a capacity to know the truth of these maxims, upon the first occasion that shall make him put together those ideas in his mind, and observe whether they agree or disagree, according as is expressed in those propositions.

5.6 . . . Since no proposition can be innate unless the *ideas* about which it is be innate, this will be to suppose all our ideas of colours, sounds, tastes, figure, etc., innate, than which there cannot be anything more opposite to reason and experience. Universal and ready assent upon hearing and understanding the terms is, I grant, a mark of self-evidence; but self-evidence, depending not on innate impressions but on something else, belongs to several propositions which nobody was yet so extravagant as to pretend to be innate.

5.7 Let us then suppose the mind to be, as we say, white paper, void of all characters, without any ideas; how comes it to be furnished? Whence comes it by that vast store which the busy and boundless fancy of man has painted on it with an almost endless variety? Whence has it all the materials of reason and knowledge? To this I answer, in one word, from *experience.* In that all our knowledge is founded, and from that it ultimately derives itself. Our observation employed either about external sensible objects, or about the internal operations of our minds perceived and reflected on by ourselves, is that which supplies our understandings with all the materials of thinking. These two are the fountains of knowledge, from whence all the ideas we have, or can naturally have, do spring.

First, our senses, conversant about particular sensible objects, do convey into the mind several distinct perceptions of things, according to those various ways wherein those objects do affect them. And thus we come by those *ideas* we have of *yellow, white, heat, cold, soft, hard, bitter, sweet,* and all those which we call sensible qualities; which when I say the senses convey into the mind, I mean, they from external objects convey into the mind what produces there those perceptions. This great source of most of the ideas we have, depending wholly upon our senses, and derived by them to the understanding, I call SENSATION.

Secondly, the other fountain from which experience furnisheth the understanding with ideas is the perception of the operations of our own mind within us, as it is employed about the ideas it has got; which operations, when the soul comes to reflect on and consider, do furnish the understanding with another set of ideas, which could not be had from things without. And such are *perception, thinking, doubting, believing, reasoning, knowing, willing,* and all the different actings of our own minds; which we being conscious of, and observing in ourselves, do from these receive into our understandings as distinct ideas as we do from bodies affecting our senses. This source of ideas every man has wholly in himself; and though it be not sense, as having nothing to do with external objects, yet it is very like it, and might properly enough be called internal sense. But as I call the other sensation, so I call this REFLECTION, the ideas it affords being such only as the mind gets by reflecting on its own operations within itself. By 'reflection' then, in the following part of this discourse, I would be understood to mean that notice which the mind takes of its own operations, and the manner of them, by reason whereof there come to be ideas of these operations in the understanding. These two, I say, viz. external material things, as the objects of SENSATION, and the operations of our own minds within, as the objects of REFLECTION, are to me the only originals from whence all our ideas take their beginnings.

5.8 I pretend not to teach, but to inquire; and therefore cannot but confess here again that external and internal sensation are the only passages that I can find of knowledge to the understanding. These alone, as far as I can discover, are the windows by which light is let into this *dark room.* For, methinks, the understanding is not much unlike a closet wholly shut from light, with only some little opening left, to let in external visible resemblances, or ideas of things without; would the pictures coming into such a dark room but stay there, and lie so orderly as to be found upon occasion, it would very much resemble the understanding of a man, in reference to all objects of sight, and the ideas of them.

5.9 If it shall be demanded then, when a man begins to have any ideas, I think the true answer is when he first has any sensation. For, since there appear not to be any ideas in the mind before the senses have conveyed any in, I conceive that ideas in the understanding are coeval with *sensation; which is such an impression or motion made in some part of the body, as produces some perception in the understanding.* It is about these impressions made on our senses by outward objects that the mind seems first to employ itself in such operations as we call perception, remembering, consideration, reasoning, etc.

The cause of all sensations—qualities of bodies.

5.10 To discover the nature of our ideas the better, and to discourse of them intelligibly, it will be convenient to distinguish them *as they are ideas or perceptions in our minds;* and *as they are modifications of matter in the bodies that cause such perceptions in us;* that so we may not think (as perhaps usually is done) that they are exactly the images and resemblances of something inherent in the subject, most of those of sensation being in the mind no more the likeness of something existing without us, than the names that stand for them are the likeness of our ideas, which yet upon hearing they are apt to excite in us.

Whatsoever the mind perceives in itself, or is the immediate object of perception, thought, or understanding, that I call *idea;* and the power to produce any idea in our mind I call *quality* of the subject wherein that power is. Thus a snowball having the power to produce in us the ideas of white, cold, and round, the powers to produce those ideas in us, as they are in the snowball, I call qualities; and as they are sensations or perceptions in our understandings, I call them ideas; which ideas, if I speak of sometimes as in the things themselves, I would be understood to mean those qualities in the objects which produce them in us.

Qualities thus considered in bodies are,

First, such as are utterly inseparable from the body, in what estate soever it be; and such as in all the alterations and changes it suffers, all the force can be used upon it, it constantly keeps; and such as sense constantly finds in every particle of matter which has bulk enough to be perceived; and the mind finds inseparable from every particle of matter, though less than to make itself singly be perceived by our senses: v.g. take a grain of wheat, divide it into two parts: each part has still solidity, extension, figure, and mobility; divide it again, and it retains still the same qualities; and so divide it on, till the parts become insensible; they must retain still each of them all those qualities. For division (which is all that a mill, or pestle, or any other body, does upon another, in reducing it to insensible parts) can never take away either solidity, extension, figure, or mobility from any body, but only makes two or more distinct separate masses of matter, of that which was but one before; all which distinct masses, reckoned as so many distinct bodies, after division, make a certain number. These I call *original* or *primary qualities* of body, which I think we may observe to produce simple ideas in us, viz. solidity, extension, figure, motion or rest, and number.

Secondly, such qualities which in truth are nothing in the objects themselves but powers to produce various sensations in us by their primary qualities, i.e. by the bulk, figure, texture, and motion of their insensible parts, as colours, sounds, tastes, etc. These I call *secondary qualities.* To these might be added a *third* sort, which are allowed to be barely powers, though they are as much real qualities in the subject as those which I, to comply with the common way of speaking, call qualities, but for distinction, secondary qualities. For the power in fire to produce a new colour, or consistency, in wax or clay, by its primary qualities, is as much a quality in fire, as the power it has to produce in me a new idea or sensation of warmth or burning, which I felt not before, by the same primary qualities, viz. the bulk, texture, and motion of its insensible parts.

5.11 The ideas of primary qualities of bodies are resemblances of them, and their patterns do really exist in the bodies themselves, but the ideas produced in us by these secondary qualities have no resemblance of them at all. There is nothing like our ideas existing in the bodies themselves. They are, in the bodies we denominate from them, only a power to produce those sensations in us; and what is sweet, blue, or warm in idea is but the certain bulk, figure, and motion of the insensible parts, in the bodies themselves, which we call so.

The different kinds of ideas.

5.12 The better to understand the nature, manner, and extent of our knowledge, one thing is carefully to be observed concerning the ideas we have; and that is, that some of them are *simple* and some *complex.*

Though the qualities that affect our senses are, in the things themselves, so united and blended that there is no separation, no distance between them, yet it is plain, the ideas they produce in the mind enter by the senses simple and unmixed. For, though the sight and touch often take in from the same object, at the same time, different ideas, as a man sees at once motion and colour, the hand feels softness and warmth in the same piece of wax; yet the simple ideas thus united in the same subject are as perfectly distinct as those that come in by different senses. The coldness and hardness which a man feels in a piece of ice being as distinct ideas in the mind as the smell and whiteness of a lily, or as the taste of sugar, and smell of a rose. And there is nothing can be plainer to a man than the clear and distinct perception he has of those simple ideas; which, being each in itself uncompounded, contains in it nothing but *one uniform appearance, or conception in the mind,* and is not distinguishable into different ideas.

These simple ideas, the materials of all our knowledge, are suggested and furnished to the mind only by those two ways above mentioned, viz. sensation and reflection. When the understanding is once stored with these simple ideas, it has the power to repeat, compare, and unite them, even to an almost infinite variety, and so can make at pleasure new complex ideas. But it is not in the power of the most exalted wit, or enlarged understanding, by any quickness or variety of thought, to *invent* or *frame* one new simple idea in the mind, not taken in by the ways before mentioned: nor can any force of the understanding *destroy* those that are there.

5.13 The better to conceive the ideas we receive from sensation, it may not be amiss for us to consider them, in reference to the different ways whereby they make their approaches to our minds, and make themselves perceivable by us.

First, then, There are some which come into our minds *by one sense only.*

Secondly, There are others that convey themselves into the mind *by more senses than one.*

Thirdly, Others that are had from *reflection only.*

Fourthly, There are some that make themselves way, and are suggested to the mind *by all the ways of sensation and reflection.*

5.14 We have hitherto considered those ideas in the reception whereof the mind is only passive, which are those simple ones received from sensation and reflection before mentioned, whereof the mind cannot make one to itself, nor have any idea which does not wholly consist of them. But as the mind is wholly passive in the reception of all its simple ideas, so it exerts several acts of its own, whereby out of its simple ideas, as the materials and foundations of the rest, the others are framed. The acts of the mind, wherein it exerts its power over its simple ideas, are chiefly these three: (1) Combining several simple ideas into one compound one; and thus all *complex ideas* are made. (2) The second is bringing two ideas, whether simple or complex, together, and setting them by one another, so as to take a view of them at once, without uniting them into one; by which way it gets all its *ideas of relations.* (3) The third is separating them from all other ideas that accompany them in their real existence; this is called *abstraction:* and thus all its *general ideas* are made.

5.15 *Complex ideas,* however compounded and decompounded, though their number be infinite, and the variety endless, wherewith they fill and entertain the thoughts of men, yet I think they may be all reduced under these three heads:

1. *Modes.*
2. *Substances.*
3. *Relations.*

First, *Modes* I call such complex ideas which, however compounded, contain not in them the supposition of subsisting by themselves, but are considered as dependences on, or affections of substances; such are the ideas signified by the words *triangle, gratitude, murder,* etc. And if in this I use the word *mode* in somewhat a different sense from its ordinary signification, I beg pardon; it being unavoidable in discourses differing from the ordinary received notions either to make new words, or to use old words in somewhat a new signification; the latter whereof, in our present case, is perhaps the more tolerable of the two.

Of these *modes* there are two sorts which deserve distinct consideration: First, there are some which are only variations, or different combinations, of the same simple idea, without the mixture of any other, as a dozen, or score; which are nothing but the ideas of so many distinct units added together, and these I call *simple modes* as being contained within the bounds of one simple idea. Secondly, there are others compounded of simple ideas of several kinds, put together to make one complex one: v.g. beauty, consisting of a certain composition of colour and figure, causing delight in the beholder; theft, which being the concealed change of the possession of anything, without the consent of the proprietor, contains, as is visible, a combination of several ideas of several kinds; and these I call *mixed modes.*

5.16 The use of words then being to stand as outward marks of our internal ideas, and those ideas being taken from particular things, if every particular idea that we take in should have a distinct name, names must be endless. To prevent this, the mind makes the particular ideas received from particular objects to become general; which is done by considering them as they are in the mind such appearances, separate from all other existences, and the circumstances of real existence, as time, place, or any other concomitant ideas. This is called ABSTRACTION, whereby ideas taken from particular beings become general representatives of all of the same kind; and their names general names, applicable to whatever exists conformable to such abstract ideas. Such precise, naked appearances in the mind, without considering how, whence, or with what others they came there, the understanding lays up (with names commonly annexed to them) as the standards to rank real existences into sorts, as they agree with these patterns, and to denominate them accordingly. Thus the same colour being observed to-day in chalk or snow, which the mind yesterday received from milk, it considers that appearance alone, makes it a representative of all of that kind; and having given it the name *whiteness,* it by that sound signifies the same quality wheresoever to be imagined or met with; and thus universals, whether ideas or terms, are made.

5.17 . . . When we nicely reflect upon them, we shall find that *general ideas* are fictions and contrivances of the mind, that carry difficulty with them, and do not so easily offer themselves as we are apt to imagine. For example, does it not require some pains and skill to form the general idea of a triangle (which is yet none of the most abstract, comprehensive, and difficult), for it must be neither oblique nor rectangle, neither equilateral, equicrural, nor scalenon; but all and none of these at once. In effect, it is something imperfect, that cannot exist; an idea wherein some parts of several different and inconsistent ideas are put together.

5.18 Words become general by being made the signs of general ideas; and ideas become general by separating from them the circumstances of time and place, and any other ideas that may determine them to this or that particular existence. By this way of abstraction they are made capable of representing more individuals than one; each of which, having in it a conformity to that abstract idea, is (as we call it) of that sort.

5.19 As a clear idea is that whereof the mind has such a full and evident perception as it does receive from an outward object operating duly on a well-disposed organ, so a *distinct* idea is that wherein the mind perceives a difference from all other; and a *confused* idea is such an one as is not sufficiently distinguishable from another, from which it ought to be different.

The objects of mathematical knowledge and the meanings of mathematical symbols—ideas of number.

5.20 Words, in their primary or immediate signification, stand for nothing but *the ideas in the mind of him that uses them,* how imperfectly soever or carelessly those ideas are collected from the things which they are supposed to represent. When a man speaks to another, it is that he may be understood; and the end of speech is that those sounds, as marks, may make known his ideas to the hearer. That then which words are the marks of are the ideas of the speaker; nor can anyone apply them as marks, immediately, to anything else but the ideas that he himself hath; for this would be to make them signs of his own conceptions, and yet apply them to other ideas; which would be to make them signs and not signs of his ideas at the same time, and so in effect to have no signification at all. Words being voluntary signs, they cannot be voluntary signs imposed by him on things he knows not. That would be to make them signs of nothing, sounds without signification.

5.21 . . . We must, I say, observe two sorts of propositions that we are capable of making:—

First, *mental,* wherein the ideas in our understandings are without the use of words put together, or separated, by the mind perceiving or judging of their agreement or disagreement.

Secondly, *verbal* propositions, which are words, the signs of our ideas, put together or separated in affirmative or negative sentences. By which way of affirming or denying, these signs made by sounds are, as it were, put together or separated one from another. So that proposition consists in joining or separating signs; and truth consists in the putting together or separating these signs, according as the things which they stand for agree or disagree.

5.22 Mathematicians abstracting their thoughts from names, and accustoming themselves to set before their minds the ideas themselves that they would consider, and not sounds instead of them, have avoided thereby a great part of that perplexity, puddering, and confusion, which has so much hindered men's progress in other parts of knowledge.

5.23 Amongst all the ideas we have, as there is none suggested to the mind by more ways, so there is none more simple, than that of *unity,* or one; it has no shadow of variety or composition in it; every object our senses are employed about, every idea in our understandings, every thought of our minds, brings this idea along with it. And therefore it is the most intimate to our thoughts, as well as it is, in its agreement to all other things, the most universal idea we have. For number applies itself to men, angels, actions, thoughts, everything that either doth exist, or can be imagined.

By repeating this idea in our minds, and adding the repetitions together, we come by the *complex* ideas of the *modes* of it. Thus, by adding one to one, we have the complex idea of a couple; by putting twelve units together, we have the complex idea of a dozen; and of a score, or a million, or any other number.

The *simple modes* of *number* are of all other the most distinct; every the least variation, which is an unit, making each combination as clearly different from that which approacheth nearest to it, as the most remote; two being as distinct from one, as two hundred; and the idea of two as distinct from the idea of three, as the magnitude of the whole earth is from that of a mite. This is not so in other simple modes, in which it is not so easy, nor perhaps possible for us to distinguish betwixt two approaching ideas, which yet are really different. For who will undertake to find a difference between the white of this paper and that of the next degree to it, or can form distinct ideas of every the least excess in extension?

The clearness and distinctness of each mode of number from all others, even those that approach nearest, makes me apt to think that demonstrations in numbers, if they are not more evident and exact than in extension, yet they are more general in their use, and more determinate in their application. Because the ideas of numbers are more precise and distinguishable than in extension, where every equality and excess are not so easy to be observed or measured, because our thoughts cannot in space arrive at any determined smallness beyond which it cannot go, as an unit; and therefore the quantity or proportion of any the least excess cannot be discovered.

By the repeating, as has been said, the idea of an unit, and joining it to another unit, we make thereof one collective idea, marked by the name 'two.' And whosoever can do this, and proceed on, still adding one more to the last collective idea which he had of any number, and give a name to it, may count, or have ideas, for several collections of units, distinguished one from another, as far as he hath a series of names for following numbers, and a memory to retain that series, with their several names. For, the several simple modes of numbers being in our minds but so many combinations of units, which have no variety, nor are capable of any other difference but more or less, names or marks for each distinct combination seem more necessary than in any other sort of ideas. For, without such names or marks, we can hardly well make use of numbers in reckoning, especially where the combination is made up of any great multitude of units; which put together, without a name or mark to distinguish that precise collection, will hardly be kept from being a heap in confusion.

This I think to be the reason why some Americans [Indians] I have spoken with (who were otherwise of quick and rational parts enough), could not, as we do, by any means count to 1000, nor had any distinct idea of that number, though they could reckon very well to 20. Because their language being scanty, and accommodated only to the few necessaries of a needy, simple life, unacquainted either with trade or mathematics, had no words in it to stand for 1000; so that when they were discoursed with of those greater numbers, they would show the hairs of their head, to express a great multitude, which they could not number; which inability, I suppose, proceeded from their want of names.

For he that will count twenty, or have any idea of that number, must know that nineteen went before, with the distinct name or sign of every one of them, as they stand marked in their order; for wherever this fails a gap is made, the chain breaks, and the progress in numbering can go no further. So that to reckon right, it is required: (1) that the mind distinguish carefully two ideas, which are different one from another only by the addition or subtraction of one unit; (2) that it retain in memory the names or marks of the several combinations, from an unit to that number, and that not confusedly, and at random, but in that exact order that the numbers follow one another. In either of which, if it trips, the whole business of numbering will be disturbed, and there will remain only the confused idea of multitude, but the ideas necessary to distinct numeration will not be attained to.

This further is observable in number that it is that which the mind makes use of in measuring all things that by us are measurable, which principally are *expansion* and *duration;* and our idea of infinity, even when applied to those, seems to be nothing but the infinity of number. For what else are our ideas of eternity and immensity, but the repeated additions of certain ideas of imagined parts of duration and expansion, with the infinity of number; in which we can come to no end of addition? For let a man collect into one sum as great a number as he pleases, this multitude, how great soever, lessens not one jot the power of adding to it, or brings him any nearer the end of the inexhaustible stock of number, where still there remains as much to be added, as if none were taken out. And this *endless addition* or *addibility* (if anyone like the word better) of numbers, so apparent to the mind, is that, I think, which gives us the clearest and most distinct idea of infinity.

5.24 I doubt not but it will be easily granted that the knowledge we have of mathematical truths is not only certain, but real knowledge, and not the bare empty vision of vain, insignificant chimeras of the brain; and yet, if we will consider, we shall find that it is only of our own ideas. The mathematician considers the truth and properties belonging to a rectangle or circle only as they are in idea in his own mind. For it is possible he never found either of them existing mathematically, i.e. precisely true, in his life. But yet the knowledge he has of any truths or properties belonging to a circle, or any other mathematical figure, are nevertheless true and certain, even of real things existing; because real things are no further concerned, nor intended to be meant by any such propositions, than as things really agree to those archetypes in his mind. Is it true of the *idea* of a triangle that its three angles are equal to two right ones? It is true also of a triangle, wherever it *really exists.* Whatever other figure exists, that is not exactly answerable to that idea of a triangle in his mind, is not at all concerned in that proposition. And therefore he is certain all his knowledge concerning such ideas is real knowledge.

Knowledge is the perception of the connection between ideas.

5.25 *Knowledge* then seems to me to be nothing but *the perception of the connexion and agreement, or disagreement and repugnancy, of any of our ideas.* In this alone it consists. Where this perception is, there is knowledge, and where it is not, there, though we may fancy, guess, or believe, yet we always come short of knowledge. For when we know that white is not black, what do we else but perceive that these two ideas do not agree? When we possess ourselves with the utmost security of the demonstration that the three angles of a triangle are equal to two right ones, what do we more but perceive that equality to two right ones does necessarily agree to, and is inseparable from, the three angles of a triangle?

But to understand a little more distinctly wherein this agreement or disagreement consists, I think we may reduce it all to these four sorts:

1. *Identity,* or *diversity.*
2. *Relation.*
3. *Co-existence,* or *necessary connexion.*
4. *Real existence.*

First, as to the first sort of agreement or disagreement, viz. *identity* or *diversity.* It is the first act of the mind, when it has any sentiments or ideas at all, to perceive its ideas; and so far as it perceives them, to know each what it is, and thereby also to perceive their difference, and that one is not another. This is so absolutely necessary that without it there could be no knowledge, no reasoning, no imagination, no distinct thoughts at all. By this the mind clearly and infallibly perceives each idea to agree with itself, and to be what it is, and all distinct ideas to disagree, i.e. the one not to be the other; and this it does without pains, labour, or deduction, but at first view, by its natural power of perception and distinction. And though men of art have reduced this into those general rules, *What is, is,* and *It is impossible for the same thing to be and not to be,* for ready application in all cases, wherein there may be occasion to reflect on it, yet it is certain that the first exercise of this faculty is about particular ideas. A man infallibly knows, as soon as ever he has them in his mind, that the ideas he calls *white* and *round* are the very ideas they are, and that they are not other ideas which he calls *red* or *square.* Nor can any maxim or proposition in the world make him know it clearer or surer than he did before, and without any such general rule. This then is the first agreement or disagreement which the mind perceives in its ideas, which it always perceives at first sight; and if there ever happen any doubt about it, it will always be found to be about the names, and not the ideas themselves, whose identity and diversity will always be perceived, as soon and as clearly as the ideas themselves are; nor can it possibly be otherwise.

Secondly, the next sort of agreement or disagreement the mind perceives in any of its ideas may, I think, be called *relative,* and is nothing but the perception of the *relation* between any two ideas, of what kind soever, whether substances, modes, or any other. For, since all distinct ideas must eternally be known not to be the same, and so be universally and constantly denied one of another, there could be no room for any positive knowledge at all, if we could not perceive any relation between our ideas, and find out the agreement or disagreement they have one with another, in several ways the mind takes of comparing them.

Mathematical knowledge is both certain and non-trivial.

5.26 Knowledge, as has been shown, consists in the perception of the agreement or disagreement of ideas. Now, where that agreement or disagreement is perceived immediately by itself, without the intervention or help of any other, there our knowledge is self-evident. This will appear to be so to any who will but consider any of those propositions which, without any proof, he assents to at first sight; for in all of them he will find that the reason of his assent is from that agreement or disagreement which the mind, by an immediate comparing them, finds in those ideas answering the affirmation or negation in the proposition.

This being so, in the next place let us consider whether this self-evidence be peculiar only to those propositions which commonly pass under the name of maxims, and have the dignity of axioms allowed them. And here it is plain that several other truths, not allowed to be axioms, partake equally with them in this self-evidence. This we shall see, if we go over these several sorts of agreement or disagreement of ideas which I have above mentioned, viz. identity, relation, co-existence, and real existence; which will discover to us that not only those few propositions which have had the credit of maxims are self-evident, but a great many, even almost an infinite number of, other propositions are such.

5.27 . . . As to the *relations of modes,* mathematicians have framed many axioms concerning that one relation of equality. As "Equals taken from equals, the remainder will be equal"; which, with the rest of that kind, however they are received for maxims by the mathematicians, and are unquestionable truths, yet, I think, that anyone who considers them will not find that they have a clearer self-evidence than these, that "One and one are equal to two"; that "If you take from the five fingers of one hand two, and from the five fingers of the other hand two, the remaining numbers will be equal." These and a thousand other such propositions may be found in numbers, which, at the very first hearing, force the assent, and carry with them an equal, if not greater, clearness than those mathematical axioms.

5.28 Before a man makes any proposition, he is supposed to understand the terms he uses in it, or else he talks like a parrot, only making a noise by imitation, and framing certain sounds, which he has learnt of others; but not as a rational creature, using them for signs of ideas which he has in his mind. The hearer also is supposed to understand the terms as the speaker uses them, or else he talks jargon, and makes an unintelligible noise. And therefore he trifles with words who makes such a proposition, which, when it is made, contains no more than one of the terms does, and which a man was supposed to know before: v.g. "A triangle hath three sides," or "Saffron is yellow." And this is no further tolerable than where a man goes to explain his terms to one who is supposed or declares himself not to understand him; and then it teaches only the signification of that word, and the use of that sign.

We can know then the truth of two sorts of propositions with perfect certainty. The one is of those trifling propositions which have a certainty in them, but it is but a verbal certainty, but not instructive. And, secondly, we can know the truth, and so may be certain in propositions, which affirm something of another, which is a necessary consequence of its precise complex idea, but not contained in it: as that the external angle of all triangles is bigger than either of the opposite internal angles. Which relation of the outward angle to either of the opposite internal angles, making no part of the complex idea signified by the name 'triangle,' this is a real truth, and conveys with it instructive real knowledge.

5.29 But since the essences of things are thought by some (and not without reason) to be wholly unknown, it may not be amiss to consider the several significations of the word 'essence.'

First, essence may be taken for the being of anything, whereby it is what it is. And thus the real internal, but generally (in substances) unknown constitution of things, whereon their discoverable qualities depend, may be called their essence. This is the proper original signification of the word, as is evident from the formation of it, *essentia,* in its primary notation, signifying properly being. And in this sense it is still used, when we speak of the essence of *particular* things, without giving them any name.

Secondly, it being evident that things are ranked under names into sorts or species only as they agree to certain abstract ideas, to which we have annexed those names, the essence of each *genus,* or sort, comes to be nothing but that abstract idea which the general, or sortal (if I may have leave so to call it from *sort,* as I do *general* from *genus*) name stands for. And this we shall find to be that which the word *essence* imports in its most familiar use. These two sorts of essences, I suppose, may not unfitly be termed, the one the *real,* the other *nominal essence.*

5.30 Essences being thus distinguished into nominal and real, we may further observe that, in the species of simple ideas and modes, they are always the same, but in substances always quite different. Thus, a figure including a space between three lines is the real as well as nominal essence of a triangle, it being not only the abstract idea to which the general name is annexed, but the very *essentia* or being of the thing itself: that foundation from which all its properties flow, and to which they are all inseparably annexed. But it is far otherwise concerning that parcel of matter which makes the ring on my finger, wherein these two essences are apparently different. For, it is the real constitution of its insensible parts, on which depend all those properties of colour, weight, fusibility, fixedness, etc., which makes it to be gold, or gives it a right to that name, which is therefore its nominal essence.

Intuitive versus demonstrative knowledge.

5.31 The different clearness of our knowledge seems to me to lie in the different way of perception the mind has of the agreement or disagreement of any of its ideas. For if we will reflect on our own ways of thinking, we shall find that sometimes the mind perceives the agreement or disagreement of two ideas *immediately by themselves,* without the intervention of any other; and this I think we may call *intuitive knowledge.* For in this the mind is at no pains of proving or examining, but perceives the truth as the eye doth light, only by being directed toward it. Thus the mind perceives that white is not black, that a circle is not a triangle, that three are more than two and equal to one and two. Such kinds of truths the mind perceives at the first sight of the ideas together by bare intuition, without the intervention of any other idea; and this kind of knowledge is the clearest and most certain that human frailty is capable of. It is on this intuition that depends all the certainty and evidence of all our knowledge; which certainty everyone finds to be so great that he cannot imagine, and therefore not require, a greater. He that demands a greater certainty than this demands he knows not what, and shows only that he has a mind to be a sceptic, without being able to be so.

The next degree of knowledge is where the mind perceives the agreement or disagreement of any ideas, but not immediately. Though, wherever the mind perceives the agreement or disagreement of any of its ideas, there be certain knowledge, yet it does not always happen that the mind sees that agreement or disagreement, which there is between them, even where it is discoverable; and in that case remains in ignorance, and at most gets no further than a probable conjecture. The reason why the mind cannot always perceive presently the agreement or disagreement of two ideas is because those ideas, concerning whose agreement or disagreement the inquiry is made, cannot by the mind be so put together as to show it. In this case then, when the mind cannot so bring its ideas together as by their immediate comparison, and as it were juxta-position or application one to another, to perceive their agreement or disagreement, it is fain, by the intervention of other ideas (one or more, as it happens) to discover the agreement or disagreement which it searches; and this is that which we call *reasoning.* Thus, the mind, being willing to know the agreement or disagreement in bigness between the three angles of a triangle and two right ones, cannot by an immediate view and comparing them do it; because the three angles of a triangle cannot be brought at once, and be compared with any one, or two, angles; and so of this the mind has no immediate, no intuitive, knowledge. In this case the mind is fain to find out some other angles, to which the three angles of a triangle have an equality; and, finding those equal to two right ones, comes to know their equality to two right ones.

5.32 This knowledge by intervening proofs, though it be certain, yet the evidence of it is not altogether so clear and bright, nor the assent so ready, as in intuitive knowledge. For, though in demonstration the mind does at last perceive the agreement or disagreement of the ideas it considers, yet it is not without pains and attention; there must be more than one transient view to find it.

5.33 If we will observe the actings of our own minds, we shall find that we reason best and clearest, when we only observe the connexion of the proof, without reducing our thoughts to any rule of syllogism. And therefore we may take notice that there are many men that reason exceeding clear and rightly, who know not how to make a syllogism. He that will look into many parts of Asia and America, will find men reason there perhaps as acutely as himself, who yet never heard of a syllogism, nor can reduce any one argument to those forms. All who have so far considered *syllogism,* as to see the reason why in three propositions laid together in one form, the conclusion will be certainly right, but in another not certainly so, I grant are certain of the conclusion they draw from the premises in the allowed *modes* and *figures.* But they who have not so far looked into those forms, are not sure by virtue of syllogism, that the conclusion certainly follows from the premises; they only take it to be so by an implicit faith in their teachers and a confidence in those forms of argumentation; but this is still but believing, not being certain. Now, if, of all mankind those who can make syllogisms are extremely few in comparison of those who cannot; and if, of those few who have been taught logic, there is but a very small number who do any more than believe that syllogisms, in the allowed *modes* and *figures* do conclude right, without knowing certainly that they do so; if syllogisms must be taken for the only proper instrument of reason and means of knowledge, it will follow, that, before Aristotle, there was not one man that did or could know anything by reason; and that, since the invention of syllogisms, there is not one of ten thousand that doth.

6

NEWTON · Introduction

Isaac Newton (1642–1727) is probably more commonly thought of today as a mathematician or physicist than as a philosopher, but during the 17th century men that today we call "physicists" were generally referred to as either "natural philosophers" or "mathematicians." The father of "modern" (Newtonian) physics invented the calculus early in his career, but his failure to publish his work on this topic until after Leibniz had published the results of his own work on the same subject led to a heated debate over who deserved the credit for being *the* inventor of the calculus. After many years, the debate was ultimately resolved when it was agreed that the two men apparently did their work independently, and that they thus deserve equal credit. In addition to his work on mathematics and physics, Newton spent much time and effort on studies in theology and alchemy.

Although he extended the application of arithmetical concepts and methods to geometry even further than had Descartes, Newton apparently had strong reservations about the justifiability of this procedure, as indicated by the following statement which appeared in one of his later books, the *Universal Arithmetick* of 1707.

> Equations are expressions of arithmetical computation, and properly have no place in Geometry, except as far as quantities truly geometrical (that is lines, surfaces, solids and proportions) may be said to be some equal to others. Multiplications, divisions, and such sort of computations, are newly introduced into Geometry, and that unwarily, and contrary to the first design of this science. . . . Therefore these two sciences ought not to be confounded. The Ancients did so industriously distinguish them from one another, that they never introduced arithmetical terms into Geometry. And the Moderns by confounding both, have lost the simplicity in which the elegancy of Geometry consists.

The traditional "Aristotelean" principles were certainly deeply etched on even the most innovative minds of the 17th century, but fortunately men such as Newton did not always allow their ideals to affect their practice.

Newton never published any works which would be considered purely philosophical in the present-day sense of the term, but his writings on physics and mathematics contain both implicit suggestions and explicit discussions of his thoughts on various ontological and epistemological issues. However, his views on these issues were confused to the point of being self-contradictory, particularly his discussions of "evanescent quantities" (see 5.7–5.8). Newton thus provides the historian with an effective counter-example to any claims that it is necessary to have an adequate philosophical foundation in order to make a significant contribution to mathematics.

Although Platonic, Cartesian, Hobbesian, and other influences can be identified in Newton's writings, the strongest influence on his thinking about mathematics was probably his own work in the physical sciences. He not only accepted the premise that the universe is governed by mathematical laws, but he also believed that mathematics itself involves the study of the physical universe—more specifically, of physical space and the quantitative properties of material objects. His description of the analytic method of mathematics (see 5.9) indicates that mathematical knowledge is to be acquired by induction and/or abstraction from sense experience. But he nowhere provided a sufficiently detailed account of the processes involved in acquiring this knowledge, and he did not even make any unequivocal statements as to whether mathematical knowledge is absolutely certain or merely probable. However, certain aspects of his theory of knowledge can be inferred by a careful reading of the following selections in which he discusses what he considered to be the fundamental objects of mathematical knowledge—space and evanescent quantities (see 5.5–5.8).

Newton apparently believed—as did many of his predecessors and contemporaries—that all we immediately perceive are ideas, and that some of these ideas (those caused by the primary qualities of things) image or picture 'real' objects in the physical world. He also explicitly rejected Hobbes' identification of ideas with images, and asserted that to know or understand something does not necessarily require having an image before the mind. Thus, he ultimately sided with Descartes and opposed Hobbes on the question of positive infinity, on the grounds that we "understand" the concept even though we don't have an image of it.

Despite his recognition of this non-imagistic process of understanding, he almost always relied on imagistic examples in his discussions of even the most abstract topics. This was undoubtedly a major source of the confusions which plagued his discussions of such difficult problems as those of infinitely large and infinitely small quantities. It was by doing as Newton taught, rather than by following his example, that mathematicians were able to make the remarkable advances they did during the following centuries. (They were probably following their own instincts rather than Newton's advice to break free from imagistic thinking.)

Although he agreed with Descartes that man possesses a *positive* concept of infinity, he denied that it was *innate,* and believed that it could somehow be derived by the mind from sense experience. He therefore could not use Descartes' proof of the existence of God from the innate idea of infinity. Instead, he used a version of the neo-Aristotelean "argument from design" (see 5.4), which is particularly compatible with the principle that all knowledge must be grounded in experience of the physical universe. But his assertion that space is the "sensorium of God" is clearly not derivable from empirical observation, and reflects his strong interest in theological questions more than it shows any clarity or consistency in his philosophical and mathematical thinking.

Although Newton influenced the thinking of many philosophers, his greatest influence was on mathematicians. Much of this influence was indirect, and resulted from his development of a tool (however crude and unrefined) in the form of the calculus which could be applied to countless significant problems having to do with man's control of the physical universe. This provided the impetus for a new "golden age" of mathematical research, which in turn generated a whole new set of problems for the philosopher of mathematics, problems which have occupied the minds of the most outstanding philosophers of the past century.

Sources

6.1 *The Mathematical Principles of Natural Philosophy* (hereafter *Principles*), translated by Andrew Motte, 1st American edition, New York, 1846, p. 77.

6.2 *Principles,* p. 79.

6.3 *Opticks,* 4th edition, London, 1730, reprinted by Dover Publications, New York, 1952, p. 370.

6.4 *Opticks,* pp. 402–404.

6.5 "De Gravitatione et Aequipondio Fluidorum" (c. 1670), translated by A. Rupert Hall and Marie Boas Hall, *Unpublished Scientific Papers of Isaac Newton,* Cambridge University Press, Cambridge, 1962, pp. 131–138.

6.6 "Letter II," in *Four Letters From Sir Isaac Newton to Doctor Bentley,* London, 1755, pp. 15–18.

6.7 *Principles,* pp. 102–103.

6.8 *Principles,* p. 261.

6.9 *Opticks,* pp. 404–405.

NEWTON · Selections

The distinction between absolute and relative space.

6.1 Absolute space, in its own nature, without regard to anything external, remains always similar and immovable. Relative space is some movable dimension or measure of the absolute spaces; which our senses determine by its position to bodies; and which is vulgarly taken for immovable space; such is the dimension of a subterraneous, an aerial, or celestial space, determined by its position in respect of the earth. Absolute and relative space are the same in figure and magnitude; but they do not remain always numerically the same. For if the earth, for instance, moves, a space of our air, which relatively and in respect of the earth remains always the same, will at one time be one part of the absolute space into which the air passes; at another time it will be another part of the same, and so, absolutely understood, it will be perpetually mutable.

Place is a part of space which a body takes up, and is according to the space, either absolute or relative. I say, a part of space; not the situation, nor the external surface of the body.

6.2 As the order of the parts of time is immutable, so also is the order of the parts of space. Suppose those parts to be moved out of their places, and they will be moved (if the expression may be allowed) out of themselves. For times and spaces are, as it were, the places as well of themselves as of all other things. All things are placed in time as to order of succession; and in space as to order of situation. It is from their essence or nature that they are places; and that the primary places of things should be movable, is absurd. These are therefore the absolute places; and translations out of those places, are the only absolute motions.

But because the parts of space cannot be seen, or distinguished from one another by our senses, therefore in their stead we use sensible measures of them. For from the positions and distances of things from any body considered as immovable, we define all places; and then with respect to such places, we estimate all motions, considering bodies as transferred from some of those places into others. And so, instead of absolute places and motions, we use relative ones; and that without any inconvenience in common affairs; but in philosophical disquisitions, we ought to abstract from our senses, and consider things themselves, distinct from what are only sensible measures of them. For it may be that there is no body really at rest, to which the places and motions of others may be referred.

Absolute space is the sensorium of God.

6.3 Is not the sensory of animals that place to which the sensitive substance is present, and into which the sensible species of things are carried through the nerves and brain, that there they may be perceived by their immediate presence to that substance? And these things being rightly dispatched, does it not appear from phenomena that there is a Being incorporeal, living, intelligent, omnipresent, who in infinite space, as it were in His Sensory, sees the things themselves intimately, and thoroughly perceives them, and comprehends them wholly by their immediate presence to Himself: of which things the images only carried through the organs of sense into our little sensoriums, are there seen and beheld by that which in us perceives and thinks. And though every true step made in this philosophy brings us not immediately to the knowledge of the first cause, yet it brings us nearer to it, and on that account is to be highly valued.

6.4 . . . All material things seem to have been composed of the hard and solid particles [i.e., atoms], variously associated in the first creation by the counsel of an intelligent agent. For it became him who created them to set them in order. And if he did so, it is unphilosophical to seek for any other origin of the world, or to pretend that it might arise out of a chaos by the mere laws of nature; though being once formed, it may continue by those laws for many ages. For while comets move in very eccentric orbs in all manner of positions, blind fate could never make all the planets move one and the same way in orbs concentric, some inconsiderable irregularities excepted, which may have risen from the mutual actions of comets and planets upon one another, and which will be apt to increase, till this system wants a reformation. Such a wonderful uniformity in the planetary system must be allowed the effect of choice. And so must the uniformity in the bodies of animals, they having generally a right and a left side shaped alike, and on either side of their bodies two legs behind, and either two arms, or two legs, or two wings before upon their shoulders, and between their shoulders a neck running down into a back-bone, and a head upon it; and in the head two ears, two eyes, a nose, a mouth, and a tongue, alike situated. Also the first contrivance of those very artificial parts of animals, the eyes, ears, brain, muscles, heart, lungs, midriff, glands, larynx, hands, wings, swimming bladders, natural spectacles, and other organs of sense and motion; and the instinct of brutes and insects, can be the effect of nothing else than the wisdom and skill of a powerful ever-living agent, who being in all places, is more able by his will to move the bodies within his boundless uniform sensorium, and thereby to form and reform the parts of the universe, than we are by our will to move the parts of our own bodies. And yet we are not to consider the world as the body of God, or the several parts thereof, as the parts of God. He is an uniform being, void of organs, members or parts, and they are his creatures subordinate to him, and subservient to his will; and he is no more the soul of them, than the soul of man is the soul of the species of things carried through the organs of sense into the place of its sensation, where it perceives them by means of its immediate presence, without the intervention of any third thing. The organs of sense are not for enabling the soul to perceive the species of things in its sensorium, but only for conveying them thither; and God has no need of such organs, he being every where present to the things themselves. And since space is divisible *in infinitum,* and matter is not necessarily in all places, it may be also allowed that God is able to create particles of matter of several sizes and figures, and in several proportions to space, and perhaps of different densities and forces, and thereby to vary the laws of nature, and make worlds of several sorts in several parts of the universe. At least, I see nothing of contradiction in all this.

The nature of absolute space.

6.5 In addition, as Descartes . . . seems to have demonstrated that body does not differ at all from extension, abstracting hardness, colour, weight, cold, heat and the remaining qualities which body can lack, so that at last there remains only its extension in length, width and depth which hence alone appertain to its essence; and as this has been taken as proved by many, and is in my view the only reason for having confidence in this opinion; lest any doubt should remain about the nature of motion, I shall reply to this argument by explaining what extension and body are, and how they differ from each other. For since the distinction of substances into thinking and extended [entities], or rather, into thoughts and extensions, is the principal foundation of Cartesian philosophy, which he contends to be even better known than mathematical demonstrations: I consider it most important to overthrow [that philosophy] as regards extension, in order to lay truer foundations of the mechanical sciences.

Perhaps now it may be expected that I should define extension as substance or accident or else nothing at all. But by no means, for it has its own manner of existence which fits neither substances nor accidents. It is not substance; on the one hand, because it is not absolute in itself, but is as it were an emanent effect of God, or a disposition of all being; on the other hand, because it is not among the proper dispositions that denote substance, namely actions, such as thoughts in the mind and motions in body. For although philosophers do not define substance as an entity that can act upon things, yet all tacitly understand this of substances, as follows from the fact that they would readily allow extension to be substance in the manner of body if only it were capable of motion and of sharing in the actions of body. And on the contrary they would hardly allow that body is substance if it could not move nor excite in the mind any sensation or perception whatever. Moreover, since we can clearly conceive extension existing without any subject, as when we may imagine spaces outside the world or places empty of body, and we believe [extension] to exist wherever we imagine there are no bodies, and we cannot believe that it would perish with the body if God should annihilate a body, it follows that [extension] does not exist as an accident inherent in some subject. And hence it is not an accident. And much less may it be said to be nothing, since it is rather something, than an accident, and approaches more nearly to the nature of substance. There is no idea of nothing, nor has nothing any properties, but we have an exceptionally clear idea of extension, abstracting the dispositions and properties of a body so that there remains only the uniform and unlimited stretching out of space in length, breadth and depth. And furthermore, many of its properties are associated with this idea; these I shall now enumerate not only to show that it is something, but what it is.

1. In all directions, space can be distinguished into parts whose common limits we usually call surfaces; and these surfaces can be distinguished in all directions into parts whose common limits we usually call lines; and again these lines can be distinguished in all directions into parts which we call points. And hence surfaces do not have depth, nor lines breadth, nor points dimension, unless you say that coterminous spaces penetrate each other as far as the depth of the surface between them, namely what I have said to be the boundary of both or the common limit; and the same applies to lines and points. Furthermore spaces are everywhere continguous to spaces, and extension is everywhere placed next to extension, and so there are everywhere common boundaries to contiguous parts; that is, there are everywhere surfaces acting as a boundary to solids on this side and that; and everywhere lines in which parts of the surfaces touch each other; and everywhere points in which the continuous parts of lines are joined together. And hence there are everywhere all kinds of figures, everywhere spheres, cubes, triangles, straight lines, everywhere circular, elliptical, parabolical and all other kinds of figures, and those of all shapes and sizes, even though they are not disclosed to sight. For the material delineation of any figure is not a new production of that figure with respect to space, but only a corporeal representation of it, so that what was formerly insensible in space now appears to the senses to exist. For thus we believe all those spaces to be spherical through which any sphere ever passes, being progressively moved from moment to moment, even though a sensible trace of the sphere no longer remains there. We firmly believe that the space was spherical before the sphere occupied it, so that it could contain the sphere; and hence as there are everywhere spaces that can adequately contain any material sphere, it is clear that space is everywhere spherical. And so of other figures. In the same way we see no material shapes in clear water, yet

there are many in it which merely introducing some colour into its parts will cause to appear in many ways. However, if the colour were introduced, it would not constitute material shapes but only cause them to be visible.

2. Space extends infinitely in all directions. For we cannot imagine any limit anywhere without at the same time imagining that there is space beyond it. And hence all straight lines, paraboloids, hyperboloids, and all cones and cylinders and other figures of the same kind continue to infinity and are bounded nowhere, even though they are crossed here and there by lines and surfaces of all kinds extending transversely, and with them form segments of figures in all directions. You may have in truth an instance of infinity; imagine any triangle whose base and one side are at rest and the other side so turns about the contiguous end of its base in the plane of the triangle that the triangle is by degrees opened at the vertex; and meanwhile take a mental note of the point where the two sides meet, if they are produced that far: it is obvious that all these points are found on the straight line along which the fixed side lies, and that they become perpetually more distant as the moving side turns further until the two sides become parallel and can no longer meet anywhere. Now, I ask, what was the distance of the last point where the sides met? It was certainly greater than any assignable distance, or rather none of the points was the last, and so the straight line in which all those meeting-points lie is in fact greater than finite. Nor can anyone say that this is infinite only in imagination, and not in fact; for if a triangle is actually drawn, its sides are always, in fact, directed towards some common point, where both would meet if produced, and therefore there is always such an actual point where the produced sides would meet, although it may be imagined to fall outside the limits of the physical universe. And so the line traced by all these points will be real, though it extends beyond all distance.

If anyone now objects that we cannot imagine that there is infinite extension, I agree. But at the same time I contend that we can understand it. We can imagine a greater extension, and then a greater one, but we understand that there exists a greater extension than any we can imagine. And here, incidentally, the faculty of understanding is clearly distinguished from imagination.

Should it be further said that we do not understand what an infinite being is, save by negating the limitations of a finite being, and that this is a negative and faulty conception, I deny this. For the limit or boundary is the restriction or negation of greater reality or existence in the limited being, and the less we conceive any being to be constrained by limits, the more we observe something to be attributed to it, that is, the more positively we conceive it. And thus by negating all limits the conception becomes positive in the highest degree. 'End' [*finis*] is a word negative as to sense, and thus 'infinity' [not-end] as it is the negation of a negation (that is, of ends) will be a word positive in the highest degree with respect to our perception and comprehension, though it seems grammatically negative. Add that positive and finite quantities of many surfaces infinite in length are accurately known to Geometers. And so I can positively and accurately determine the solid quantities of many solids infinite in length and breadth and compare them to given finite solids. But this is irrelevant here.

If Descartes now says that extension is not infinite but rather indefinite, he should be corrected by the grammarians. For the word 'indefinite' is never applied to that which actually is, but always relates to a future possibility signifying only something which is not yet determined and definite. Thus before God had decreed anything about the creation of the world (if there was ever a time when he had not), the quantity of matter, the number of the stars and all other things were indefinite; once the world was created they were defined. Thus matter is indefinitely divisible, but is always divided either finitely or infinitely. Thus an indefinite line is one whose future length is still undetermined. And so an indefinite space is one whose future magnitude is not yet determined; for indeed that which actually is, is not to be defined, but does either have limits or not and so is either finite or infinite. Nor is it an objection that he takes space to be indefinite in relation to ourselves; that is, we simply do not know its limits and are not absolutely sure that there are none. This is because although we are ignorant beings God at least understands that there are no limits not merely indefinitely but certainly and positively, and although we negatively imagine it to transcend all limits, yet we positively and most certainly understand that it does so. But I see what Descartes feared, namely that if he should consider space infinite, it would perhaps become God because of the perfection of infinity. But by no means, for infinity is not perfection except when it is an attribute of perfect things. Infinity of intellect, power, happiness and so forth is the height of perfection; but infinity of ignorance, impotence, wretchedness and so on is the height of imperfection; and infinity of extension is so far perfect as that which is extended.

3. The parts of space are motionless. If they moved, it would have to be said either that the motion of each part is a translation from the vicinity of other contiguous parts, as Descartes defined the motion of bodies; and that this is absurd has been sufficiently shown; or that it is a translation out of space into space, that is out of itself, unless perhaps it is said that two spaces everywhere coincide, a moving one and a motionless one. Moreover the immobility of space will be best exemplified by duration. For just as the parts of duration derive their individuality from their order, so that (for example) if yesterday could change places with today, and become the latter of the two, it would lose its individuality and would no longer be yesterday, but today; so the parts of space derive their character from their positions, so that if any two could change their positions, they would change their character at the same time and each would be converted numerically into the other. The parts of

duration and space are only understood to be the same as they really are because of their mutual order and position; nor do they have any hint of individuality apart from that order and position which consequently cannot be altered.

4. Space is a disposition of being *qua* being. No being exists or can exist which is not related to space in some way. God is everywhere, created minds are somewhere, and body is in the space that it occupies; and whatever is neither everywhere nor anywhere does not exist. And hence it follows that space is an effect arising from the first existence of being, because when any being is postulated, space is postulated. And the same may be asserted of duration: for certainly both are dispositions of being or attributes according to which we denominate quantitatively the presence and duration of any existing individual thing. So the quantity of the existence of God was eternal, in relation to duration, and infinite in relation to the space in which he is present; and the quantity of the existence of a created thing was as great, in relation to duration, as the duration since the beginning of its existence, and in relation to the size of its presence as great as the space belonging to it.

Moreover, lest anyone should for this reason imagine God to be like a body, extended and made of divisible parts, it should be known that spaces themselves are not actually divisible, and furthermore, that any being has a manner proper to itself of being in spaces. For thus there is a very different relationship between space and body, and space and duration. For we do not ascribe various durations to the different parts of space, but say that all endure together. The moment of duration is the same at Rome and at London, on the Earth and on the stars, and throughout all the heavens. And just as we understand any moment of duration to be diffused throughout all spaces, according to its kind, without any thought of its parts, so it is no more contradictory that mind also, according to its kind, can be diffused through space without any thought of its parts.

5. The positions, distances and local motions of bodies are to be referred to the parts of space. And this appears from the properties of space enumerated as 1. and 4. above, and will be more manifest if you conceive that there are vacuities scattered between the particles, or if you pay heed to what I have formerly said about motion. To that it may be further added that in space there is no force of any kind which might impede or assist or in any way change the motions of bodies. And hence projectiles describe straight lines with a uniform motion unless they meet with an impediment from some other source. But more of this later.

6. Lastly, space is eternal in duration and immutable in nature, and this because it is the emanent effect of an eternal and immutable being. If ever space had not existed, God at that time would have been nowhere; and hence he either created space later (in which he was not himself), or else, which is not less repugnant to reason, he created his own ubiquity. Next, although we

can possibly imagine that there is nothing in space, yet we cannot think that space does not exist, just as we cannot think that there is no duration, even though it would be possible to suppose that nothing whatever endures. This is manifest from the spaces beyond the world, which we must suppose to exist (since we imagine the world to be finite), although they are neither revealed to us by God, nor known from the senses, nor does their existence depend upon that of the spaces within the world. But it is usually believed that these spaces are nothing; yet indeed they are true spaces. Although space may be empty of body, nevertheless it is not in itself a void; and *something* is there, because spaces are there, although nothing more than that. Yet in truth it must be acknowledged that space is no more space where the world is, than where no world is, unless perchance you say that when God created the world in this space he at the same time created space in itself, or that if God should annihilate the world in this space, he would also annihilate the space in it. Whatever has more reality in one space than in another space must belong to body rather than to space; the same thing will appear more clearly if we lay aside that puerile and jejune prejudice according to which extension is inherent in bodies like an accident in a subject without which it cannot actually exist.

A 'proof' that not all infinites are equal.

6.6 But you argue, in the next paragraph of your letter, that every particle of matter in an infinite space, has an infinite quantity of matter on all sides, and by consequence an infinite attraction every way, and therefore must rest in equilibrio, because all infinites are equal. Yet you suspect a paralogism in this argument; and I conceive the paralogism lies in the position, that all infinites are equal. The generality of mankind consider infinites no other ways than indefinitely; and in this sense, they say all infinites are equal; though they would speak more truly if they should say, they are neither equal nor unequal, nor have any certain difference or proportion one to another. In this sense therefore, no conclusions can be drawn from them, about the equality, proportions, or differences of things, and they that attempt to do it usually fall into paralogisms. So when men argue against the infinite divisibility of magnitude, by saying, that if an inch may be divided into an infinite number of parts, the sum of those parts will be an inch; and if a foot may be divided into an infinite number of parts, the sum of those parts must be a foot, and therefore since all infinites are equal, those sums must be equal, that is, an inch equal to a foot.

The falseness of the conclusion shows an error in the premises, and the error lies in the position, that all infinites are equal. There is therefore another way of considering infinites used by mathematicians, and that is, under certain definite restrictions and limitations, whereby infinites are determined to have certain differences or proportions to one another. Thus Dr. Wallis considers them in his *Arithmetica Infinitorum*, where by the various proportions of infinite sums, he gathers the various proportions of infinite magnitudes: which way of arguing is generally allowed by mathematicians, and yet would not be good were all infinites equal. According to the same way of considering infinites, a mathematician would tell you, that though there be an infinite number of infinite little parts in an inch, yet there is twelve times that number of such parts in a foot, that is, the infinite number of those parts in a foot is not equal to, but twelve times bigger than the infinite number of them in an inch. And so a mathematician will tell you, that if a body stood in equilibrio between any two equal and contrary attracting infinite forces; and if to either of these forces you add any new finite attracting force, that new force, how little soever, will destroy their equilibrium, and put the body into the same motion into which it would put it were those two contrary equal forces but finite, or even none at all; so that in this case the two equal infinites by the addition of a finite to either of them, become unequal in our ways of reckoning; and after these ways we must reckon, if from the considerations of infinites we would always draw true conclusions.

The problem of the ultimate 'building-blocks' of mathematics and geometry.

6.7 Those things which have been demonstrated of curved lines, and the surfaces which they comprehend, may be easily applied to the curved surfaces and contents of solids. These Lemmas are premised to avoid the tediousness of deducing involved demonstrations *ad absurdum*, according to the method of the ancient geometers. For demonstrations are shorter by the method of indivisibles: but because the hypothesis of indivisibles seems somewhat harsh, and therefore that method is reckoned less geometrical, I chose rather to reduce the demonstrations of the following propositions to the first and last sums and ratios of nascent and evanescent quantities, that is, to the limits of those sums and ratios; and so to premise, as short as I could, the demonstrations of those limits. For hereby the same thing is performed as by the method of indivisibles; and now those principles being demonstrated, we may use them with more safety. Therefore if hereafter I should happen to consider quantities as made up of particles, or should use little curved lines for right ones, I would not be understood to mean indivisibles, but evanescent divisible quantities; not the sums and ratios of determinate parts, but always the limits of sums and ratios; and that the force of such demonstrations always depends on the method laid down in the foregoing Lemmas.

Perhaps it may be objected, that there is no ultimate proportion of evanescent quantities; because the proportion, before the quantities have vanished, is not the ultimate, and when they are vanished, is none. But by the same argument, it may be alleged that a body arriving at a certain place, and there stopping, has no ultimate velocity: because the velocity, before the body comes to the place, is not its ultimate velocity; when it has arrived, there is none. But the answer is easy; for by the ultimate velocity is meant that with which the body is moved, neither before it arrives at its last place and the motion ceases, nor after, but at the very instant it arrives; that is, that velocity with which the body arrives at its last place, and with which the motion ceases. And in like manner, by the ultimate ratio of evanescent quantities is to be understood the ratio of the quantities not before they vanish, nor afterwards, but with which they vanish. In like manner the first ratio of nascent quantities is that with which they begin to be. And the first or last sum is that with which they begin and cease to be (or to be augmented or diminished). There is a limit which the velocity at the end of the motion may attain, but not exceed. This is the ultimate velocity. And there is the like limit in all quantities and proportions that begin and cease to be. And since such limits are certain and definite, to determine the same is a problem strictly geometrical. But whatever is geometrical we may use in determining and demonstrating any other thing that is likewise geometrical.

It may also be objected, that if the ultimate ratios of evanescent quantities are given, their ultimate magnitudes will be also given: and so all quantities will consist of indivisibles, which is contrary to what *Euclid* has demonstrated concerning incommensurables, in the tenth Book of his *Elements.* But this objection is founded on a false supposition. For those ultimate ratios with which quantities vanish are not truly the ratios of ultimate quantities, but limits towards which the ratios of quantities decreasing without limit do always converge; and to which they approach nearer than by any given difference, but never go beyond, nor in effect attain to, till the quantities are diminished *in infinitum.* This thing will appear more evident in quantities infinitely great. If two quantities, whose difference is given, be augmented *in infinitum,* the ultimate ratio of these quantities will be given, to wit, the ratio of equality; but it does not from thence follow, that the ultimate or greatest quantities themselves, whose ratio that is, will be given. Therefore if in what follows, for the sake of being more easily understood, I should happen to mention quantities as least, or evanescent, or ultimate, you are not to suppose that quantities of any determinate magnitude are meant, but such as are conceived to be always diminished without end.

6.8 I call any quantity a *genitum* which is not made by addition or subtraction of divers parts, but is generated or produced in arithmetic by the multiplication, division, or extraction of the root of any terms whatsoever; in geometry by the invention of contents and sides, or of the extremes and means of proportionals. Quantities of this kind are products, quotients, roots, rectangles, squares, cubes, square and cubic sides, and the like. These quantities I here consider as variable and indetermined, and increasing or decreasing, as it were, by a perpetual motion or flux; and I understand their momentary increments or decrements by the name of moments; so that the increments may be esteemed as added or affirmative moments; and the decrements as subtracted or negative ones. But take care not to look upon finite particles as such. Finite particles are not moments, but the very quantities generated by the moments. We are to conceive them as the just nascent principles of finite magnitudes.

The mathematical method—analysis versus synthesis.

6.9 As in mathematics, so in natural philosophy, the investigation of difficult things by the method of analysis, ought ever to precede the method of composition. This analysis consists in making experiments and observations, and in drawing general conclusions from them by induction, and admitting of no objections against the conclusions, but such as are taken from experiments, or other certain truths. For hypotheses are not to be regarded in experimental philosophy. And although the arguing from experiments and observations by induction be no demonstration of general conclusions; yet it is the best way of arguing which the nature of things admits of, and may be looked upon as so much the stronger, by how much the induction is more general. And if no exception occur from phenomena, the conclusion may be pronounced generally. But if at any time afterwards any exception shall occur from experiments, it may then begin to be pronounced with such exceptions as occur. By this way of analysis we may proceed from compounds to ingredients, and from motions to the forces producing them; and in general, from effects to their causes, and from particular causes to more general ones, till the argument end in the most general. This is the method of analysis: and the synthesis consists in assuming the causes discovered and established as principles, and by them explaining the phenomena proceeding from them, and proving the explanations.

7

LEIBNIZ · Introduction

Gottfried Wilhelm Leibniz (1646–1716) was a mathematician of the caliber of Descartes and Newton, and he is also well known for his work as a philosopher, historian, lawyer, diplomat, scientist, logician, and theologian. E. T. Bell in his *Men of Mathematics* described Leibniz as a "spectacular exception" to the folk adage "Jack of all trades, master of none," because his work in any *one* of these many areas would have been sufficient to establish a reputation for him as an extraordinary intellect. In addition to his work on the development of the calculus, Leibniz the mathematician laid the groundwork for the field of combinatorial analysis, which ultimately led to the development of Boolean algebra and mathematical logic.

Leibniz was a prolific writer, and it has been estimated that when his collected works are finally published they will fill forty or more large volumes. He was constantly coming up with new ideas, and throughout his career continued to add to and revise his earlier work. Thus it is difficult, if not impossible, to present a concise and accurate statement of *the* philosophy of Leibniz. The passages in this section represent mainly those elements of his thinking which had the greatest impact on the work of other philosophers and mathematicians, and for the most part they also represent concepts and theories to which he adhered for most of his career. It was his epistemology which was most influential on later thinkers, at least in part because Leibniz presented it in a generally more consistent and less confused manner than his ontology. Thus, even though his ontology and epistemology are closely tied in most of his writings, the following selections emphasize his epistemological ideas.

Leibniz considered his philosophy to be a refinement and extension of Descartes' philosophy, although he was also influenced by Hobbes and others. Instead of focusing his attention on the Cartesian distinction between mental and physical substances, Leibniz began with what he considered to be a more fundamental distinction—that between simple and compound substances. He defined a simple substance as one which is essentially indivisible, complete, and active. He reasoned that the only thing which would satisfy this definition is some sort of point. He argued that it could not be a physical point, because physical points are infinitely divisible; and it could not be a mathematical point, since mathematical points are not 'real.' He then postulated the existence of metaphysical points, called "monads," which satisfy his definition of "simple substance," and he asserted that the universe is composed of an infinite number of these monads—including God, the "Supreme Monad" and creator of all the others.

In so far as a monad must be completely self-sufficient and self-contained (with the exception of its original dependence on God for its creation), it can not receive or transmit any signals, since this involves adding to or subtracting from that entity. It follows that an individual monad's perception of the "external" world is illusory; all of its percepts (or ideas) must have been contained in it from the moment of its creation, and it becomes conscious of these ideas in a sequence pre-determined by God at its creation (since not even God can act on it after it has come to exist). It follows that *all* knowledge must be innate (see 7.10), a position much more radical than that of Plato (for whom only knowledge of the Forms is innate). Since God is not a deceiver, and since he makes it *appear* to the perceiving monad that things exist externally, such correlates of perceptions must exist.

Leibniz borrowed more from Descartes' theory of knowledge, including the confusion between psychological and logical criteria of certainty, than from his ontology. He made a sharp distinction between necessary and contingent truths, between "truths of reason" and "truths of fact" (see 7.4–7.8), on the basis of essentially logical criteria (perhaps borrowed from Hobbes—see 4.27). He characterized truths of reason as being identity statements of the form "A is A," whose denial is self-contradictory. Truths of fact, on the other hand, are such that their denial is conceivable, and their truth or falsity can be ascertained only by appealing to non-logical facts and/or to the "principle of sufficient reason." This distinction can also be expressed in terms of the relation between the subject and predicate terms of a sentence; in a necessary proposition the concept of the predicate is "contained in" the concept of the subject, and in a contingent proposition it is distinct from the subject-concept. Significant use was made of this kind of distinction by David Hume in his development of a radically different philosophy, and it provided Kant with the basis for his "analytic-synthetic" distinction which (with various refinements) led to important developments in many areas of philosophy and is still central to much work in philosophy today.

The most radical and controversial aspects of Leibniz' philosophy appear when his theory of monads is brought together with his theory of necessary truth. In so far as the subject term of a proposition refers to a monad or simple substance, and since all past, present, and future states and properties of the monad are contained in it from the moment of its creation, every true statement about each particular monad must *necessarily* be true. The distinction between necessary and contingent knowledge thus seems to be reduced to a psychological rather than a logical distinction. (Leibniz was apparently quite uncomfortable about this conclusion because he did not mention it explicitly in any of the writings whose publication he authorized during his lifetime.)

Although he was reluctant to assert that *all* true knowledge consists of necessary truths, Leibniz explicitly stated in many places that all *mathematical* truths are truths of reason, and that they can be reduced to or derived from the fundamental laws of logic—the laws of identity and non-contradiction. He also suggested that a new symbolic language (which he called the "Universal Characteristic") could be constructed in such a way that all true statements (including those of mathematics) could be expressed in this symbolism in such a way as to facilitate their derivation from the laws of logic. Leibniz himself never carried out this project, but his description of it was sufficient to provide the impetus for some of the most significant work in the areas of mathematics, logic, and the philosophy of mathematics during the second half of the 19th and the first half of the 20th centuries. The brilliant work of Gottlob Frege, Bertrand Russell, and others fell short of the Leibnizian ideal of reducing all mathematics to logic. But the insights which were gained and the new concepts and techniques (including symbolic logic) which were developed in pursuit of this goal have led to the opening of whole new areas of inquiry in both philosophy and mathematics, and to new ways of approaching some of the most difficult of the traditional problems. Leibniz could not have envisioned much of the recent work in philosophy of mathematics, but he did much to make it possible.

Sources

All selections are from *The Philosophical Works of Leibnitz,* translated by George Martin Duncan, New Haven, The Tuttle, Morehouse, & Taylor Company, 1908 (abbreviated to D); *New Essays Concerning Human Understanding,* translated by Alfred Gideon Langley, La Salle, The Open Court Publishing Company, 1949 (abbreviated to L); or *Leibniz—Basic Writings,* translated by George R. Montgomery, La Salle, The Open Court Publishing Company, 1968 (abbreviated to M).

7.1 *Discourse on Metaphysics* (hereafter *Discourse*), XXIV, XXV, (M, pp. 41–43)
7.2 *On the Supersensible in Knowledge* (hereafter *Supersensible*), (D, pp. 158–160)
7.3 *Discourse,* XXVII, (M, p. 46)
7.4 *Letter to M. Coste,* (D, p. 259)
7.5 *New Essays,* II, i, (D, pp. 236–237)
7.6 *On Descartes' "Principles",* (D, p. 49)
7.7 *Monadology,* 29–36, (D, pp. 312–313)
7.8 *Reflections on Locke's "Essay",* (D, pp. 100–101)
7.9 *On Wisdom,* (D, pp. 296–297)

7.10 *Discourse*, XXVI, (M, pp. 44–45)
7.11 *Supersensible*, (D, pp. 163–164)
7.12 *New Essays*, II, i, 2, (D, p. 204)
7.13 *New Essays*, I, i, 10–11, (D, p. 199)
7.14 *New Essays*, Preface, (D, p. 175)
7.15 *Discourse*, XXVIII, (M, pp. 46–47)
7.16 *Discourse*, XXIX, (M, p. 48)
7.17 *Supersensible*, (D, pp. 160–161)
7.18 *Supersensible*, (D, pp. 161–162)
7.19 *Letters to Clarke*, 2nd Paper, (D, pp. 330–331)
7.20 *New Essays*, I, i, 4, (D, p. 197)
7.21 *New Essays*, I, i, 5, (D, p. 197)
7.22 *On Knowledge, Truth, and Ideas*, (D, pp. 29–30)
7.23 *New Essays*, IV, vii, 10, (L, p. 472)
7.24 *On Descartes' "Principles"*, (D, pp. 48–49)
7.25 *New Essays*, II, xiii, 17, (D, p. 214)
7.26 *Letters to Clarke*, 3rd Paper, (D, pp. 334–335)
7.27 *On Descartes' "Principles"*, (D, p. 53)
7.28 *New Essays*, II, xvii, 1, 3, (D, pp. 215–216)
7.29 *Reflections on Locke's "Essay"*, (D, p. 103)
7.30 *Reply to M. Foucher*, (D, pp. 70–71)
7.31 *Remarks on Letter from Arnauld*, (M, pp. 116–117)
7.32 *Letter to Arnauld*, July, 1686, (M, pp. 121–122)
7.33 *Discourse*, XIII, (M, pp. 19–20)
7.34 *Discourse*, XIII, (M, pp. 22–23)

LEIBNIZ · Selections

The objects of all knowledge—ideas and concepts.

7.1 In order to understand better the nature of ideas it is necessary to touch somewhat upon the various kinds of knowledge. When I am able to recognize a thing among others, without being able to say in what its differences or characteristics consist, the knowledge is confused. Sometimes indeed we may know clearly, that is without being in the slightest doubt, that a poem or a picture is well or badly done because there is in it an "I know not what" which satisfies or shocks us. Such knowledge is not yet distinct. It is when I am able to explain the peculiarities which a thing has, that the knowledge is called distinct. Such is the knowledge of an assayer who discerns the true gold from the false by means of certain proofs or marks which make up the definition of gold. But distinct knowledge has degrees, because ordinarily the conceptions which enter into the definitions will themselves have need of definition, and are only known confusedly. When at length everything which enters into a definition or into distinct knowledge is known distinctly, even back to the primitive conceptions, I call that knowledge adequate. When my mind understands at once and distinctly all the primitive ingredients of a conception, then we have intuitive knowledge. This is extremely rare as most human knowledge is only confused or indeed assumed. It is well also to distinguish nominal from real definition. I call a definition nominal when there is doubt whether an exact conception of it is possible; as for instance, when I say that an endless screw is a line in three dimensional space whose parts are congruent or fall one upon another. Now although this is one of the reciprocal properties of an endless screw, he who did not know from elsewhere what an endless screw was could doubt if such a line were possible, because the other lines whose ends are congruent (there are only two: the circumference of a circle and the straight line) are plane figures, that is to say they can be described *in plano.* This instance enables us to see that any reciprocal property can serve as a nominal definition, but when the property brings us to see the possibility of a thing it makes the definition real, and as long as one has only a nominal definition he cannot be sure of the consequences which he draws, because if it conceals a contradiction or an impossibility he would be able to draw the opposite conclusions. That is why truths do not depend upon names and are not arbitrary, as some of our new philosophers think. There is also a considerable difference among real definitions, for when the possibility proves itself only by experience, as in the definition of quicksilver, whose possibility we know because such a body, which is both an extremely heavy fluid and quite volatile, actually exists, the definition is merely real and nothing more. If, however, the proof of the possibility is *a priori*, the definition is not only real but also causal as for instance when it contains the possible generation of a thing. Finally when the definition, without assuming anything which requires a proof *a priori* of its possibility, carries the analysis clear to the primitive conception, the definition is perfect or essential.

Now it is manifest that we have no idea of a conception when it is impossible. And in case the knowledge, where we have the idea of it, is only assumed, we do not visualize it because such a conception is known only in like manner as conceptions internally impossible. And if it be in fact possible, it is not by this kind of knowledge that we learn its possibility. For instance, when I am thinking of a thousand or of a chiliagon, I frequently do it without contemplating the idea. Even if I say a thousand is ten times a hundred, I frequently do not trouble to think what ten and a hundred are, because I assume that I know, and I do not consider it necessary to stop just at present to conceive of them. Therefore it may well happen, as it in fact does happen often enough, that I am mistaken in regard to a conception which I assume that I understand, although it is an impossible truth or at least is incompatible with others with which I join it, and whether I am mistaken or not, this way of assuming our knowledge remains the same. It is, then, only when our knowledge is clear in regard to confused conceptions, and when it is intuitive in regard to those which are distinct, that we see its entire idea.

7.2 The purpose of nominal definitions is to give sufficient marks by which the thing may be recognized; for example, assayers have marks by which they distinguish gold from every other metal, and even if a man had never seen gold these signs might be taught him so that he would infallibly recognize it if he should some day meet with it. But it is not the same with these sensible qualities; and marks to recognize blue, for example, could not be given if we had never seen it. So that blue is its own mark, and in order that a man may know what blue is it must necessarily be shown to him.

It is for this reason that we are accustomed to say that the *notions* of these qualities are *clear,* for they serve to recognize them; but that these same notions are not *distinct,* because we cannot distinguish or develop that which they include. It is an *I know not what* of which we are conscious, but for which we cannot account. Whereas we can make another understand what a thing is of which we have some description or nominal definition, even although we should not have the thing itself at hand to show him. However we must do the senses the justice to say that, in addition to these occult qualities, they make us know other qualities which are more manifest and which furnish more distinct notions. And these are those which we ascribe to the *common sense,* because there is no external sense to which they are particularly attached and belong. And here definitions of the terms or words employed may be given. Such is the idea of *numbers,* which is found equally in sounds, colors, and touches. It is thus that we perceive also *figures,* which are common to colors and to touches, but which we do not notice in sounds. Although it is true that in order to conceive distinctly numbers and even figures, and to form sciences of them, we must come to something which the senses cannot furnish, and which the understanding adds to the senses.

As therefore our soul compares (for example) the numbers and figures which are in colors with the numbers and figures which are found by touch, there must be an *internal sense,* in which the perceptions of these different external senses are found united. This is what is called *imagination,* which comprises at once the *notions of the particular senses,* which are *clear* but *confused,* and the *notions of the common sense,* which are clear and distinct. And these clear and distinct ideas which are subject to the imagination are the objects of the *mathematical sciences,* namely of arithmetic and geometry, which are *pure* mathematical sciences, and of the application of these sciences to nature, forming mixed mathematics. It is evident also that particular sensible qualities are susceptible of explanations and of reasonings only in so far as they involve what is common to the objects of several external senses, and belong to the internal sense. For those who try to explain sensible qualities distinctly always have recourse to the ideas of mathematics, and these ideas always involve *size* or multitude of parts. It is true that the mathematical sciences would not be demonstrative, and would consist in a simple induction or observation, which would never assure us of the perfect generality of the truths there found, if something higher and which intelligence alone can furnish did not come to the aid of *imagination* and the *senses.*

There are, therefore, objects of still other nature, which are not included at all in what is observed in the objects of the senses in particular or in common, and which consequently are not objects of the imagination either. Thus besides the *sensible* and *imageable,* there is that which is purely *intelligible,* as being the *object of the understanding alone,* and such is the object of my thought when I think of myself.

This thought of the *Ego,* which informs me of sensible objects, and of my own action resulting therefrom, adds something to the objects of the senses. To think a color and to observe that one thinks it, are two very different thoughts, as different as the color is from the Ego which thinks it. And as I conceive that other beings may also have the right to say I, or that it could be said for them, it is through this that I conceive what is called *substance* in general, and it is also the consideration of the Ego itself which furnishes other *metaphysical* notions, such as cause, effect, action, similarity, etc., and even those of *logic* and of *ethics.* Thus it can be said that there is nothing in the understanding which does not come from the senses, except the understanding itself, or that which understands.

There are then three grades of notions: the *sensible only,* which are the objects appropriate to each sense in particular; *the sensible and at the same time intelligible,* which pertain to the common sense; and the *intelligible only,* which belong to the understanding. The first and the second are both imageable, but the third are above the imagination. The second and third are intelligible and distinct; but the first are confused, although they are clear or recognizable.

7.3 . . . We may say that knowledge is received from without through the medium of the senses because certain exterior things contain or express more particularly the causes which determine us to certain thoughts. Because in the ordinary uses of life we attribute to the soul only that which belongs to it most manifestly and particularly, and there is no advantage in going further. When, however, we are dealing with the exactness of metaphysical truths, it is important to recognize the powers and independence of the soul which extend infinitely further than is commonly supposed. In order, therefore, to avoid misunderstandings it would be well to choose separate terms for the two. These expressions which are in the soul whether one is conceiving of them or not may be called ideas, while those which one conceives of or constructs may be called conceptions, *conceptus.* But whatever terms are used, it is always false to say that all our conceptions come from the so-called external senses, because those conceptions which I have of myself and of my thoughts, and consequently of being, of substance, of action, of identity, and of many others came from an inner experience.

The two basic kinds of knowledge—truths of reason and truths of fact.

7.4 A truth is *necessary* when the opposite implies contradiction, and when it is not necessary it is called *contingent.* That God exists, that all right angles are equal, etc., are necessary truths; but that I myself exist, and that there are bodies in nature which show an angle actually right, are contingent truths. For the whole universe might be otherwise; time, space, and matter being absolutely indifferent to motion and forms. And God has chosen among an infinite number of possibles what he judged most fit. But since he has chosen, it must be affirmed that everything is comprised in his choice and that nothing could be changed, since he has once for all foreseen and regulated all; he who could not regulate things piece-meal and by fits and starts. Therefore the sins and evils which he has judged proper to permit for greater goods, are comprised in his choice.

7.5 *Primitive* truths, which are known by *intuition,* are of two kinds, like the *derivative.* They are either truths of *reason,* or truths of *fact.* Truths of reason are necessary, and those of fact are contingent. Primitive truths of *reason* are those which I call by the general name of *identical,* because it seems that they do nothing but repeat the same thing without giving us any information. They are affirmative or negative. . . .

As respects *primitive truths of fact,* they are the immediate internal experiences of an *immediateness of feeling.* And here it is that the first truth of the Cartesians or of St. Augustine: *I think, hence I am,* that is, *I am a thing which thinks,* holds good. But it should be known that just as the identicals are general or particular, and that the one class is as clear as the other (since it is just as clear to say that *A is A,* as to say that *a thing is what it is*), so it is also with first truths of fact. For not only is it clear to me immediately that *I think;* but it is just as clear to me that *I have different thoughts;* that sometimes *I think of A,* and that sometimes *I think of B,* etc. Thus the Cartesian principle is good, but it is not the only one of its kind. You see by this that all *primitive truths* of reason or of fact have this in common, that they cannot be proved by anything more certain.

7.6 *I think therefore I am* is well remarked by Descartes to be among the first truths. But it was but just that he should not neglect others equal to this. In general, therefore, it may be said: Truths are either of fact or of reason. The first of the truths of reason is, as Aristotle rightly observed, the principle of *contradiction* or, what amounts to the same thing, of *identity*. First truths of fact are as many as the immediate perceptions, or those of consciousness, so to speak. Moreover not only am I conscious of my thinking but also of my thoughts. Nor is it more certain that I think than that this or that is thought by me. Thus first truths of fact may not inconveniently be traced back to these two, *I think*, and *Various things are thought by me*. Whence it follows not only that I am but also that I am affected in various ways.

7.7 The knowledge of necessary and eternal truths is what distinguishes us from mere animals and furnishes us with *reason* and the sciences, raising us to a knowledge of ourselves and of God. This is what we call in us the rational soul or *spirit*.

It is also by the knowledge of necessary truths, and by their abstractions, that we rise to *acts of reflection*, which make us think of that which calls itself "I," and to observe that this or that is within *us:* and it is thus that, in thinking of ourselves, we think of being, of substance, simple or compound, of the immaterial and of God himself, conceiving that what is limited in us is in him without limits. And these reflective acts furnish the principal objects of our reasonings.

Our reasonings are founded on *two great principles*, that of *contradiction*, in virtue of which we judge that to be *false* which involves contradiction, and that *true*, which is opposed or contradictory to the false.

And that of *sufficient reason*, in virtue of which we hold that no fact can be real or existent, no statement true, unless there be a sufficient reason why it is so and not otherwise, although most often these reasons cannot be known to us.

There are also two kinds of *truths*, those of *reasoning* and those of *fact*. Truths of reasoning are necessary and their opposite is impossible, and those of *fact* are contingent and their opposite is possible. When a truth is necessary its reason can be found by analysis, resolving it into more simple ideas and truths until we reach those which are primitive.

It is thus that mathematicians by analysis reduce speculative *theorems* and practical *canons* to *definitions*, *axioms* and *postulates*.

And there are finally simple ideas, definitions of which cannot be given; there are also axioms and postulates, in a word, *primary principles*, which cannot be proved, and indeed need no proof; and these are *identical propositions*, whose opposite involves an express contradiction.

But there must also be a *sufficient reason* for *contingent truths*, or those *of fact*,—that is, for the sequence of things diffused through the universe of created objects—where the resolution into particular reasons might run into a detail without limits, on account of the immense variety of the things in nature and the division of bodies *ad infinitum*. There is an infinity of figures and of movements, present and past, which enter into the efficient cause of my present writing, and there is an infinity of slight inclinations and dispositions, past and present, of my soul, which enter into the final cause.

7.8 My opinion is, then, that nothing ought to be taken as primitive principles except *experiences* and the axiom of *identity,* or, what is the same thing, *contradiction,* which is primitive, since otherwise there would be no difference between truth and falsehood; and since all researches would cease at the start if to say yes or no were indifferent. We cannot, therefore, prevent ourselves from assuming this principle as soon as we wish to reason. All other truths are capable of proof, and I highly esteem the method of Euclid, who without stopping at what would be thought to be sufficiently proved by the so-called ideas, has proved, for example, that in a triangle one side is always less than the other two together. Yet Euclid was right in taking some axioms for granted, not as if they were truly primitive and undemonstrable, but because he would have come to a standstill if he had wished to draw conclusions only after an accurate discussion of principles. Thus he judged it proper to content himself with having pushed the proofs up to this small number of propositions, so that it can be said that if they are true, all that he says is also true. He has left to others the trouble of demonstrating further these principles themselves, which, besides, are already justified by experience; but in these matters this does not satisfy us. This is why Appolonius, Proclus and others have taken the trouble to demonstrate some of Euclid's axioms. This manner of proceeding ought to be imitated by philosophers in order to arrive finally at some established positions, even if they be but provisional, after the way of which I have just spoken.

7.9 THE ART OF DISCOVERY consists in the following maxims:

1. In order to know a thing we must consider all the requisites of that thing, that is to say, all that which suffices to distinguish it from every other thing. This is what is called definition, nature, reciprocal property.

2. Having once found a means of distinguishing it from every other thing, this same first rule must be applied to the consideration of each condition or requisite which enters into this means, and all the requisites of each requisite must be considered. And this is what I call true *analysis* or distribution of the difficulty into several parts.

3. When we have pushed the analysis to the end, that is to say, when we have considered the requisites which enter into the consideration of the thing proposed and even the requisites of the requisites, and when we have finally come to the consideration of some natures which are understood only through themselves, which are without requisites and which need nothing outside of themselves in order to be conceived, we have reached a *perfect knowledge* of the thing proposed.

4. When the thing deserves it, we must try to have this perfect knowledge present in the mind all at once, and this is done by repeating the analysis several times until it seems to us that we see the whole of it at a single glance of the mind. And for this result a certain order in repetition must be observed.

5. The mark of perfect knowledge is when nothing presents itself in the thing in question for which we cannot account and when there is no conjuncture the outcome of which we cannot *predict* beforehand. It is very difficult to carry through an analysis of things, but it is not so difficult to complete the analysis of truths of which we have need. Because the analysis of a truth is completed when its demonstration has been found, and it is not always necessary to complete the analysis of the subject or predicate in order to find the demonstration of a proposition. Most often the beginning of the analysis of a thing suffices for the analysis or perfect knowledge of the truth which we know of the thing.

All knowledge is innate and dependent on God.

7.10 In order to see clearly what an idea is, we must guard ourselves against a misunderstanding. Many regard the idea as the form or the differentiation of our thinking, and according to this opinion we have the idea in our mind, in so far as we are thinking of it, and each separate time that we think of it anew we have another idea although similar to the preceding one. Some, however, take the idea as the immediate object of thought, or as a permanent form which remains even when we are no longer contemplating it. As a matter of fact our soul has the power of representing to itself any form or nature whenever the occasion comes for thinking about it, and I think that this activity of our soul is, so far as it expresses some nature, form or essence, properly the idea of the thing. This is in us, and is always in us, whether we are thinking of it or no. (Our soul expresses God and the universe and all essences as well as all existences.) This position is in accord with my principles that naturally nothing enters into our minds from outside.

It is a bad habit we have of thinking as though our minds receive certain messengers, as it were, or as if they had doors or windows. We have in our minds all those forms for all periods of time because the mind at every moment expresses all its future thoughts and already thinks confusedly of all that of which it will ever think distinctly. Nothing can be taught us of which we have not already in our minds the idea. This idea is as it were the material out of which the thought will form itself. This is what Plato has excellently brought out of his doctrine of reminiscence, a doctrine which contains a great deal of truth, provided that it is properly understood and purged of the error of pre-existence, and provided that one does not conceive of the soul as having already known and thought at some other time what it learns and thinks now. Plato has also confirmed his position by a beautiful experiment. He introduces a small boy, whom he leads by short steps, to extremely difficult truths of geometry bearing on incommensurables, all this without teaching the boy anything, merely drawing out replies by a well arranged series of questions. This shows that the soul virtually knows those things, and needs only to be reminded (animadverted) to recognize the truths. Consequently it possesses at least the idea upon which those truths depend. We may say even that it already possesses those truths, if we consider them as the relations of the ideas.

7.11 In truth there are *experiments* which succeed numberless times and ordinarily, and yet it is found in some extraordinary cases that there are *instances* where the experiment does not succeed. . . . for example, we ordinarily find that two lines which continually approach each other finally meet, and many people will almost swear that this could never be otherwise. And nevertheless geometry furnishes us with extraordinary lines, which are for this reason called *asymptotes,* which prolonged *ad infinitum* continally approach each other, and nevertheless never meet.

This consideration shows also that there is a *light born with us.* For since the senses and inductions could never teach us truths which are thoroughly universal, nor that which is absolutely necessary, but only that which is, and that which is found in particular examples; and since we nevertheless know necessary and universal truths of the sciences, a privilege which we have above the brutes; it follows that we have derived these truths in part from what is within us. Thus we may lead a child to these by simple interrogations, after the manner of Socrates, without telling him anything, and without making him experiment at all upon the truth of what is asked him. And this could very easily be practiced in numbers and other similar matters.

I agree, nevertheless, that in the present state the external senses are necessary to us for thinking, and that, if we had none, we could not think. But that which is necessary for something does not for all that constitute its essence. Air is necessary for life, but our life is something else than air. The senses furnish us the matter for reasoning, and we never have thoughts so abstract that something from the senses is not mingled therewith; but reasoning requires something else in addition to what is from the senses.

7.12 Experience is, I admit, necessary in order that the soul be determined to such or such thoughts, and in order that it take notice of the ideas which are in us; but by what means can experience and the senses give ideas? Has the soul windows? does it resemble tablets? is it like wax? It is evident that all who think of the soul thus, make it at bottom corporeal. This axiom received among the philosophers, will be opposed to me, that there is nothing in the soul which does not come from the senses. But the soul itself and its affections must be excepted.

7.13 If the mind had only the simple capacity of receiving knowledge or passive power for it, as indeterminate as that which the wax has for receiving figures, and the blank tablet for receiving letters, it would not be the source of necessary truths, as I have just shown it to be; for it is incontestable that the senses do not suffice to show their necessity, and that thus the mind has a disposition (as much active as passive) to draw them itself from its depths; although the senses are necessary in order to give it the occasion and attention for this, and to carry it to some rather than to others. You see therefore, that these people, otherwise very able, who are of a different opinion, seem not to have sufficiently meditated on the consequences of the difference which there is between necessary or eternal truths and the truths of experience, as I have already remarked, and as all our discussion shows. The original proof of necessary truths comes from the understanding alone, and the other truths come from experiences or from the observations of the senses. Our mind is capable of knowing both, but it is the source of the former; and whatever number of particular experiences we may have of a universal truth, we could not be assured of it forever by induction, without knowing its necessity through the reason.

It is the particular relation of the human mind to these truths which renders the exercise of the faculty easy and natural as respects them, and which causes them to be called innate. It is not, therefore, a naked faculty which consists in the mere possibility of understanding them; it is a disposition, an aptitude, a preformation, which determines our soul and which brings it about that they may be derived from it. Just as there is a difference between the figures which are given to the stone or marble indifferently, and those which its veins already mark out, or are disposed to mark out, if the workman profits by them.

The intellectual ideas, which are the source of necessary truths, do not come from the senses; and you recognize that there are ideas which are due to the reflection of the mind when it reflects upon itself.

7.14 Reflection is nothing else than attention to what is in us, and the senses do not give us that which we already carry with us. This being so, can it be denied that there is much that is innate in our mind, since we are innate, so to say, in ourselves? and that there is in us ourselves, being, unity, substance, duration, change, action, perception, pleasure, and a thousand other objects of our intellectual ideas? And these objects being immediate to our understanding and always present (although they cannot be always perceived on account of our distractions and wants), why be astonished that we say that these ideas, with all which depends on them, are innate in us? I have made use also of the comparison of a block of marble which has veins, rather than of a block of marble wholly even, or of blank tablets, that is to say, of what is called among philosophers *tabula rasa.* For if the soul resembled these blank tablets, truths would be in us as the figure of Hercules is in marble when the marble is entirely indifferent toward receiving this figure or some other. But if there were veins in the block which should mark out the figure of Hercules rather than other figures, the block would be more determined thereto, and Hercules would be in it as in some sort innate, although it would be necessary to labor in order to discover these veins and to cleanse them by polishing and by cutting away that which prevents them from appearing. It is thus that ideas and truths are innate in us, as inclinations, dispositions, habits, or natural capacities. . . .

7.15 In the strictly metaphysical sense no external cause acts upon us excepting God alone, and he is in immediate relation with us only by virtue of our continual dependence upon him. Whence it follows that there is absolutely no other external object which comes into contact with our souls and directly excites perceptions in us. We have in our souls ideas of everything, only because of the continual action of God upon us, that is to say, because every effect expresses its cause and therefore the essences of our souls are certain expressions, imitations or images of the divine essence, divine thought and divine will, including all the ideas which are there contained. We may say, therefore, that God is for us the only immediate external object, and that we see things through him. For example, when we see the sun or the stars, it is God who gives to us and preserves in us the ideas and whenever our senses are affected according to his own laws in a certain manner, it is he, who by his continual concurrence, determines our thinking. God is the sun and the light of souls. . . .

7.16 It is indeed inconceivable that the soul should think using the ideas of something else. The soul when it thinks of anything must be affected effectively in a certain manner, and it must needs have in itself in advance not only the passive capacity of being thus affected, a capacity already wholly determined, but it must have besides an active power by virtue of which it has always had in its nature the marks of the future production of this thought, and the disposition to produce it at its proper time. All of this shows that the soul already includes the idea which is comprised in any particular thought.

Mathematical knowledge is innate and necessary, based on the principle of contradiction.

7.17 *Being* itself and *truth* are not known wholly through the senses; for it would not be impossible for a creature to have long and orderly dreams, resembling our *life,* of such a sort that everything which it thought it perceived through the senses would be but mere *appearances.* There must therefore be something beyond the senses, which distinguishes the true from the apparent.

But the truth of the demonstrative sciences is exempt from these doubts, and must even serve for judging of the truth of sensible things. For as able philosophers, ancient and modern, have already well remarked:—if all that I should think that I see should be but a dream, it would always be true that I who think while dreaming, would be something, and would actually think in many ways, for which there must always be some reason.

Thus what the ancient Platonists have observed is very true, and is very worthy of being considered, that the existence of sensible things and particularly of the *Ego* which thinks and which is called spirit or soul, is incomparably more sure than the existence of sensible things: and that thus it would not be impossible, speaking with metaphysical rigor, that there should be at bottom only these intelligible substances, and that sensible things should be but appearances. While on the other hand our lack of attention makes us take sensible things for the only true things. It is well also to observe that if I should discover any demonstrative truth, mathematical or other, while dreaming (as might in fact be), it would be just as certain as if I had been awake. This shows us how intelligible truth is independent of the truth or of the existence outside of us of sensible and material things.

This conception of *being* and of *truth* is found therefore in the Ego and in the understanding, rather than in the external senses and in the perception of exterior objects. There we find also what it is to affirm, to deny, to doubt, to will, to act. But above all we find there the *force of the consequences* of reasoning, which are a part of what is called the *natural light.*

7.18 It is by this *natural light* that the *axioms* of mathematics are recognized; for example, that *if from two equal things the same quantity be taken away the things which remain are equal;* likewise that *if in a balance everything is equal on the one side and on the other, neither will incline,* a thing which we foresee without ever having experienced it. It is upon such foundations that we construct arithmetic, geometry, mechanics and the other demonstrative sciences; in which, in truth, the senses are very necessary, in order to have certain ideas of sensible things, and experiments are necessary to establish certain facts, and even useful to verify reasonings as by a kind proof. But the force of the demonstrations depends upon intelligible notions and truths, which alone are capable of making us discern what is necessary, and which, in the conjectural sciences, are even capable of determining demonstratively the degree of probability upon certain given suppositions, in order that we may choose rationally among opposite appearances, the one which is greatest. Nevertheless this part of the art of reasoning has not yet been cultivated as much as it ought to be.

But to return to *necessary truths,* it is generally true that we know them only by this natural light, and not at all by the experiences of the senses. For the senses can very well make known, in some sort, what is, but they cannot make known what *ought to be* or could not be otherwise.

7.19 The great foundation of mathematics is the *principle of contradiction or identity,* that is, that a proposition cannot be *true and false* at the same time; and that therefore *A* is *A*, and cannot be *not A*. This single principle is sufficient to demonstrate every part of arithmetic and geometry, that is, all *mathematical* principles. But in order to proceed from *mathematics* to *natural philosophy,* another principle is requisite, as I have observed in my *Theodicœa:* I mean, *the principle of a sufficient reason, viz:* that nothing happens without a *reason* why it should be so, rather than *otherwise.* And therefore *Archimedes* being desirous to proceed from *mathematics* to *natural philosophy,* in his book *De Aequilibrio,* was obliged to make use of a particular case of the great principle of *a sufficient reason.* He takes it for granted, that if there be a *balance,* in which every thing is alike on both sides, and if equal weights are hung on the two ends of that balance, the whole will be at rest. 'Tis because no reason can be given, why one side should weigh down, rather than the other. Now, by that single principle, *viz:* that *there ought to be a sufficient reason why things should be so, and not otherwise,* one may demonstrate the being of a *God,* and all the other parts of *metaphysics* or *natural theology;* and even, in some measure, those principles of *natural philosophy,* that are independent upon *mathematics:* I mean, the *dynamic* principles, or the principles *of force.*

7.20 To reply again to what you say against the general approbation given to the two great speculative principles, [i.e. the principle of contradiction and the principle of sufficient reason], which are nevertheless the best established, I may say to you that even if they were not known, they would none the less be innate, because they are recognized as soon as heard; but I will add further, that at bottom everyone knows them and makes use at every moment of the principle of contradiction (for example) without examining it distinctly, and there is no barbarian, who, in a matter which he considers serious, would not be shocked at the conduct of a liar who contradicts himself. Thus these maxims are employed without being expressly considered. And it is very much so that we have virtually in mind the propositions suppressed in enthymemes, which are set aside not only externally, but also in our thought.

7.21 We have an infinity of knowledge of which we are not always conscious, not even when we have need of it. It is for memory to retain it and for reminiscence to represent it to us, as it often does, but not always when needed. This is very well called remembrance (*subvenire*), for reminiscence requires some help. And it must be that in this multiplicity of our knowledge we are determined by something to renew one portion rather than another, since it is impossible to think distinctly and at once of *all that we know.* . . .

In a sense it must be said that all arithmetic and all geometry are innate and are in us virtually, so that they may be found there if we consider attentively and arrange what is already in the mind, without making use of any truth learned by experience or by the tradition of others, as Plato has shown in a dialogue, where he introduces Socrates leading a child to abstract truths by mere questions, without telling him anything. We may therefore invent these sciences in our libraries and even with closed eyes, without learning by sight or even by touch, the truths which we need; although it is true that we would not consider the ideas in question if we had never seen or touched anything.

7.22 As for composite notions where each of the component marks is sometimes clearly known, although in a confused way, as gravity, color, aqua fortis, which form a part of those [the marks] of gold, it follows that such a knowledge of gold is distinct without always being *adequate.* But when all the elements of a distinct notion are themselves also known distinctly, or when its analysis is complete, the idea is *adequate.* I do not know that men can give a perfect example of this, although the knowledge of numbers approaches it very nearly. It very often happens, nevertheless, especially in a long analysis, that we do not perceive the whole nature of the object at one time, but substitute in place of the things, signs, the explanation of which, in any present thought, we are accustomed for the sake of abbreviation to omit, knowing or believing that we can give it; thus when I think a chiliogon, or polygon with a thousand equal sides, I do not always consider the nature of a side, of equality, and of the number thousand (or of the cube of ten); but these words, the sense of which presents itself to my mind in an obscure, or at least imperfect manner, take the place to me of the ideas which I have of them, because my memory attests to me that I know the signification of these words, and that their explanation is not now necessary for any judgment. I am accustomed to call this thought *blind* or again *symbolical;* and we make use of it in algebra, in arithmetic and almost everywhere. And assuredly when a question is very complex, we cannot embrace in thought at the same time all the elementary notions which compose it; but when this can be done, or at least as far as this can be done, I call this knowledge *intuitive.* We can only have an intuitive knowledge of a distinct, primitive notion, as most often we have only a symbolical knowledge of composite ideas.

From this it clearly follows that even of the things which we know distinctly, we only conceive the ideas in as far as they form the object of intuitive thought.

7.23 That two and two are four is not a truth at once immediate, supposing that *four* signifies three and one. We can then demonstrate it, and in this way:—

Definitions.—

(1) *Two* is one and one.
(2) *Three* is two and one.
(3) *Four* is three and one.

Axiom.—Putting equal things in their place, the equality remains.

Demonstration.— 2 and 2 is 2 and 1 and 1 (by def. 1) .... 2 + 2
2 and 1 and 1 is 3 and 1 (by def. 2) ..... 2 + 1 + 1
3 and 1 is 4 (by def. 3) .... 3 + 1
4

Then (by the axiom) 2 and 2 is 4. Which was to be demonstrated. I might, instead of saying that 2 and 2 is 2 and 1 and 1, say that 2 and 2 is equal to 2 and 1 and 1, and thus with the others. But it may be understood throughout in order to shorten the process; and that, in virtue of another axiom which states that a thing is equal to itself, or that what is the same, is equal.

7.24 We cannot otherwise doubt of mathematical demonstrations except as error may be feared in the reckoning of arithmeticians. This cannot be remedied except by examining the reckoning often, or by different reckonings, confirming proofs being added. This weakness of the human mind, arising from want of attention and of memory, cannot be perfectly removed, and what is adduced by Descartes as a remedy is useless. The same thing suffices as guarantee in other departments which suffices in mathematics; indeed all reasoning, even the Cartesian, however proved or accurate, will yet be subject to this doubt, whatever may finally be thought of any powerful deceptive genius or of the difference between sleep and wakefulness.

Space is relational, not absolute.

7.25 Some have believed that God is the place of things. Lessius and Guerike, if I am not mistaken, were of this opinion; but then place contains something more than we attribute to space which we strip of all action; and in this way it is no more a substance than time, and if it has parts it could not be God. It is a relation, an order, not only among existing things, but also among possible things as they may exist. But its truth and reality is founded in God, like all the eternal truths.

It is best then to say that space is an order, but that God is its source.

7.26 As for my own opinion, I have said more than once, that I hold *space* to be something *merely* relative, as *time* is; that I hold it to be an order of *coexistences,* as *time* is an *order of successions.* For *space* denotes, in terms of possibility, *an order* of things which exist at the same time, considered as existing *together;* without inquiring into their particular manner of existing. And when many things are seen *together,* one perceives *that order of things among themselves.*

I have many demonstrations, to confute the fancy of those who take *space* to be a *substance,* or at least an absolute *being.* But I shall only use, at the present, one demonstration, which the author here gives me occasion to insist upon. I say then, that if *space* was an absolute *being,* there would something happen, for which it would be impossible there should be a *sufficient reason.*

Which is against my Axiom. And I can prove it thus. *Space* is something absolutely *uniform;* and, without the things placed in it, *one point* of space does not absolutely differ in any respect whatsoever from *another point* of space. Now from hence it follows, (supposing space to be something in itself, besides the *order of bodies among themselves,*) that 'tis impossible there should be a *reason,* why God, preserving the same situations of bodies among themselves, should have placed them in space after *one certain particular manner,* and not *otherwise;* why everything was not placed the *quite contrary way,* for instance, by changing *east* into *west.* But if space is nothing else, but that *order* or *relation;* and is nothing at all without bodies, but the possibiility of placing them; then those two states, the *one* such as it now is, the *other* supposed to be the quite contrary way, would not at all differ from one another. *Their difference* therefore is only to be found in our *chimerical* supposition of the *reality* of space in itself. But in truth the *one* would exactly be the same thing as the *other,* they being absolutely *indiscernible;* and consequently there is no room to enquire after a reason of the preference of the one to the other.

The infinitely large and the infinitely small.

7.27 Although we are finite, we may nevertheless know many things concerning the infinite, as concerning asymptotic lines or those which produced *ad infinitum* always approach but never meet; concerning infinite spaces not greater than a finite length as respects area; concerning the last members of series which are infinite. Otherwise we should know nothing certain concerning God either. Moreover it is one thing to know something of a thing, another to comprehend the thing, that is to have in our power whatever lies hidden in the thing.

7.28 Take a straight line and prolong it in such a way that it is double the first. Now it is clear that the second, being perfectly similar to the first, can be doubled in the same way in order to give a third, which is also similar to the preceding; and the same *ratio* always holding it will never be possible to stop; thus the line can be prolonged *ad infinitum;* in such a way that the consideration of the infinite comes from that of similarity or of the same *ratio,* and its origin is the same as that of universal and necessary truths. This shows how what gives completion to the conception of this idea is found in us and could not come from the experiences of the senses; just as necessary truths could not be proved by induction nor by the senses. The idea of the *absolute* is in us internally, like that of being. These absolutes are nothing but the attributes of God and it can be said that they are no less the source of ideas than God is himself the principle of beings. The idea of the absolute in relation to space, is no other than that of the immensity of God, and so of the others. But we deceive ourselves in wishing to imagine an absolute space, which would be an infinite whole, composed of parts. There is no such thing. It is a notion which involves a contradiction, and these infinite wholes, and their opposites, the infinitely minutes, are only admissible in the calculations of geometers, just like the imaginary roots of algebra.

I believe that we have a positive idea [of infinity], and this idea will be true provided it is not conceived as an infinite whole but as an absolute or attribute without limits, which is the case as regards the *eternity* in the necessity of the existence of God, without depending on parts and without forming the notion by an addition of times. From this is also seen, as I have already said, that the origin of the notion of the infinite comes from the same source as that of necessary truths.

7.29 I believe, in truth, that, strictly speaking, it may be said that there is no space, no time and no number which is infinite, but that it is only true that however great may be a space, a time or a number, there is always another larger than it, *ad infinitum;* and that thus the true infinite is not found in a whole made up of parts. It is none the less, however, found elsewhere; namely, in the *Absolute,* which is without parts and which has influence upon compound things because they result from the limitation of the absolute. Hence the *positive infinite* being nothing else than the absolute, it may be said that there is in this sense a positive idea of the infinite, and that it is anterior to that of the finite.

7.30 As regards *indivisibles,* when by that word is understood simple extremities of time or of line, new extremities could not be conceived in them, nor actual nor potential parts. Thus points are neither large nor small, and there needs no leap to pass them. However, although there are such *indivisibles* everywhere, the continuum is not composed of them, as the objections of the sceptics appear to suppose. In my opinion these objections have nothing insurmountable about them, as will be found by reducing them to form. Gregory of St. Vincent has well shown by the calculations even of divisibility *ad infinitum,* the place where Achilles ought to overtake the tortoise which precedes him, according to the proportion of velocities. Thus geometry serves to dissipate these apparent difficulties.

I am so much in favor of the actual infinite that instead of admitting that nature abhors it, as is commonly said, I hold that it affects it everywhere in order better to mark the perfections of its author. So I believe that there is no part of matter which is not, I do not say divisible, but actually divided; and consequently the least particle must be regarded as a world full of an infinity of different creatures.

ALL *true propositions are analytic, i.e. the concept of the predicate is contained in the concept of the subject.*

7.31 . . . In order to determine the concept of an individual substance it is good to consult the concept which I have of myself, just as the specific concept of the sphere must be consulted in order to determine its properties. Nevertheless, there is a great difference in the two cases for the concept of myself and of any other individual substance, is infinitely more extended and more difficult to understand than is a specific concept like that of a sphere which is only incomplete. It is not sufficient that I feel myself as a substance which thinks; I must also distinctly conceive whatever distinguishes me from all other spirits. But of this I have only a confused experience.

Therefore, although it is easy to determine that the number of feet in the diameter is not involved in the concept of the sphere in general, it is not so easy to decide if the journey which I intend to make is involved in my concept; otherwise, it would be as easy for us to become prophets as to be Geometers. I am uncertain whether I will make the journey but I am not uncertain that, whether I make it or no, I will always be myself. Such human previsions are not the same as distinct notions or distinct knowledge. They appear to us undetermined because the evidences or marks which are found in our substance are not recognizable by us. Very much as those who regard sensations merely, ridicule one who says that the slightest movement is communicated as far as matter extends, because experience alone could not demonstrate this to them. When, however, they consider the nature of motion and matter they are convinced of it. It is the same here when the confused experience, which one has of his individual concept in particular, is consulted. He does not take care to notice this interconnection of events, but, when he considers general and distinct notions which enter into them, he finds the connection. In fact, when I consult the conception which I have of all true propositions, I find that every necessary or contingent predicate, every past, present, or future, predicate, is involved in the concept of the subject, and I ask no more.

7.32 I will say one word in regard to the cause for the difference which there is here between concepts of space and those of individual substances, rather in relation to the divine will than in relation to the simple understanding. This difference is because the most abstract specific concepts embrace only necessary or eternal truths which do not depend upon the decrees of God (whatever the Cartesians may say about this) but the concepts of individual substances which are complete, and sufficient to identify entirely their subjects and which involve consequently truths that are contingent or of fact, namely, individual circumstances of time, of space, etc.—such substances, I say, should also involve in their concept taken as possible, the free decrees or will of God, likewise taken as possible, because these free decrees are the principle sources for existences or facts while essences are in the divine understanding before his will is taken into consideration.

Not all certain *truths are* necessary.

7.33 We have said that the concept of an individual substance includes once for all everything which can ever happen to it and that in considering this concept one will be able to see everything which can truly be said concerning the individual, just as we are able to see in the nature of a circle all the properties which can be derived from it. But does it not seem that in this way the difference between contingent and necessary truths will be destroyed, that there will be no place for human liberty, and that an absolute fatality will rule as well over all our actions as over all the rest of the events of the world? To this I reply that a distinction must be made between that which is certain and that which is necessary. Every one grants that future contingencies are assured since God foresees them, but we do not say just because of that that they are necessary. But it will be objected, that if any conclusion can be deduced infallibly from some definition or concept, it is necessary; and now since we have maintained that everything which is to happen to anyone is already virtually included in his nature or concept, as all the properties are contained in the definition of a circle, therefore, the difficulty still remains. In order to meet the objection completely, I say that the connection or sequence is of two kinds; the one, absolutely necessary, whose contrary implies contradiction, occurs in the eternal verities like the truths of geometry; the other is necessary only *ex hypothesi,* and so to speak by accident, and in itself it is contingent since the contrary is not implied. This latter sequence is not founded upon ideas wholly pure and upon the pure understanding of God, but upon his free decrees and upon the processes of the universe.

7.34 . . . A demonstration of some predicate as belonging to Caesar is not as absolute as are those of numbers or of geometry, but this predicate supposes a sequence of things which God has shown by his free will. This sequence is based on the first free decree of God which was to do always that which is the most perfect and upon the decree which God made following the first one, regarding human nature, which is that men should always do, although freely, that which appears to be the best. Now every truth which is founded upon this kind of decree is contingent, although certain, for the decrees of God do not change the possibilities of things and, as I have already said, although God assuredly chooses the best, this does not prevent that which is less perfect from being possible in itself. Although it will never happen, it is not its impossibility but its imperfection which causes him to reject it. Now nothing is necessitated whose opposite is possible. One will then be in a position to satisfy these kinds of difficulties, however great they may appear (and in fact they have not been less vexing to all other thinkers who have ever treated this matter), provided that he considers well that all contingent propositions have reasons why they are thus, rather than otherwise, or indeed (what is the same thing) that they have proof *a priori* of their truth, which render them certain and show that the connection of the subject and predicate in these propositions has its basis in the nature of the one and of the other, but he must further remember that such contingent propositions have not the demonstrations of necessity, since their reasons are founded only on the principle of contingency or of the existence of things, that is to say, upon that which is, or which appears to be the best among several things equally possible. Necessary truths, on the other hand, are founded upon the principle of contradiction, and upon the possibility or impossibility of the essences themselves, without regard here to the free will of God or of creatures.

8

BERKELEY · Introduction

George Berkeley (1685–1753), Bishop of Cloyne, studied and wrote on a wide range of topics, including mechanics, optics, linguistics, ethics, and metaphysics, as well as religion. His understanding of and insights into these other areas were generally quite sharp and clear: for example, his criticisms of the concepts of absolute space, time, and motion in Newtonian physics are today recognized to have been anticipatory of arguments used by Einstein almost two hundred years later. Berkeley was also an extremely competent mathematician, although he did not make any original positive contributions to the discipline. He was familiar with the most recent developments in the mathematics of his day, and he initiated one of the most heated debates in the history of mathematics with his tract entitled *The Analyst*, in which he pointed out certain difficulties in both Newton's and Leibniz' formulations of the calculus.

Berkeley was influenced by many of his predecessors and contemporaries, including the French philosophers Pierre Bayle, a sceptic, and Nicolas Malebranche, a Cartesian who argued that mental and physical substances could not possibly interact. The most immediate influence on his philosophical thinking was John Locke, whose recently published *Essay* was a basic text during Berkeley's student days at Trinity College, Dublin. Berkeley borrowed much from Locke, but he also differed sharply with him on a number of crucial points, several of which are brought out particularly well in his discussions of the philosophy of mathematics.

Just as Locke had devoted the first of the four books of his *Essay* to a critique of Descartes' concept of *innate* ideas, Berkeley devoted the lengthy introduction to his best known work—*The Principles of Human Knowledge*—to a critique of Locke's concept of *abstract general* ideas (or at least what he took to be Locke's concept). He interpreted Locke's use of the term "idea" to mean "image" or "picture"; although Locke's discussions of the concepts were indeed vague and ambiguous (see 5.16—5.18), most Berkeley scholars concede that Berkeley's interpretation is not completely fair. However, whether his critical arguments are fair to Locke or not, Berkeley's positive discussion of the meanings of abstract general terms can stand by itself, and it provides the foundation for his philosophy of geometry (see 8.21—8.24).

Like Hobbes and Locke, Berkeley believed that all ideas must be traceable ultimately to experience; that is, he rejected any and every kind of innate idea and/or pre-knowledge. But he differed with them—in fact, with *all* of his predecessors—in so far as he denied that the meaning of every word is a specific idea. He believed that it was this assumption which led Locke (and others) to assert that abstract general ideas *must* exist as the significates or denotata of abstract general nouns). Berkeley shared Hobbes' theory that abstract general terms, in so far as they signify any ideas at all, signify collections of particular ideas. But he rejected Hobbes' unquestioning assumption—and in this he was several centuries ahead of his time—that if there is no particular idea corresponding to some abstract general noun, then that word is meaningless. Berkeley argued that words can be meaningful if they serve some function or have some use even if they do not (or even if they cannot) denote or signify any ideas. In asserting this, he anticipated by over 200 years the 20th century school of philosophers whose motto is "Don't ask for the meaning of a word or statement, ask for its use" (see 8.7—8.12). Although his ultimate aim in formulating this theory of language was to demonstrate that *religious* discourse can be meaningful even though certain basic terms (such as "grace") do not denote any ideas (see 8.10), it also led him to formulate a philosophy of mathematics which was more than one hundred years ahead of its time. This case illustrates clearly that significant, though unexpected, results are sometimes obtained when an exceptional intellect is working on apparently unrelated problems.

The full significance of Berkeley's theory of language and philosophy of mathematics was overlooked by his contemporaries and immediate successors in part because it was so radical and too far ahead of its time. But more importantly, it was overshadowed by another element of Berkeley's philosophy which he himself proclaimed to be his major contribution—his immaterialist thesis, i.e., the rejection of the concept of material substance. As Berkeley intended, it was this aspect of his writings on which most other philosophers were to focus their attention during the next two hundred years.

Berkeley's arguments against material substance are both numerous and complex. In part they borrow from other philosophers' criticisms of Descartes, such as Malebranche's demonstrations of the impossibility of interaction between mental and physical substance; but they go far beyond all previous arguments. Berkeley attempted to show not only that there is no idea denoted by "material substance," but also that attempts to define the term involve internal inconsistencies. He also argued that the distinction between ideas caused by primary and secondary qualities is untenable, and concluded that all ideas are of the type which Locke characterised as "secondary," that is, all ideas are in some way "products of the mind or of mental activity" rather than of any independently existing non-mental substance. Thus, according to Berkeley, no ideas are copies or representations of external material substances; those ideas which are not produced by one's own mind and which appear to come from "outside" are produced by another mind, in most cases by the infinite mind—God.

This central core of Berkeley's ontology is reflected in several significant ways in his philosophy of mathematics, particularly in his completely original distinction between arithmetic and geometry. It will be remembered that from the time of Aristotle there were two basic differences which were assumed to hold between arithmetic and geometry—methodological (arithmetic uses the analytic method, geometry the synthetic) and ontological (arithmetic is the study of the discrete, geometry of the continuous). All of Berkeley's predecessors assumed that the signs or symbols used in both arithmetic and geometry signify or denote other ideas or concepts or things (e.g., numbers, lines, etc.); they differ only in their accounts of the nature and origin of these ideas. Berkeley broke sharply with this tradition by denying that the symbols of *arithmetic* denote anything at all. Whereas Locke and others had maintained that we have clear and distinct ideas of unity and other numbers, Berkeley insisted that there is no idea or thing which is signified by the symbol "1" or by any other numeral (see 8.13—8.18). His argument here is similar to his argument against physical substance where he denied that there is anything corresponding to or denoted by the basic term "material substance." In maintaining that arithmetic is essentially the manipulation of uninterpreted or non-signifying marks according to some set of rules, Berkeley was clearly anticipating the philosophy of mathematics which became known as "formalism" at the end of the 19th century and which is one of the most widely accepted theories today.

Although he insisted that the term "material substance" is totally meaningless and insignificant, Berkeley did not attempt to deny that arithmetical statements are meaningful even though the individual symbols lack denotata. In this case he appealed to his own theory of language, and argued that since arithmetic is obviously very useful for performing certain tasks, arithmetical statements must be meaningful—and their meaning is their use (see 9.20). In doing this, Berkeley was anticipating the twentieth century philosophy of mathematics known as "instrumentalism." As indicated earlier, it is doubtful that the developers of the modern theories of formalism and instrumentalism were aware of Berkeley's theory, or that he exerted any kind of a direct influence on the development of these theories; he was simply too far ahead of his time.

It does not appear that Berkeley ever considered extending his theory of the nature of arithmetic to the field of geometry. It seemed clear to him that we possess certain ideas which are adequate for geometry, although he firmly rejected the possibility of ideas such as Locke's abstract general ideas of triangles, or even the ideas of dimensionless points and lines without breadth. A careful reading of sections 8.21 to 8.24 will reveal significant similarities to Hobbes' philosophy of geometry, and many of these ideas and arguments are repeated in John Stuart Mill's discussions of geometry as well as in various twentieth century treatments of the topic. It is ironic that the least original portion of Berkeley's philosophy of mathematics apparently had the greatest influence on his successors.

Sources

All selections are from the standard edition of the *Works of George Berkeley*, edited by A. A. Luce and T. E. Jessop, Thomas Nelson and Sons Ltd., London, 1948. The following abbreviations are used:

TV — *An Essay towards a New Theory of Vison* (the numerals refer to the paragraph numbering).
P — *A Treatise Concerning the Principles of Human Knowledge* (the numerals refer to the paragraph numbering).
A — *Alciphron or the Minute Philosopher* (numerals refer to the chapter and section numbers).

8.1	P, 1	**8.13**	P, 119
8.2	P, 89	**8.14**	P, 13
8.3	P, Introduction, 7	**8.15**	P, 120
8.4	P, Introduction, 9	**8.16**	P, 12
8.5	TV, 122–125	**8.17**	TV, 109
8.6	P, Introduction, 10	**8.18**	A, VII, 5
8.7	P, Introduction, 19–20	**8.19**	A, VII, 12
8.8	A, VII, 5	**8.20**	P, 121–122
8.9	A, VII, 6	**8.21**	P, Introduction, 12
8.10	A, VII, 6–7	**8.22**	P, Introduction, 18
8.11	A, VII, 11	**8.23**	P, Introduction, 15–16
8.12	A, VII, 13–15	**8.24**	P, 124–131

BERKELEY · Selections

The objects of all human knowledge—ideas and notions.

8.1 It is evident to anyone who takes a survey of the objects of human knowledge, that they are either *ideas* actually imprinted on the senses; or else such as are perceived by attending to the passions and operations of the mind; or lastly, *ideas* formed by help of memory and imagination—either compounding, dividing, or barely representing those originally perceived in the aforesaid ways.—By sight I have the ideas of light and colours, with their several degrees and variations. By touch I perceive hard and soft, heat and cold, motion and resistance, and of all these more and less either as to quantity or degree. Smelling furnishes me with odours; the palate with tastes; and hearing conveys sounds to the mind in all their variety of tone and composition.—And as several of these are observed to accompany each other, they come to be marked by one name, and so to be reputed as one *thing*.

8.2 *Thing* or *being* is the most general name of all: it comprehends under it two kinds entirely distinct and heterogeneous, and which have nothing common but the name, viz. *spirits* and *ideas*. The former are active, indivisible substances: the latter are inert, fleeting, or dependent beings, which subsist not by themselves, but are supported by, or exist in minds or spiritual substances. We comprehend our own existence by inward feeling or Reflection, and that of other spirits by Reason.—We may be said to have some knowledge or *notion* of our own minds, of spirits and active beings—whereof in a strict sense we have not ideas. In like manner, we know and have a *notion* of relations between things or ideas—which relations are distinct from the ideas or things related, inasmuch as the latter may be perceived by us without our perceiving the former. To me it seems that *ideas, spirits,* and *relations* are all, in their respective kinds, the object of human knowledge and subject of discourse, and that the term *idea* would be improperly extended to signify *everything* we know or have any notion of.

The argument against abstract general ideas.

8.3 It is agreed on all hands that the qualities or modes of things do never really exist each of them apart by itself, and separated from all others, but are mixed, as it were, and blended together, several in the same object. But, we are told, the mind being able to consider each quality singly, or abstracted from those other qualities with which it is united, does by that means frame to itself abstract ideas. For example, there is perceived by sight an object extended, colored, and moved: this mixed or compound idea the mind resolving into its simple, constituent parts, and viewing each by itself, exclusive of the rest, does frame the abstract ideas of extension, color, and motion. Not that it is possible for color or motion to exist without extension; but only that the mind can frame to itself by *abstraction* the idea of color exclusive of extension, and of motion exclusive of both color and extension.

8.4 And as the mind frames to itself abstract ideas of qualities or modes, so does it, by the same precision or mental separation, attain abstract ideas of the more compounded beings which include several coexistent qualities. For example, the mind having observed that Peter, James, and John resemble each other in certain common agreements of shape and other qualities, leaves out of the complex or compounded idea it has of Peter, James, and any other particular man, that which is peculiar to each, retaining only what is common to all, and so makes an abstract idea wherein all the particulars equally partake; abstracting entirely from and cutting off all those circumstances and differences which might determine it to any particular existence. And after this manner it is said we come by the abstract idea of man, or, if you please, humanity, or human nature; wherein it is true there is included color, because there is no man but has some color, but then it can be neither white, nor black, nor any particular color, because there is no one particular color wherein all men partake. So likewise there is included stature, but then it is neither tall stature, nor low stature, nor yet middle stature, but something abstracted from all these. And so of the rest.

8.5 . . . I am apt to think that when men speak of extension as being an idea common to two senses, it is with a secret supposition that we can single out extension from all other tangible and visible qualities, and form thereof an abstract idea, which idea they will have common both to sight and touch. We are therefore to understand by extension in abstract, an idea of extension—for instance, a line or surface entirely stripped of all other sensible qualities and circumstances that might determine it to any particular existence; it is neither black, nor white, nor red, nor has it any color at all, or any tangible quality whatsoever, and consequently it is of no finite determinate magnitude; for that which bounds or distinguishes one extension from another is some quality or circumstance wherein they disagree.

Now, I do not find that I can perceive, imagine, or anywise frame in my mind such an abstract idea as is here spoken of. A line or surface which is neither black, nor white, nor blue, nor yellow, etc., nor long, nor short, nor rough, nor smooth, nor square, nor round, etc., is perfectly incomprehensible. This I am sure of as to myself; how far the faculties of other men may reach they best can tell.

It is commonly said that the object of geometry is abstract extension. But geometry contemplates figures: now, figure is the termination of magnitude; but we have shown that extension in abstract has no finite determinate magnitude; whence it clearly follows that it can have no figure, and consequently is not the object of geometry. It is indeed a tenet, as well of the modern as of the ancient philosophers, that all general truths are concerning universal abstract ideas; without which, we are told, there could be no science, no demonstration of any general proposition in geometry. But it were no hard matter, did I think it necessary to my present purpose, to show that propositions and demonstrations in geometry might be universal, though they who make them never think of abstract general ideas of triangles or circles.

After reiterated endeavors to apprehend the general idea of a triangle, I have found it altogether incomprehensible. And surely, if anyone were able to introduce that idea into my mind, it must be the author of the *Essay Concerning Human Understanding*: he, who has so far distinguished himself from the generality of writers by the clearness and significancy of what he says. Let us therefore see how this celebrated author describes the general or abstract idea of a triangle. "It must be," says he, "neither oblique nor rectangular, neither equilateral, equicrural, nor scalenum; but all and none of these at once. In effect it is somewhat imperfect that cannot exist; an idea, wherein some parts of several different and inconsistent ideas are put together." (*Essay On Human Understanding*, Bk. IV, chap. 7, sec. 9 [sec. 5.17 of this book].) This is the idea which he thinks needful for the enlargement of knowledge, which is the subject of mathematical demonstration, and without which we could never come to know any general proposition concerning triangles. That author acknowledges it does "require some pains and skill to form this general idea of a triangle." (*Ibid.*) But, had he called to mind what he says in another place, to wit, "that ideas of mixed modes wherein any inconsistent ideas are put together cannot so much as exist in the mind, i.e., be conceived" (*vide* Bk. III, chap. 10, sec. 33, ibid.)—I say, had this occurred to his thoughts, it is not improbable he would have owned it above all the pains and skill he was master of, to form the above-mentioned idea of a triangle, which is made up of manifest staring contradictions. That a man who laid so great a stress on clear and determinate ideas should nevertheless talk at this rate seems very surprising. But the wonder will lessen if it be considered that the source whence this opinion flows is the prolific womb which has brought forth innumerable errors and difficulties, in all parts of philosophy, and in all the sciences. But this matter, taken in its full extent, were a subject too comprehensive to be insisted on in this place. And so much for extension in abstract.

8.6 Whether others have this wonderful faculty of abstracting their ideas, they best can tell; for myself, I find indeed I have a faculty of imagining, or representing to myself, the ideas of those particular things I have perceived, and of variously compounding and dividing them. I can imagine a man with two heads, or the upper parts of a man joined to the body of a horse. I can consider the hand, the eye, the nose, each by itself abstracted or separated from the rest of the body. But then whatever hand or eye I imagine, it must have some particular shape and color. Likewise the idea of man that I frame to myself must be either of a white, or a black, or a tawny, a straight, or a crooked, a tall, or a low, or a middle-sized man. I cannot by any effort of thought conceive the abstract idea above described. And it is equally impossible for me to form the abstract idea of motion distinct from the body moving, and which is neither swift nor slow, curvilinear nor rectilinear; and the like may be said of all other abstract general ideas whatsoever. To be plain, I own myself able to abstract in one sense, as when I consider some particular parts or qualities separated from others, with which, though they are united in some object, yet it is possible they may really exist without them. But I deny that I can abstract from one another, or conceive separately, those qualities which it is impossible should exist so separated; or that I can frame a general notion, by abstracting from particulars in the manner aforesaid—which last are the two proper acceptations of 'abstraction.' And there are grounds to think most men will acknowledge themselves to be in my case.

Words may be meaningful even if they do not signify ideas or things.

8.7 But to give a farther account how words came to produce the doctrine of abstract ideas, it must be observed that it is a received opinion that language has no other end but the communicating our ideas, and that every significant name stands for an idea. This being so and it being withal certain that names which yet are not thought altogether insignificant do not always mark out particular conceivable ideas, it is straightway concluded that they stand for abstract notions. That there are many names in use among speculative men which do not always suggest to others determinate, particular ideas, or in truth anything at all, is what nobody will deny. And a little attention will discover that it is not necessary (even in the strictest reasonings) significant names which stand for ideas should, every time they are used, excite in the understanding the ideas they are made to stand for; in reading and discoursing, names being for the most part used as letters are in algebra, in which, though a particular quantity be marked by each letter, yet to proceed right it is not requisite that in every step each letter suggest to your thoughts that particular quantity it was appointed to stand for.

Besides, the communicating of ideas marked by words is not the chief and only end of language, as is commonly supposed. There are other ends, as the raising of some passion, the exciting to or deterring from an action, the putting the mind in some particular disposition; to which the former is in many cases barely subservient, and sometimes entirely omitted, when these can be obtained without it, as I think does not unfrequently happen in the familiar use of language. I entreat the reader to reflect with himself, and see if it doth not often happen, either in hearing or reading a discourse, that the passions of fear, love, hatred, admiration, disdain, and the like, arise immediately in his mind upon the perception of certain words, without any ideas coming between. At first, indeed, the words might have occasioned ideas that were fitting to produce those emotions; but, if I mistake not, it will be found that when language is once grown familiar, the hearing of the sounds or sight of the characters is oft immediately attended with those passions which at first were wont to be produced by the intervention of ideas that are now quite omitted. May we not, for example, be affected with the promise of a *good thing,* though we have not an idea of what it is? Or is not the being threatened with danger sufficient to excite a dread, though we think not of any particular evil likely to befall us, nor yet frame to ourselves an idea of danger in abstract? If anyone shall join ever so little reflection of his own to what has been said, I believe that it will evidently appear to him that general names are often used in the propriety of language without the speaker's designing them for marks of ideas in his own, which he would have them raise in the mind of the hearer. Even proper names themselves do not seem always spoken with a design to bring into our view the ideas of those individuals that are supposed to be marked by them. For example, when a schoolman tells me "Aristotle hath said it," all I conceive he means by it is to dispose me to embrace his opinion with the deference and submission which custom has annexed to that name. And this effect is often so instantly produced in the minds of those who are accustomed to resign their judgment to authority of that philosopher, as it is impossible any idea either of his person, writings, or reputation should go before. Innumerable examples of this kind may be given but why should I insist on those things which everyone's experience will, I doubt not, plentifully suggest unto him?

8.8 . . . Words may not be insignificant, although they should not, every time they are used, excite the ideas they signify in our minds; it being sufficient that we have it in our power to substitute things or ideas for their signs when there is occasion. It seems also to follow that there may be another use of words besides that of marking and suggesting distinct ideas, to wit, the influencing our conduct and actions, which may be done either by forming rules for us to act by, or by raising certain passions, dispositions, and emotions in our minds. A discourse, therefore, that directs how to act or excites to the doing or forbearance of an action may, it seems, be useful and significant, although the words whereof it is composed should not bring each a distinct idea into our minds.

[The following is a portion of a dialogue between Alciphron and Euphranor. Euphranor is speaking for Berkeley.]

8.9 EUPHRANOR. . . . Let me entreat you, Alciphron, be not amused by terms: lay aside the *word* force, and exclude every other thing from your thoughts, and then see what precise *idea* you have of force.

ALCIPHRON. Force is that in bodies which produceth motion and other sensible effects.

EUPHRANOR. It is then something distinct from those effects?

ALCIPHRON. It is.

EUPHRANOR. Be pleased now to exclude the consideration of its subject and effects, and contemplate force itself in its own precise idea.

ALCIPHRON. I profess I find it no such easy matter.

EUPHRANOR. Take your own advice, and shut your eyes to assist your meditation.—Upon this, Alciphron, having closed his eyes and mused a few minutes, declared he could make nothing of it.

And that, replied EUPHRANOR, which it seems neither you nor I can frame an idea of, by your own remark of men's minds and faculties being made much alike, we may suppose others have no more an idea of than we.

ALCIPHRON. We may.

8.10 EUPHRANOR. . . . Upon the whole, therefore, may we not pronounce that—excluding body, time, space, motion, and all its sensible measures and effects—we shall find it as difficult to form an idea of force as of grace?

ALCIPHRON. I do not know what to think of it.

EUPHRANOR. And yet, I presume, you allow there are very evident propositions or theorems relating to force, which contain useful truths: for instance, that a body with conjunct forces describes the diagonal of a parallelogram in the same time that it would the sides with separate. Is not this a principle of very extensive use? Doth not the doctrine of the composition and resolution of forces depend upon it, and, in consequence thereof, numberless rules and theorems directing men how to act, and explaining phenomena throughout the Mechanics and mathematical philosophy? And if, by considering this doctrine of force, men arrive at the knowledge of many inventions in Mechanics, and are taught to frame engines, by means of which things difficult and otherwise impossible may be performed; and if the same doctrine which is so beneficial here below serveth also as a key to discover the nature of the celestial motions; shall we deny that it is of use, either in practice or speculation, because we have no distinct idea of force? Or that which we admit with regard to force, upon what pretence can we deny concerning grace? If there are queries, disputes, perplexities, diversity of notions and opinions about the one, so there are about the other also: if we can form no precise distinct idea of the one, so neither can we of the other. Ought we not therefore, by a parity of reason, to conclude there may be possibly diverse true and useful propositions concerning the one as well as the other?

. . . Although terms are signs, yet having granted that those signs may be significant, though they should not suggest ideas represented by them, provided they serve to regulate and influence our wills, passions, or conduct, you have consequently granted that the mind of man may assent to propositions containing such terms, when it is so directed or affected by them, notwithstanding it should not perceive distinct ideas marked by those terms.

8.11 Science and faith agree in this, that they both imply an assent of the mind: and, as the nature of the first is most clear and evident, it should be first considered in order to cast a light on the other. To trace things from their original, it seems that the human mind, naturally furnished with the ideas of things particular and concrete, and being designed, not for the bare intuition of ideas, but for action and operation about them, and pursuing her own happiness therein, stands in need of certain general rules or theorems to direct her operations in this pursuit; the supplying which want is the true, original, reasonable end of studying the arts and sciences. Now, these rules being general, it follows that they are not to be obtained by the mere consideration of the original ideas, or particular things, but by the means of marks and signs, which, being so far forth universal, become the immediate instruments and materials of science. It is not, therefore, by mere contemplation of particular things, and much less of their abstract general ideas, that the mind makes her progress, but by an apposite choice and skilful management of signs: for instance, force and number, taken in concrete, with their adjuncts, subjects, and signs, are what every one knows; and considered in abstract, so as making precise ideas of themselves, they are what nobody can comprehend. That their abstract nature, therefore, is not the foundation of science is plain: and that barely considering their ideas in concrete is not the method to advance in the respective sciences is what every one that reflects may see; nothing being more evident than that one who can neither write nor read, in common use understands the meaning of numeral words as well as the best philosopher or mathematician.

8.12 I am inclined to think the doctrine of signs a point of great importance and general extent, which, if duly considered, would cast no small light upon things, and afford a just and genuine solution of many difficulties.

Thus much, upon the whole, may be said of all signs:—that they do not always suggest ideas signified to the mind: that when they suggest ideas, they are not general abstract ideas: that they have other uses besides barely standing for and exhibiting ideas, such as raising proper emotions, producing certain dispositions or habits of mind, and directing our actions in pursuit of that happiness which is the ultimate end and design, the primary spring and motive, that sets rational agents at work: that signs may imply or suggest the relations of things; which relations, habitudes or proportions, as they cannot be by us understood but by the help of signs, so being thereby expressed and confuted, they direct and enable us to act with regard to things: that the true end of speech, reason, science, faith, assent, in all its different degrees, is not merely, or principally, or always, the imparting or acquiring of ideas, but rather something of an active operative nature, tending to a conceived good: which may sometimes be obtained, *not only although the ideas marked are not offered to the mind, but even although there should be no possibility of offering or exhibiting any such idea to the mind:* * for instance, the algebraic mark which denotes the root of

* [Italics mine. Ed.]

a negative square, hath its use in logistic operations, although it be impossible to form an idea of any such quantity. And what is true of algebraic signs is also true of words or language, modern algebra being in fact a more short, apposite, and artificial sort of language, and it being possible to express by words at length, though less conveniently, all the steps of an algebraical process. And it must be confessed that even the mathematical sciences themselves, which above all others are reckoned the most clear and certain, if they are considered, not as instruments to direct our practise, but as speculations to employ our curiosity, will be found to fall short in many instances of those clear and distinct ideas which, it seems, the minute philosophers of this age, whether knowingly or ignorantly, expect and insist upon in the mysteries of religion.

Be the science or subject what it will, whensoever men quit particulars for generalities, things concrete for abstractions, when they forsake practical views, and the useful purposes of knowledge for barren speculation, considering means and instruments as ultimate ends, and labouring to obtain precise ideas which they suppose indiscriminately annexed to all terms, they will be sure to embarrass themselves with difficulties and disputes. Such are those which have sprung up in geometry about the nature of the angle of contact, the doctrine of proportions, of indivisibles, infinitesimals, and divers other points; notwithstanding all which, that science is very rightly esteemed an excellent and useful one, and is really found to be so in many occasions of human life, wherein it governs and directs the actions of men, so that by the aid or influence thereof those operations become just and accurate which would otherwise be faulty and uncertain.

[End of the excerpt from the dialogue.]

Arithmetic is not the study of numbers or ideas of numbers.

8.13 Arithmetic has been thought to have for its object abstract ideas of *number;* of which to understand the properties and mutual habitudes, is supposed no mean part of speculative knowledge. The opinion of the pure and intellectual nature of numbers in abstract hath made them in esteem with those philosophers who seem to have affected an uncommon fineness and elevation of thought. It hath set a price on the most trifling numerical speculations which in practice are of no use, but serve only for amusement; and hath therefore so far infected the minds of some, that they have dreamed of mighty mysteries involved in numbers, and attempted the explication of natural things by them. But, if we inquire into our own thoughts, and consider what hath been premised, we may perhaps entertain a low opinion of those high flights and abstractions, and look on all inquiries, about numbers only as so many *difficiles nugae,* so far as they are not subservient to practise, and promote the benefit of life.

8.14 Unity I know some will have to be a simple or uncompounded idea, accompanying all other ideas into the mind. That I have any such idea answering the word *unity* I do not find; and if I had, methinks I could not miss finding it: on the contrary, it should be the most familiar to my understanding, since it is said to accompany all other ideas, and to be perceived by all the ways of sensation and reflexion. To say no more, it is an *abstract* idea.

8.15 . . . Number being defined a 'collection of units,' we may conclude that, if there be no such thing as unity or unit in abstract, there are no ideas of number in abstract denoted by the numeral names and figures. The theories therefore in arithmetic, if they are abstracted from the names and figures, as likewise from all use and practise, as well as from the particular things numbered, can be supposed to have nothing at all for their object; hence we may see how entirely the science of numbers is subordinate to practise, and how jejune and trifling it becomes when considered as a matter of mere speculation.

8.16 That number is entirely the creature of the mind, even though the other qualities be allowed to exist without, will be evident to whoever considers that the same thing bears a different denomination of number as the mind views it with different respects. Thus, the same extension is one, or three, or thirty-six, according as the mind considers it with reference to a yard, a foot, or an inch. Number is so visibly relative, and dependent on men's understanding, that it is strange to think how anyone should give it an absolute existence without the mind. We say one book, one page, one line, etc.; all these are equally units, though some contain several of the others. And in each instance, it is plain, the unit relates to some particular combination of ideas arbitrarily put together by the mind.

8.17 But, for a fuller illustration of this matter, it ought to be considered that number (however some may reckon it amongst the primary qualities) is nothing fixed and settled, really existing in things themselves. It is entirely the creature of the mind, considering either an idea by itself or any combination of ideas to which it gives one name, and so makes it pass for a unit. According as the mind variously combines its ideas, the unit varies; and as the unit, so the number, which is only a collection of units, also varies. We call a window "one," a chimney "one"; and yet a house, in which there are many windows and many chimneys, has an equal right to be called "one"; and many houses go to the making of one city. In these and the like instances, it is evident the unit constantly relates to the particular drafts the mind makes of its ideas, to which it affixes names, and wherein it includes more or less, as best suits its own ends and purposes. Whatever, therefore, the mind considers as one, that is a unit. Every combination of ideas is considered as one thing by the mind, and in token thereof is marked by one name. Now, this naming and combining together of ideas is perfectly arbitrary, and done by the mind in such sort as experience shows it to be most convenient—without which our ideas had never been collected into such sundry distinct combinations as they now are.

[The following is another excerpt from the dialogue between Alciphron and Euphranor.]

8.18 ALCIPHRON. Will you not allow then that the mind can abstract?

EUPHRANOR. I do not deny it may abstract in a certain sense, inasmuch as those things that can really exist, or be really perceived asunder, may be conceived asunder, or abstracted one from the other; for instance, a man's head from his body, colour from motion, figure from weight. But it will not thence follow that the mind can frame abstract general ideas, which appear to be impossible.

ALCIPHRON. And yet it is a current opinion that every substantive name marks out and exhibits to the mind one distinct idea separate from all others.

EUPHRANOR. Pray, Alciphron, is not the word *number* such a substantive name?

ALCIPHRON. It is.

EUPHRANOR. Do but try now whether you can frame an idea of number in abstract, exclusive of all signs, words, and things numbered. I profess for my own part I cannot.

ALCIPHRON. Can it be so hard a matter to form a simple idea of number, the object of a most evident demonstrable science? Hold, let me see if I can't abstract the idea of number from the numerical names and characters, and all particular numerable things.—Upon which Alciphron paused a while, and then said, To confess the truth I do not find that I can.

EUPHRANOR. But, though it seems neither you nor I can form distinct simple ideas of number, we can nevertheless make a very proper and significant use of numeral names. They direct us in the disposition and management of our affairs, and are of such necessary use that we should not know how to do without them. And yet, if other men's faculties may be judged of by mine, to attain a precise simple abstract idea of number is as difficult as to comprehend any mystery in religion.

8.19 But here lies the difference: the one who understands the notation of numbers, by means thereof is able to express briefly and distinctly all the variety and degrees of number, and to perform with ease and despatch several arithmetical operations by the help of general rules. Of all which operations as the use in human life is very evident, so it is no less evident that the performing them depends on the aptness of the notation. If we suppose rude mankind without the use of language, it may be presumed they would be ignorant of arithmetic. But the use of names, by the repetition whereof in a certain order they might express endless degrees of number, would be the first step towards that science. The next step would be to devise proper marks of a permanent nature, and visible to the eye, the kind and order whereof must be chose with judgment, and accommodated to the names. Which marking or notation would, in proportion as it was apt and regular, facilitate the invention and application of general rules to assist the mind in reasoning and judging, in extending, recording, and communicating its knowledge about numbers: in which theory and operations, the mind is immediately occupied about the signs or notes, by mediation of which it is directed to act about things, or number in concrete (as the logicians call it), without ever considering the simple, abstract, intellectual, general idea of number. The signs, indeed, do in their use imply relations or proportions of things; but these relations are not abstract general ideas, being founded in particular things, and not making of themselves distinct ideas to the mind, exclusive of the particular ideas and the signs. I imagine one need not think much to be convinced that the science of arithmetic, in its rise, operations, rules, and theorems, is altogether conversant about the artificial use of signs, names, and characters. These names and characters are universal, inasmuch as they are signs. The names are referred to things, the characters to names, and both to operation. The names being few, and proceeding by a certain analogy, the characters will be more useful, the simpler they are, and the more aptly they express this analogy. Hence the old notation by letters was more useful than words written at length; and the modern notation by figures, expressing the progression or analogy of the names by their simple places, is much preferable to that, for ease and expedition, as the invention of algebraical symbols is to this, for extensive and general use. As arithmetic and algebra are sciences of great clearness, certainty, and extent, which are immediately conversant about signs, upon the skilful use and management whereof they entirely depend, so a little attention to them may possibly help us to judge of the progress of the mind in other sciences, which, though differing in nature, design, and object, may yet agree in the general methods of proof and inquiry.

[End of the passage from the dialogue.]

8.20 However, since there may be some who, deluded by the specious show of discovering abstracted verities, waste their time in arithmetical theorems and problems which have not any use, it will not be amiss if we more fully consider and expose the vanity of that pretense; and this will plainly appear by taking a view of arithmetic in its infancy, and observing what it was that originally put men on the study of that science, and to what scope they directed it. It is natural to think that at first, men, for ease of memory and help of computation, made use of counters, or in writing of single strokes, points, or the like, each whereof was made to signify an unit, i.e., some one thing of whatever kind they had occasion to reckon. Afterwards they found out the more compendious ways of making one character stand in place of several strokes or points. And, lastly, the notation of the Arabians or Indians came into use, wherein, by the repetition of a few characters or figures, and varying the significance of each figure according to the place it obtains, all numbers may be most aptly expressed; which seems to have been done in imitation of language, so that an exact anology is observed betwixt the notation by figures and names, the nine simple figures answering the nine first numeral names and places in the former, corresponding to denominations in the latter. And agreeably to those conditions of the simple and local value of figures, were contrived methods of finding, from the given figures or marks of the parts, what figures and how placed are proper to denote the whole, or *vice versa.* And having found the sought figures, the same rule or analogy being observed throughout, it is easy to read them into words; and so the number becomes perfectly known. For then the number of any particular things is said to be known, when we know the name or figures (with their due arrangement) that according to the standing analogy belong to them. For, these signs being known, we can by the operations of arithmetic know the signs of any part of the particular sums signified by them; and, thus computing in signs (because of the connection established betwixt them and the distinct multitudes of things whereof one is taken for an unit), we may be able rightly to sum up, divide, and proportion the things themselves that we intend to number.

In arithmetic, therefore, we regard not the *things,* but the *signs,* which nevertheless are not regarded for their own sake, but because they direct us how to act with relation to things, and dispose rightly of them. Now, agreeably to what we have before observed of words in general it happens here likewise that abstract ideas are thought to be signified by numeral names or characters, while they do not suggest ideas of particular things to our minds. I shall not at present enter into a more particular dissertation on this subject, but only observe that it is evident from what hath been said, those things which pass for abstract truths and theorems concerning numbers, are in reality conversant about no object distinct from particular numeral things, except only names and characters, which originally came to be considered on no other account but their being signs, or capable to represent aptly whatever particular things men had need to compute. Whence it follows that to study them for their sake would be just as wise, and to as good purpose as if a man, neglecting the true use or original intention and subserviency of language, should spend his time in impertinent criticisms upon words, or reasonings and controversies purely verbal.

Geometry is the study of particular ideas of particular figures.

8.21 By observing how ideas become general we may the better judge how words are made so. And here it is to be noted that I do not deny absolutely there are general ideas, but only that there are any *abstract* general ideas. . . . Now, if we will annex a meaning to our words, and speak only of what we can conceive, I believe we shall acknowledge that an idea which considered in itself is particular, becomes general by being made to represent or stand for all other particular ideas of the same sort. To make this plain by an example, suppose a geometrician is demonstrating the method of cutting a line in two equal parts. He draws, for instance, a black line of an inch in length: this, which in itself is a particular line, is nevertheless with regard to its signification general, since, as it is there used, it represents all particular lines whatsoever; so that what is demonstrated of it is demonstrated of all lines, or, in other words, of a line in general. And, as that particular line becomes general by being made a sign, so the *name* 'line,' which taken absolutely is particular, by being a sign is made general. And as the former owes its generality not to its being the sign of an abstract or general line, but of all particular right lines that may possibly exist, so the latter must be thought to derive its generality from the same cause, namely, the various particular lines which it indifferently denotes.

8.22 To this it will be objected that every name that has a definition is thereby restrained to one certain signification. For example, a triangle is defined to be 'a plane surface comprehended by three right lines,' by which that name is limited to denote one certain idea and no other. To which I answer that in the definition it is not said whether the surface be great or small, black or white, nor whether the sides are long or short, equal or unequal, nor with what angles they are inclined to each other; in all which there may be great variety, and consequently there is no one settled idea which limits the signification of the word triangle. It is one thing to keep a name constantly to the same definition, and another to make it stand everywhere for the same idea; the one is necessary, the other useless and impracticable.

8.23 It is, I know, a point much insisted on, that all knowledge and demonstration are about universal notions, to which I fully agree; but then it doth not appear to me that those notions are formed by abstraction in the manner premised: *universality* so far as I can comprehend, not consisting in the absolute, *positive* nature or conception of anything, but in the *relation* it bears to the particulars signified or represented by it; by virtue whereof it is that things, names, or notions, being in their own nature *particular,* are rendered *universal.* Thus when I demonstrate any proposition concerning triangles, it is to be supposed that I have in view the universal idea of a triangle; which ought not to be understood as if I could frame an idea of a triangle which was neither equilateral, nor scalenon, nor equicrual; but only that the particular triangle I consider, whether of this or that sort it matters not, doth equally stand for and represent all rectilinear triangles whatsoever, and is in that sense *universal*. All which seems very plain and not to include any difficulty in it.

But here it will be demanded how we can know any proposition to be true of all particular triangles, except we have first seen it demonstrated of the abstract idea of a triangle which equally agrees to all? For because a property may be demonstrated to agree to some one particular triangle, it will not thence follow that it equally belongs to any other triangle which in all respects is not the same with it. For example, having demonstrated that the three angles of an isosceles rectangular triangle are equal to two right ones, I cannot therefore conclude this affection agrees to all other triangles which have neither a right angle nor two equal sides. It seems therefore that, to be certain this proposition is universally true, we must either make a particular demonstration for every particular triangle, which is impossible, or once for all demonstrate it of the abstract idea of a triangle, in which all the particulars do indifferently partake and by which they are all equally represented. To which I answer, that, though the idea I have in view whilst I make the demonstration be, for instance, that of an isosceles rectangular triangle whose sides are of a determinate length, I may nevertheless be certain it extends to all other rectilinear triangles, of what sort or bigness soever. And that because neither the right angle nor the equality nor determinate length of the sides are at all concerned in the demonstration. It is true the diagram I have in view includes all these particulars, but then there is not the least mention made of them in the proof of the proposition. It is not said the three angles are equal to two right ones, because one of them is a right angle, or because the sides comprehending it are of the same length. Which sufficiently shows that the right angle might have been oblique and the sides unequal, and for all that the demonstration have held good. And for this reason it is that I conclude that to be true of any obliquangular or scalenon which I had demonstrated of a particular right-angled equicrual triangle, and not because I demonstrated the proposition of the abstract idea of a triangle. And here it must be acknowledged that a man may consider a figure merely as triangular, without attending to the particular qualities of the angles, or relations of the sides. So far he may abstract; but this will never prove that he can frame an abstract, general, inconsistent idea of a triangle. In like manner we may consider Peter so far forth as man, or so far forth as animal, without framing the forementioned abstract idea, either of man or of animal, inasmuch as all that is perceived is not considered.

8.24 Every particular finite extension which may possibly be the object of our thought is an *idea* existing only in the mind, and consequently each part thereof must be perceived. If, therefore, I cannot perceive innumerable parts in any finite extension that I consider, it is certain they are not contained in it; but it is evident that I cannot distinguish innumerable parts in any particular line, surface, or solid, which I either perceive by sense, or figure to myself in my mind: wherefore I conclude they are not contained in it. Nothing can be plainer to me than that the extensions I have in view are no other than my own ideas; and it is no less plain that I cannot resolve any one of my ideas into an infinite number of other ideas, that is, that they are not infinitely divisible. If by finite extension be meant something distinct from a finite idea, I declare I do not know what that is, and so cannot affirm or deny anything of it. But if the terms 'extension,' 'parts,' etc., are taken in any sense conceivable, that is, for ideas, then to say a finite quantity or extension consists of parts infinite in number is so manifest a contradiction, that everyone at first sight acknowledges it to be so; and it is impossible it should ever gain the assent of any reasonable creature who is not brought to it by gentle and slow degrees, as a converted Gentile to the belief of transubstantiation. Ancient and rooted prejudices do often pass into principles; and those propositions which once obtain the force and credit of a *principle,* are not only themselves, but likewise whatever is deducible from them, thought privileged from all examination. And there is no absurdity so gross, which, by this means, the mind of man may not be prepared to swallow.

He whose understanding is possessed with the doctrine of abstract general ideas may be persuaded that (whatever be thought of the ideas of sense) extension in *abstract* is infinitely divisible. And one who thinks the objects of sense exist without the mind will perhaps in virtue thereof be brought to admit that a line but an inch long may contain innumerable parts—really existing, though too small to be discerned. These errors are grafted as well in the minds of geometricians as of other men, and have a like influence on their reasonings; and it were no difficult thing to shew how the arguments from geometry made use of to support the infinite divisibility of extension are bottomed on them. At present we shall only observe in general whence it is the mathematicians are all so fond and tenacious of that doctrine.

It hath been observed in another place that the theorems and demonstrations in geometry are conversant about universal ideas; where it is explained in what sense this ought to be understood, to wit, the particular lines and figures included in the diagram are supposed to stand for innumerable others of different sizes; or, in other words, the geometer considers them abstracting from their magnitude—which does not imply that he forms an abstract idea, but only that he cares not what the particular magnitude is, whether great or small, but looks on that as a thing different to the demonstration. Hence it follows that a line in the scheme but an inch long must be spoken of as though it contained ten thousand parts, since it is regarded not in itself, but as it is universal; and it is universal only in its signification, whereby it represents innumerable lines greater than itself, in which may be distinguished ten thousand parts or more, though there may not be above an inch in it. After this manner, the properties of the lines signified are (by a very usual figure) transferred to the sign, and thence, through mistake, thought to appertain to it considered in its own nature.

Because there is no number of parts so great but it is possible there may be a line containing more, the inch-line is said to contain parts more than any assignable number; which is true, not of the inch taken absolutely, but only for the things signified by it. But men, not retaining that distinction in their thoughts, slide into a belief that the small particular line described on paper contains in itself parts innumerable. There is no such thing as the ten-thousandth part of an inch; but there is of a mile or diameter of the earth, which may be signified by that inch. When therefore I delineate a triangle on paper, and take one side not above an inch, for example, in lengths to be the radius, this I consider as divided into ten thousand or an hundred thousand parts or more; for, though the ten-thousandth part of that line considered in itself is nothing at all, and consequently may be neglected without an error or inconveniency, yet these described lines, being only marks standing for greater quantities, whereof it may be the ten-thousandth part is very considerable, it follows that, to prevent notable errors in practice, the radius must be taken of ten thousand parts or more.

From what hath been said the reason is plain why, to the end any theorem become universal in its use, it is necessary we speak of the lines described on paper as though they contained parts which really they do not. In doing of which, if we examine the matter thoroughly, we shall perhaps discover that we cannot conceive an inch itself as consisting of, or being divisible into, a thousand parts, but only some other line which is far greater than an inch, and represented by it; and that when we say a line is infinitely divisible, we must mean a line which is infinitely great. What we have here observed seems to be the chief cause why, to suppose the infinite divisibility of finite extension hath been thought necessary in geometry.

The several absurdities and contradictions which flowed from this false principle might, one would think, have been esteemed so many demonstrations against it. But, by I know not what logic, it is held that proofs *a posteriori* are not to be admitted against propositions relating to infinity, as though it were not impossible even for an infinite mind to reconcile contradictions; or as if anything absurd and repugnant could have a necessary connection with truth or flow from it. But whoever considers the weakness of this pretense will think it was contrived on purpose to humor the laziness of the mind which had rather acquiesce in an indolent scepticism than be at the pains to go through with a severe examination of those principles it hath ever embraced for true.

Of late the speculations about infinites have run so high, and grown to such strange notions, as have occasioned no scruples and disputes among the geometers of the present age. Some there are of great note who, not content with holding that finite lines may be divided into an infinite number of parts, do yet farther maintain that each of those infinitesimals is itself subdivisible into an infinity of other parts or infinitesimals of a second order, and so on *ad infinitum*. These, I say, assert there are infinitesimals of infinitesimals of infinitesimals, etc., without ever coming to an end! so that according to them an inch does not barely contain an infinite number of parts, but an infinity of an infinity of an infinity *ad infinitum* of parts. Others there be who hold all orders of infinitesimals below the first to be nothing at all; thinking it with good reason absurd to imagine there is any positive quantity or part of extension which, though multiplied infinitely, can never equal the smallest given extension. And yet on the other hand it seems no less absurd to think the square, cube or other power of a positive real root, should itself be nothing at all; which they who hold infinitesimals of the first order, denying all of the subsequent orders, are obliged to maintain.

Have we not therefore reason to conclude they are *both* in the wrong, and that there is in effect no such thing as parts infinitely small, or an infinite number of parts contained in any finite quantity? But you will say that if this doctrine obtains it will follow the very foundations of geometry are destroyed, and those great men who have raised that science to so astonishing a height, have been all the while building a castle in the air. To this it may be replied that whatever is useful in geometry, and promotes the benefit of human life, does still remain firm and unshaken on our principles: that science considered as practical will rather receive advantage than any prejudice from what has been said. But to set this in a due light, and shew how lines and figures may be measured, and their properties investigated, without supposing finite extension to be infinitely divisible, may be the proper business of another place. For the rest, though it should follow that some of the more intricate and subtle parts of Speculative Mathematics may be pared off without any prejudice to truth, yet I do not see what damage will be thence derived to mankind.

9

HUME · Introduction

David Hume (1711–1776) probably had less formal training and interest in mathematics than any other individual represented in this volume. However, he recognized that in order to deal adequately with the philosophical issues with which he was primarily concerned—including ethics, philosophy of religion, and philosophy of the natural sciences (particularly the problem of induction) —it was necessary to provide some account of the nature of mathematical knowledge. Unfortunately, his discussions of the nature of mathematical knowledge are not only extremely brief, but also vague and ambiguous, and perhaps even inconsistent. The following collection of Hume's statements on the philosophy of mathematics is included in this volume because of the influence which they have had on subsequent philosophies of mathematics, and also because of the intrinsic interest of some of the specific points on which Hume touched. As with all the selections in this volume, the reader can learn much from a thoughtful analysis of even the mistakes of the great thinkers of the past.

Hume's philosophy, with its emphasis on epistemological rather than ontological questions, resembles the theories of many of his predecessors in a variety of ways, but it also contains several strikingly original features. One of the fundamental assumptions underlying all of Hume's philosophical work is the traditional view that in order for any statement to be meaningful or significant each word in it must denote or signify something. His distinction between impressions and ideas (see 9.3) in terms of force or vivacity is so vague and subjective as to be of little if any intrinsic value; however, when coupled with his assertion that every idea must be ultimately grounded in a particular impression(see 9.5), it provides a basis (however weak) for his rejection of innate ideas, abstract general ideas, and a variety of specific ideas such as the ideas of causal connections and of all kinds of substance (including God, material substance, mental substance, etc.). On Hume's account, the absence of an idea—for example, of material substance—implies only the meaninglessness of the term "material substance." Given the meaninglessness of the subject term, the statement "Material substance does not exist" can be neither true nor false, but is simply meaningless. This is in contrast to statements such as

"Unicorns do not exist," for it is possible to construct an idea of a unicorn out of the impressions of horses and horns, which makes the term "unicorn" meaningful by Hume's criterion. In denying that there are any ideas at all corresponding to terms such as "material substance," "God," and "mental substance," Hume was not merely denying the existence of material substance, etc., but was attempting to eliminate the possibility of even raising in any significant way such questions as "Does material substance exist?" (Hume also provided an interesting account of the meaning of the term "exists," but since it is only of secondary interest to his philosophy of mathematics and is quite complex, it can not be discussed here.)

Hume's use of the ideational theory of meaning has significant ramifications for his theory of knowledge, as well as for his ontology. Hume ultimately concluded that in so far as the term "mind" could have any meaning at all, it could only refer to a "bundle" of ideas, and not to anything which underlies or contains or perceives or is conscious of the ideas—because he believed that there is no impression or idea of such a being or entity and thus that no such being or entity itself could be said to exist. But if the mind is merely a bundle of ideas, and if the things of which the mind is conscious or which it knows are those very same ideas, then the knower and the thing-known must be one and the same. To say that "Tom knows that $2 + 2 = 4$" would seem to be nothing more than to say that the bundle or set of ideas known as "Tom" contains the idea "$2 + 2 = 4$" as well as some other idea corresponding to the term "knows" which is related to the idea "$2 + 2 = 4$" in some way. It is impossible here to go into all of the difficulties generated by this theory, although it should be pointed out that many philosophers believe that Hume's discussions of these problems are more accurately characterizable as psychological than as philosophical.

The preceding paragraphs provide at best a rough sketch of those aspects of Hume's general philosophy which have some bearing on his discussions of the nature of mathematical knowledge. As the following selections show, he was primarily concerned with two aspects of the philosophy of mathematics: first, he wanted to ascertain what ideas (and ultimately what impressions) are signified by mathematical symbols, that is, to determine their meanings; second, he was concerned with identifying the nature of the relations among the ideas in order to determine the nature of the necessity which he assumed is essential to mathematical knowledge.

Since there can be no doubt that mathematical discourse is significant and meaningful, and since Hume adhered rigidly to the ideational theory of meaning, it is not suprising that he concluded that the objects of mathematical knowledge are the *ideas* of number, space, etc. which are ultimately traceable to sense impressions as their source. This last statement might require some qualification depending on one's interpretation of an enigmatic passage in later work, the *Enquiry Concerning Human Understanding,* where he asserted that: "Though there never were a circle or triangle in nature, the truths demonstrated by Euclid would for ever retain their certainty and evidence" (see 9.11). This statement is strikingly similar to one of Locke's assertions about mathematical knowledge (see 5.24), and it does not violate Hume's demand that all ideas are traceable to impressions, in so far as a constructivist account similar to Locke's can be provided. Unfortunately, however, Hume not only did not provide any such account, but in his first work, the *Treatise on Human Nature,* he even seems to be denying that such an account is possible (see 9.9 and 9.15). There is disagreement among scholars as to whether or not these passages are truly inconsistent, whether they represent a change of mind or merely a further development of Hume's thinking, and whether or not Hume was making a conscious distinction between pure and applied geometry. Because of the brevity of Hume's discussion of these matters, it is doubtful that these questions will ever be answered to everyone's satisfaction, although it might be agreed that these passages reflect a confusion resulting from Hume's cursory treatment of the problems of the philosophy of mathematics in general.

With regard to the question of the nature of the necessity of mathematical truths, Hume argued that this necessity results only from the nature of the ideas denoted by the terms of a proposition and the relations between them. His discussions of this matter are somewhat more detailed and complex in the *Treatise* (see 9.14—9.15) than in the *Enquiry* (see 9.11), and again the two accounts can be interpreted in different ways—even as being inconsistent with one another. However, it is the account in the later work, the *Enquiry,* which Hume himself wished to have taken as his ultimate statement on this topic, and it is this account which in fact did have the greatest influence on subsequent philosophers of mathematics. Thus, any apparent inconsistencies between passages in the *Treatise* and the *Enquiry* can be resolved by discounting the earlier discussions in the *Treatise.*

It is difficult to assess the extent of Hume's influence on later work on the problems of the philosophy of mathematics. Kant was aware of the similarities between Hume's concept of truths of relations of ideas and Leibniz' definition of truths of reason (see 7.4—7.6); he incorporated both definitions into his concept of analyticity (see 10.5), although he argued against both Leibniz and Hume that mathematical truths are not of this kind. Mill also argued that mathematical truths are not truths of relations of ideas (see 11.23), but scholars have doubts that this was a conscious attack on Hume. Numerous other philosophers have made explicit references to Hume in their discussions of the philosophy of mathematics, but it would appear that many of them have been referring to interpretations of his writings which are questionable or simply mistaken. Part of this confusion has resulted from the previously noted ambiguity and vagueness inherent in Hume's own writings. But some modern writers have simply forgotten or ignored the fact that he adhered rigidly to the ideational theory of meaning, and that therefore his concept of the truth of relations of ideas cannot be viewed as even anticipative of the modern formalist theories which deny that mathematical axioms signify or denote anything at all. However, in so far as philosophers have read such possibilities into Hume's theory, his influence can be said to have been even greater than he had intended, in fact even greater than he ever could have anticipated.

Sources

All selections are from *An Enquiry Concerning Human Understanding,* Open Court Publishing Company, La Salle, 1966 (abbreviated to E), or *A Treatise of Human Nature,* reprinted from the original edition and edited by L. A. Selby-Bigge, Clarendon Press, Oxford, 1888 (abbreviated to T).

9.1 T, p. xix
9.2 T, pp. xix–xx
9.3 T, pp. 1–2
9.4 T, pp. 3–4
9.5 E, pp. 20–21
9.6 T, pp. 17–19
9.7 T, p. 22
9.8 T, pp. 30–31
9.9 T, pp. 49–51
9.10 E, p. 184
9.11 E, pp. 24–25
9.12 E, pp. 64–65
9.13 E, pp. 182–183
9.14 T, pp. 13–14
9.15 T, pp. 69–73
9.16 T, pp. 200–201
9.17 T, pp. 180–181
9.18 T, pp. 26–27
9.19 T, pp. 29–30

HUME · Selections

The psychological foundation of all knowledge.

9.1 It is evident, that all the sciences have a relation, greater or less, to human nature; and that however wide any of them may seem to run from it, they still return back by one passage or another. Even *Mathematics, Natural Philosophy, and Natural Religion*, are in some measure dependent on the science of Man; since they lie under the cognizance of men, and are judged of by their powers and faculties. It is impossible to tell what changes and improvements we might make in these sciences were we thoroughly acquainted with the extent and force of human understanding, and could explain the nature of the ideas we employ, and of the operations we perform in our reasonings.

9.2 If therefore the sciences of Mathematics, Natural Philosophy, and Natural Religion, have such a dependence on the knowledge of man, what may be expected in the other sciences, whose connection with human nature is more close and intimate? The sole end of logic is to explain the principles and operations of our reasoning faculty, and the nature of our ideas: morals and criticism regard our tastes and sentiments: and politics consider men as united in society, and dependent on each other. In these four sciences of *Logic, Morals, Criticism, and Politics*, is comprehended almost everything, which it can any way import us to be acquainted with, or which can tend either to the improvement or ornament of the human mind.

Here then is the only expedient, from which we can hope for success in our philosophical researches, to leave the tedious lingering method, which we have hitherto followed, and instead of taking now and then a castle or village on the frontier, to march up directly to the capital or center of these sciences, to human nature itself; which being once masters of, we may every where else hope for an easy victory. From this station we may extend our conquests over all those sciences, which more intimately concern human life, and may afterwards proceed at leisure to discover more fully those, which are the objects of pure curiosity. There is no question of importance, whose decision is not comprized in the science of man; and there is none, which can be decided with any certainty, before we become acquainted with that science. In pretending therefore to explain the principles of human nature, we in effect propose a complete system of the sciences, built on a foundation almost entirely new, and the only one upon which they can stand with any security.

And as the science of man is the only solid foundation for the other sciences, so the only solid foundation we can give to this science itself must be laid on experience and observation.

The objects of all knowledge—impressions and ideas.

9.3 All perceptions of the human mind resolve themselves into two distinct kinds, which I shall call Impressions and Ideas. The difference between these consists in the degrees of force and liveliness with which they strike upon the mind, and make their way into our thought or consciousness. Those perceptions, which enter with most force and violence, we may name *impressions;* and under this name I comprehend all our sensations, passions and emotions, as they make their first appearance in the soul. By *ideas* I mean the faint images of these in thinking and reasoning; such as, for instance, are all the perceptions excited by the present discourse, excepting only, those which arise from the sight and touch, and excepting the immediate pleasure or uneasiness it may occasion. I believe it will not be very necessary to employ many words in explaining this distinction. Every one of himself will readily perceive the difference between feeling and thinking. The common degrees of these are easily distinguished; though it is not impossible but in particular instances they may very nearly approach to each other. Thus in sleep, in a fever, in madness, or in any very violent emotions of soul, our ideas may approach to our impressions: As on the other hand it sometimes happens, that our impressions are so faint and low, that we cannot distinguish them from our ideas. But notwithstanding this near resemblance in a few instances, they are in general so very different, that no-one can make a scruple to rank them under distinct heads, and assign to each a peculiar name to mark the difference.

There is another division of our perceptions, which it will be convenient to observe, and which extends itself both to our impressions and ideas. This division is into Simple and Complex. Simple perceptions or impressions and ideas are such as admit of no distinction nor separation. The complex are the contrary to these, and may be distinguished into parts. Though a particular colour, taste, and smell are qualities all united together in this apple, it is easy to perceive they are not the same, but are at least distinguishable from each other.

9.4 . . . I venture to affirm, that the rule here holds without any exception, that every simple idea has a simple impression, which resembles it; and every simple impression a correspondent idea. That idea of red, which we form in the dark, and that impression, which strikes our eyes in sunshine, differ only in degree, not in nature. That the case is the same with all our simple impressions and ideas, it is impossible to prove by a particular enumeration of them. Every one may satisfy himself in this point by running over as many as he pleases. But if any one should deny this universal resemblance, I know no way of convincing him, but by desiring him to show a simple impression, that has not a correspondent idea, or a simple idea, that has not a correspondent impression. If he does not answer this challenge, as it is certain he cannot, we may from his silence and our own observation establish our conclusion.

Thus we find that all simple ideas and impressions resemble each other; and as the complex are formed from them, we may affirm in general, that these two species of perception are exactly correspondent.

9.5 Here, therefore, is a proposition, which not only seems, in itself, simple and intelligible; but, if a proper use were made of it, might render every dispute equally intelligible, and banish all that jargon, which has so long taken possession of metaphysical reasonings, and drawn disgrace upon them. All ideas, especially abstract ones, are naturally faint and obscure: the mind has but a slender hold of them: they are apt to be confounded with other resembling ideas; and when we have often employed any term, though without a distinct meaning, we are apt to imagine it has a determinate idea annexed to it. On the contrary, all impressions, that is, all sensations, either outward or inward, are strong and vivid: the limits between them are more exactly determined: nor is it easy to fall into any error or mistake with regard to them. When we entertain, therefore, any suspicion that a philosophical term is employed without any meaning or idea (as is but too frequent), we need but enquire, *from what impression is that supposed idea derived?* And if it be impossible to assign any, this will serve to confirm our suspicion. By bringing ideas into so clear a light we may reasonably hope to remove all dispute, which may arise, concerning their nature and reality.

Hume's account of abstract ideas.

9.6 A very material question has been started concerning *abstract* or *general ideas, whether they be general or particular in the mind's conception of them.* A great philosopher [Berkeley] has disputed the received opinion in this particular, and has asserted, that all general ideas are nothing but particular ones, annexed to a certain term, which gives them a more extensive signification, and makes them recall upon occasion other individuals, which are similar to them. As I look upon this to be one of the greatest and most valuable discoveries that has been made of late years in the republic of letters, I shall here endeavour to confirm it by some arguments which I hope will put it beyond all doubt and controversy.

It is evident, that in forming most of our general ideas, if not all of them, we abstract from every particular degree of quantity and quality, and that an object ceases not to be of any particular species on account of every small alteration in its extension, duration and other properties. It may therefore be thought, that here is a plain dilemma, that decides concerning the nature of those abstract ideas, which have afforded so much speculation to philosophers. The abstract idea of a man represents men of all sizes and all qualities; which it is concluded it cannot do, but either by representing at once all possible sizes and all possible qualities, or by representing no particular one at all. Now it having been esteemed absurd to defend the former proposition, as implying an infinite capacity in the mind, it has been commonly inferred in favour of the latter; and our abstract ideas have been supposed to represent no particular degree either of quantity or quality. But that this inference is erroneous, I shall endeavour to make appear, *first,* by proving, that it is utterly impossible to conceive any quantity or quality, without forming a precise notion of its degrees: and *secondly* by showing, that though the capacity of the mind be not infinite, yet we can at once form a notion of all possible degrees of quantity and quality, in such a manner at least, as, however imperfect, may serve all the purposes of reflection and conversation.

To begin with the first proposition, *that the mind cannot form any notion of quantity or quality without forming a precise notion of degrees of each;* we may prove this by the three following arguments. First, we have observed, that whatever objects are different are distinguishable, and that whatever objects are distinguishable are separable by the thought and imagination. And we may here add, that these propositions are equally true in the *inverse,* and that whatever objects are separate are also distinguishable, and that whatever objects are distinguishable are also different. For how is it possible we can separate what is not distinguishable, or distinguish what is not different? In order therefore to know, whether abstraction implies a separation, we need only consider it in this view, and examine, whether all the circumstances, which we abstract from in our general ideas, be such as are distinguishable and different from those, which we retain as essential parts of them. But it is evident at first sight, that the precise length of a line is not different nor distinguishable from the line itself; nor the precise degree of any quality from the quality. These ideas, therefore, admit no more of separation than they do of distinction and difference. They are consequently conjoined with each other in the conception; and the general idea of a line, notwithstanding all our abstractions and refinements, has in its appearance in the mind a precise degree of quantity and quality; however it may be made to represent others, which have different degrees of both.

Secondly, it is confessed, that no object can appear to the senses; or in other words, that no impression can become present to the mind, without being determined in its degrees both of quantity and quality. The confusion in which impressions are sometimes involved, proceeds only from their faintness and unsteadiness, not from any capacity in the mind to receive any impression, which in its real existence has no particular degree nor proportion. That is a contradiction in terms; and even implies the flattest of all contradictions, *viz.* that it is possible for the same thing both to be and not to be.

Now since all ideas are derived from impressions and are nothing but copies and representations of them, whatever is true of the one must be acknowledged concerning the other. Impressions and ideas differ only in their strength and vivacity. The foregoing conclusion is not founded on any particular degree of vivacity. It cannot therefore be affected by any variation in that particular. An idea is a weaker impression; and as a strong impression must necessarily have a determinate quantity and quality, the case must be the same with its copy or representative.

Thirdly, it is a principle generally received in philosophy, that everything in nature is individual, and that it is utterly absurd to suppose a triangle really existent, which has no precise proportion of sides and angles. If this therefore be absurd in *fact and reality,* it must also be absurd *in idea;* since nothing of which we can form a clear and distinct idea is absurd and impossible. But to form the idea of an object, and to form an idea simply is the same thing; the reference of the idea to an object being an extraneous denomination, of which in itself it bears no mark or character.

9.7 That we may fix the meaning of the word, figure, we may revolve in our mind the ideas of circles, squares, parallelograms, triangles of different sizes and proportions, and may not rest on one image or idea. However this may be, it is certain *that* we form the idea of individuals, whenever we use any general term; *that* we seldom or never can exhaust these individuals; and *that* those, which remain, are only represented by means of that habit, by which we recall them, whenever any present occasion requires it. This then is the nature of our abstract ideas and general terms; and it is after this manner we account for the foregoing paradox, *that some ideas are particular in their nature, but general in their representation.* A particular idea becomes general by being annexed to a general term; that is, to a term, which from a customary conjunction has a relation to many other particular ideas, and readily recalls them in the imagination.

The basic ideas of mathematics and geometry.

9.8 It is evident, that existence in itself belongs only to unity, and it is never applicable to number, but on account of the units, of which the number is composed. Twenty men may be said to exist; but it is only because one, two, three, four, etc. are existent; and if you deny the existence of the latter, that of the former falls of course. It is therefore utterly absurd to suppose any number to exist, and yet deny the existence of units; and as extension is always a number, according to the common sentiment of metaphysicians, and never resolves itself into any unit or indivisible quantity, it follows, that extension can never at all exist. It is in vain to reply, that any determinate quantity of extension is an unit; but such-a-one as admits of an infinite number of fractions, and is inexhaustible in its sub-divisions. For by the same rule these twenty men *may be considered as an unit.* The whole globe of the earth, nay the whole universe *may be considered as an unit.* That term of unity is merely a fictitious denomination, which the mind may apply to any quantity of objects it collects together; nor can such an unity any more exist alone than number can, as being in reality a true number. But the unity, which can exist alone, and whose existence is necessary to that of all number, is of another kind, and must be perfectly indivisible, and incapable of being resolved into any lesser unity.

9.9 It is true, mathematicians pretend they give an exact definition of a right line, when they say, *it is the shortest way between two points.* But in the first place, I observe, that this is more properly the discovery of one of the properties of a right line, than a just definition of it. For I ask any one, if upon mention of a right line he thinks not immediately on such a particular appearance, and if it is not by accident only that he considers this property? A right line can be comprehended alone; but this definition is unintelligible without a comparison with other lines, which we conceive to be more extended. In common life it is established as a maxim, that the straightest way is always the shortest; which would be as absurd as to say, the shortest way is always the shortest, if our idea of a right line was not different from that of the shortest way between two points.

Secondly, I repeat what I have already established, that we have no precise idea of equality and inequality, shorter and longer, more than of a right line or a curve; and consequently that the one can never afford us a perfect standard for the other. An exact idea can never be built on such as are loose and undeterminate.

The idea of a *plain surface* is as little susceptible of a precise standard as that of a right line; nor have we any other means of distinguishing such a surface, than its general appearance. It is in vain, that mathematicians represent a plain surface as produced by the flowing of a right line. It will immediately be objected, that our idea of a surface is as independent of this method of forming a surface, as our idea of an ellipse is that of a cone; that the idea of a right line is no more precise than that of a plain surface; that a right line may flow irregularly, and by that means form a figure quite different from a plane; and that therefore we must suppose it to flow along two right lines, parallel to each other, and on the same plane; which is a description, that explains a thing by itself, and returns in a circle.

It appears, then, that the ideas which are most essential to geometry, *viz.* those of equality and inequality, of a right line and a plain surface, are far from being exact and determinate, according to our common method of conceiving them. Not only we are incapable of telling, if the case be in any degree doubtful, when such particular figures are equal; when such a line is a right one, and such a surface a plain one; but we can form no idea of that proportion or of these figures, which is firm and invariable. Our appeal is still to the weak and fallible judgment, which we make from the appearance of the objects, and correct by a compass or common measure; and if we join the supposition of any farther correction, it is of such-a-one as is either useless or imaginary. In vain should we have recourse to the common topic, and employ the supposition of a deity, whose omnipotence may enable him to form a perfect geometrical figure, and describe a right line without any curve or inflection. As the ultimate standard of these figures is derived from nothing but the senses and imagination, it is absurd to talk of any perfection beyond what these faculties can judge of; since the true perfection of any thing consists in its conformity to its standard.

9.10 When we run over libraries, persuaded of these principles, what havoc must we make? If we take in our hand any volume; of divinity or school metaphysics, for instance; let us ask, *Does it contain any abstract reasoning concerning quantity or number?* No. *Does it contain any experimental reasoning concerning matters of* *fact and existence?* No. Commit it then to the flames: for it can contain nothing but sophistry and illusion.

The two kinds of knowldge—of Relations of Ideas and Matters of Fact.

9.11 All the objects of human reason or enquiry may naturally be divided into two kinds, to wit, *Relations of Ideas,* and *Matters of Fact.* Of the first kind are the sciences of Geometry, Algebra, and Arithmetic; and in short, every affirmation which is either intuitively or demonstratively certain. *That the square of the hypothenuse is equal to the squares of the two sides,* is a proposition which expresses a relation between these figures. *That three times five is equal to half of thirty,* expresses a relation between these numbers. Propositions of this kind are discoverable by the mere operation of thought, without dependence on what is anywhere existent in the universe. Though there never were a circle or triangle in nature, the truths demonstrated by Euclid would for ever retain their certainty and evidence.

Matters of fact, which are the second objects of human reason, are not ascertained in the same manner; nor is our evidence of their truth, however great, of a like nature with the foregoing. The contrary of every matter of fact is still possible; because it can never imply a contradiction, and is conceived by the mind with the same facility and distinctness, as if ever so conformable to reality. *That the sun will not rise tomorrow* is no less intelligible a proposition, and implies no more contradiction than the affirmation, *that it will rise.* We should in vain, therefore, attempt to demonstrate its falsehood. Were it demonstratively false, it would imply a contradiction, and could never be distinctly conceived by the mind.

9.12 The great advantage of the mathematical sciences above the moral consists in this, that the ideas of the former, being sensible, are always clear and determinate, the smallest distinction between them is immediately perceptible, and the same terms are still expressive of the same ideas, without ambiguity or variation. An oval is never mistaken for a circle, nor an hyperbola for an ellipsis. The isosceles and scalenum are distinguished by boundaries more exact than vice and virtue, right and wrong. If any term be defined in geometry, the mind readily, of itself, substitutes, on all occasions, the definition for the term defined: Or even when no definition is employed the object itself may be presented to the senses and by that means be steadily and clearly apprehended. But the finer sentiments of the mind, the operations of the understanding, the various agitations of the passions, though really in themselves distinct, easily escape us, when surveyed by reflection; nor is it in our power to recall the original object, as often as we have occasion to contemplate it. Ambiguity, by this means, is gradually introduced into our reasonings: Similar objects are readily taken to be the same: And the conclusion becomes at last very wide of the premises.

One may safely, however, affirm, that, if we consider these sciences in a proper light, their advantages and disadvantages nearly compensate each other, and reduce both of them to a state of equality. If the mind, with greater facility, retains the ideas of geometry clear and determinate, it must carry on a much longer and more intricate chain of reasoning, and compare ideas much wider of each other, in order to reach the abstruser truths of that science. And if the moral ideas are apt, without extreme care, to fall into obscurity and confusion, the inferences are always much shorter in these disquisitions, and the intermediate steps, which lead to the conclusion, much fewer than in the sciences which treat of quantity and number. In reality, there is scarcely a proposition in Euclid so simple, as not to consist of more parts, than are to be found in any moral reasoning which runs not into chimera and conceit. Where we trace the principles of the human mind through a few steps, we may be very well satisfied with our progress; considering how soon nature throws a bar to all our enquiries concerning causes, and reduces us to an acknowledgment of our ignorance. The chief obstacle, therefore, to our improvement in the moral or metaphysical sciences is the obscurity of the ideas, and ambiguity of the terms. The principal difficulty in the mathematics is the length of inferences and compass of thought, requisite to the forming of any conclusion.

9.13 It seems to me, that the only objects of the abstract science or of demonstration are quantity and number, and that all attempts to extend this more perfect species of knowledge beyond these bounds are mere sophistry and illusion. As the component parts of quantity and number are entirely similar, their relations become intricate and involved; and nothing can be more curious, as well as useful, than to trace, by a variety of mediums, their equality or inequality, through their different appearances. But as all other ideas are clearly distinct and different from each other, we can never advance farther, by our utmost scrutiny, than to observe this diversity, and, by an obvious reflection, pronounce one thing not to be another. Or if there be any difficulty in these decisions, it proceeds entirely from the undeterminate meaning of words, which is corrected by juster definitions. That *the square of the hypothenuse is equal to the squares of the other two sides,* cannot be known, let the terms be ever so exactly defined, without a train of reasoning and enquiry. But to convince us of this proposition, that *where there is no property, there can be no injustice,* it is only necessary to define the terms, and explain injustice to be a violation of property. This proposition is, indeed, nothing but a more imperfect definition. It is the same case with all those pretended syllogistical reasonings, which may be found in every other branch of learning, except the sciences of quantity and number; and these may safely, I think, be pronounced the only proper objects of knowledge and demonstration.

All other enquiries of men regard only matter of fact and existence; and these are evidently incapable of demonstration. Whatever *is* may *not be.* No negation of a fact can involve a contradiction. The non-existence of any being, without exception, is as clear and distinct an idea as its existence. The proposition, which affirms it not to be, however false, is no less conceivable and intelligible, than that which affirms it to be. The case is different with the sciences, properly so called. Every proposition, which is not true, is there confused and unintelligible. That the cube root of 64 is equal to the half of ten, is a false proposition, and can never be distinctly conceived. But that Caesar, or the angel Gabriel, or any being never existed, may be a false proposition, but still is perfectly conceivable, and implies no contradiction.

Mathematical knowledge is of the philosophical relation of quantity.

9.14 The word Relation is commonly used in two senses considerably different from each other. Either for that quality, by which two ideas are connected together in the imagination, and the one naturally introduces the other; or for that particular circumstance, in which, even upon the arbitrary union of two ideas in the fancy, we may think proper to compare them. In common language the former is always the sense, in which we used the word, relation; and it is only in philosophy, that we extend it to mean any particular subject of comparison, without a connecting principle. Thus distance will be allowed by philosophers to be a true relation, because we acquire an idea of it by the comparing of objects: But in a common way we say, *that nothing can be more distant than such or such things from each other, nothing can have less relation;* as if distance and relation were incompatible.

9.15 There are seven different kinds of philosophical relation, *viz. resemblance, identity, relations of time and place, proportion in quantity or number, degrees in any quality, contrariety, and causation.* These relations may be divided into two classes; into such as depend entirely on the ideas, which we compare together, and such as may be changed without any change in the ideas. It is from the idea of a triangle, that we discover the relation of equality, which its three angles bear to two right ones; and this relation is invariable, as long as our idea remains the same. On the contrary, the relations of *contiguity* and *distance* between two objects may be changed merely by an alteration of their place, without any change on the objects themselves or on their ideas; and the place depends on a hundred different accidents, which cannot be foreseen by the mind. It is the same case with *identity* and *causation.* Two objects, though perfectly resembling each other, and even appearing in the same place at different times, may be numerically different: And as the power, by which one object produces another, is never discoverable merely from their idea, it is evident *cause* and *effect* are relations, of which we receive information from experience, and not from any abstract reasoning or reflection. There is no single phaenomenon, even the most simple, which can be accounted for from the qualities of the objects, as they appear to us; or which we could foresee without the help of our memory and experience.

It appears, therefore, that of these seven philosophical relations, there remain only four, which depending solely upon ideas, can be the objects of knowledge and certainty. These four are *resemblance, contrariety, degrees in quality, and proportions in quantity or number.* Three of these relations are discoverable at first sight, and fall more properly under the province of intuition than demonstration. When any objects *resemble* each other, the resemblance will at first strike the eye, or rather the mind; and seldom requires a second examination. The case is the same with *contrariety,* and with the *degrees* of any *quality.* No one can once doubt but existence and non-existence destroy each other, and are perfectly incompatible and contrary. And though it be impossible to judge exactly of the degrees of any quality, such as colour, taste, heat, cold, when the difference between them is very small; yet it is easy to decide, that any of them is superior or inferior to another, when their difference is considerable. And this decision we always pronounce at first sight, without any enquiry or reasoning.

We might proceed, after the same manner, in fixing the *proportions* of *quantity* or *number,* and might at one view observe a superiority or inferiority between any numbers, or figures; especially where the difference is very great and remarkable. As to equality or any exact proportion, we can only guess at it from a single consideration; except in very short numbers, or very limited portions of extension; which are comprehended in an instant, and where we perceive an impossibility of falling into any considerable error. In all other cases we must settle the proportions with some liberty, or proceed in a more *artificial* manner.

I have already observed that geometry, or the *art,* by which we fix the proportions of figures; though it much excels, both in universality and exactness, the loose judgments of the senses and imagination; yet never attains a perfect precision and exactness. Its first principles are still drawn from the general appearance of the objects; and that appearance can never afford us any security, when we examine the prodigious minuteness of which nature is susceptible. Our ideas seem to give a perfect assurance, that no two right lines can have a common segment; but if we consider these ideas, we shall find, that they always suppose a sensible inclination of the two lines, and that where the angle they form is extremely small, we have no standard of a right line so precise, as to assure us of the truth of this proposition. It is the same case with most of the primary decisions of the mathematics.

There remain, therefore, algebra and arithmetic as the only sciences, in which we can carry on a chain of reasoning to any degree of intricacy, and yet preserve a perfect exactness and certainty. We are possessed of a precise standard, by which we can judge of the equality and proportion of numbers; and according as they correspond or not to that standard, we determine their relations, without any possibility of error. When two numbers are so combined, as that the one has always an unit answering to every unit of the other, we pronounce them equal; and it is for want of such a standard of equality in extension that geometry can scarce be esteemed a perfect and infallible science.

But here it may not be amiss to obviate a difficulty, which may arise from my asserting, that though geometry falls short of that perfect precision and certainty, which are peculiar to arithmetic and algebra, yet it excels the imperfect judgments of our senses and imagination. The reason why I impute any defect to geometry, is, because its original and fundamental principles are derived merely from appearances; and it may perhaps be imagined, that this defect must always attend it, and keep it from ever reaching a greater exactness in the comparison of objects or ideas, than what our eye or imagination alone is able to attain. I own that this defect so far attends it, as to keep it from ever aspiring to a full certainty: But since these fundamental principles depend on the easiest and least deceitful appearances, they bestow on their consequences a degree of exactness, of which these consequences are singly incapable. It is impossible for the eye to determine the angles of a chiliagon to be equal to 1996 right angles, or make any conjecture, that approaches this proportion; but when it determines, that right lines cannot concur; that we cannot draw more than one right line between two given points; its mistakes can never be of any consequence. And this is the nature and use of geometry, to run us up to such appearances, as, by reason of their simplicity, cannot lead us into any considerable error.

I shall here take occasion to propose a second observation concerning our demonstrative reasonings, which is suggested by the same subject of the mathematics. It is usual with mathematicians, to pretend, that those ideas, which are their objects, are of so refined and spiritual a nature, that they fall not under the conception of the fancy, but must be comprehended by a pure and intellectual view, of which the superior faculties of the soul are alone capable. The same notion runs through most parts of philosophy, and is principally made use of to explain our abstract ideas, and to show how we can form an idea of a triangle, for instance, which shall neither be an isosceles nor scalenum, nor be confined to any particular length and proportion of sides. It is easy to see, why philosophers are so fond of this notion of some spiritual and refined perceptions; since by that means they cover many of their absurdities, and may refuse to submit to the decisions of clear ideas, by appealing to such as are obscure and uncertain. But to destroy this artifice, we need but reflect on that principle so often insisted on, *that all our ideas are copied from our impressions.* For from thence we may immediately conclude, that since all impressions are clear and precise, the ideas, which are copied from them, must be of the same nature, and can never, but from our fault, contain any thing so dark and intricate. An idea is by its very nature weaker and fainter than an impression; but being in every other respect the same, cannot imply any very great mystery. If its weakness render it obscure, it is our business to remedy that defect, as much as possible, by keeping the idea steady and precise; and till we have done so, it is in vain to pretend to reasoning and philosophy.

This is all I think necessary to observe concerning those four relations, which are the foundation of science; but as to the other three, which depend not upon the idea, and may be absent or present even while *that* remains the same, it will be proper to explain them more particularly. These three relations are *identity, the situations in time and place, and causation.*

Unity, number, and identity.

9.16 As to the principle of individuation; we may observe, that the view of any one object is not sufficient to convey the idea of identity. For in that proposition, *an object is the same with itself,* if the idea expressed by the word, *object,* were no ways distinguished from that meant by *itself;* we really should mean nothing, nor would the proposition contain a predicate and a subject, which however are implied in this affirmation. One single object conveys the idea of unity, not that of identity.

On the other hand, a multiplicity of objects can never convey this idea, however resembling they may be supposed. The mind always pronounces the one not to be the other, and considers them as forming two, three, or any determinate number of objects, whose existences are entirely distinct and independent.

Since then both number and unity are incompatible with the relation of identity, it must lie in something that is neither of them. But to tell the truth, at first sight this seems utterly impossible. Between unity and number there can be no medium; no more than between existence and non-existence. After one object is supposed to exist, we must either suppose another also to exist; in which case we have the idea of number: Or we must suppose it not to exist; in which case the first object remains at unity.

To remove this difficulty, let us have recourse to the idea of time or duration. I have already observed, that time, in a strict sense, implies succession, and that when we apply its idea to any unchangeable object, it is only by a fiction of the imagination, by which the unchangeable object is supposed to participate of the changes of the co-existent objects, and in particular of that of our perceptions. This fiction of the imagination almost universally takes place; and it is by means of it, that a single object, placed before us, and surveyed for any time without our discovering in it any interruption or variation, is able to give us a notion of identity. For when we consider any two points of this time, we may place them in different lights: We may either survey them at the very same instant; in which case they give us the idea of number, both by themselves and by the object; which must be multiplied, in order to be conceived at once, as existent in these two different points of time: Or on the other hand, we may trace the succession of time by a like succession of ideas, and conceiving first one moment, along with the object then existent, imagine afterwards a change in the time without any *variation* or *interruption* in the object; in which case it gives us the idea of unity. Here then is an idea, which is a medium between unity and number; or more properly speaking, is either of them, according to the view, in which we take it: And this idea we call that of identity. We cannot, in any propriety of speech, say, that an object is the same with itself, unless we mean, that the object existent at one time is the same with itself existent at another. By this means we make a difference, between the idea meant by the word, *object,* and that meant by *itself,* without going the length of number, and at the same time without restraining ourselves to a strict and absolute unity.

Thus the principle of individuation is nothing but the *invariableness* and *uninterruptedness* of any object, through a supposed variation of time, by which the mind can trace it in the different periods of its existence, without any break of the view, and without being obliged to form the idea of multiplicity or number.

The sense in which mathematical knowledge cannot be certain.

9.17 In all demonstrative sciences the rules are certain and infallible; but when we apply them, our fallible and uncertain faculties are very apt to depart from them, and fall into error. We must, therefore, in every reasoning form a new judgment, as a check or control on our first judgment or belief; and must enlarge our view to comprehend a kind of history of all the instances, wherein our understanding has deceived us, compared with those, wherein its testimony was just and true. Our reason must be considered as a kind of cause, of which truth is the natural effect; but such-a-one as by the irruption of other causes, and by the inconstancy of our mental powers, may frequently be prevented. By this means all knowledge degenerates into probability; and this probability is greater or less, according to our experience of the veracity or deceitfulness of our understanding, and according to the simplicity or intricacy of the question.

There is no Algebraist nor Mathematician so expert in his science, as to place entire confidence in any truth immediately upon his discovery of it, or regard it as any thing, but a mere probability. Every time he runs over his proofs, his confidence increases; but still more by the approbation of his friends; and is raised to its utmost perfection by the universal assent and applauses of the learned world. Now it is evident, that this gradual increase of assurance is nothing but the addition of new probabilities, and is derived from the constant union of causes and effects, according to past experience and observation.

In accounts of any length or importance, Merchants seldom trust to the infallible certainty of numbers for their security; but by the artificial structure of the accounts, produce a probability beyond what is derived from the skill and experience of the accountant. For that is plainly of itself some degree of probability; though uncertain and variable, according to the degrees of his experience and length of the account. Now as none will maintain, that our assurance in a long numeration exceeds probability, I may safely affirm, that there scarce is any proposition concerning numbers, of which we can have a fuller security. For it is easily possible, by gradually diminishing the numbers, to reduce the longest series of addition to the most simple question, which can be formed, to an addition of two single numbers; and upon this supposition we shall find it impracticable to shew the precise limits of knowledge and of probability, or discover that particular number, at which the one ends and the other begins. But knowledge and probability are of such contrary and disagreeing natures, that they cannot well run insensibly into each other, and that because they will not divide, but must be either entirely present, or entirely absent. Besides, if any single addition were certain, every one would be so, and consequently the whole or total sum; unless the whole can be different from all its parts. I had almost said, that this was certain; but I reflect, that it must reduce *itself*, as well as every other reasoning, and from knowledge degenerate into probability.

Finite extensions are not infinitely divisible.

9.18 It is universally allowed, that the capacity of the mind is limited, and can never attain a full and adequate conception of infinity: And though it were not allowed, it would be sufficiently evident from the plainest observation and experience. It is also obvious, that whatever is capable of being divided *in infinitum,* must consist of an infinite number of parts, and that it is impossible to set any bounds to the number of parts without setting bounds at the same time to the division. It requires scarce any induction to conclude from hence, that the *idea,* which we form of any finite quantity, is not infinitely divisible, but that by proper distinctions and separations we may run up this idea to inferior ones, which will be perfectly simple and indivisible. In rejecting the infinite capacity of the mind, we suppose it may arrive at an end in the division of its ideas; nor are there any possible means of evading the evidence of this conclusion.

It is therefore certain, that the imagination reaches a *minimum,* and may raise up to itself an idea, of which it cannot conceive any sub-division, and which cannot be diminished without a total annihilation. When you tell me of the thousandth and ten thousandth part of a grain of sand, I have a distinct idea of these numbers and of their different proportions; but the images, which I form in my mind to represent the things themselves, are nothing different from each other, nor inferior to that image, by which I represent the grain of sand itself, which is supposed so vastly to exceed them. What consists of parts is distinguishable into them, and what is distinguishable is separable. But whatever we may imagine of the thing, the idea of a grain of sand is not distinguishable, nor separable into twenty, much less into a thousand, ten thousand, or an infinite number of different ideas.

9.19 Wherever ideas are adequate representations of objects, the relations, contradictions and agreements of the ideas are all applicable to the objects; and this we may in general observe to be the foundation of all human knowledge. But our ideas are adequate representations of the most minute parts of extension; and through whatever divisions and subdivisions we may suppose these parts to be arrived at, they can never become inferior to some ideas, which we form. The plain consequence is that whatever *appears* impossible and contradictory upon the comparison of these ideas, must be *really* impossible and contradictory, without any farther excuse or evasion.

Everything capable of being infinitely divided contains an infinite number of parts; otherwise the division would be stopped short by the indivisible parts, which we should immediately arrive at. If therefore any finite extension be infinitely divisible, it can be no contradiction to suppose, that a finite extension contains an infinite number of parts: And *vice versa* if it be a contradiction to suppose, that a finite extension contains an infinite number of parts, no finite extension can be infinitely divisible. But that this latter supposition is absurd, I easily convince myself by the consideration of my clear ideas. I first take the least idea I can form of a part of extension, and being certain that there is nothing more minute than this idea, I conclude, that whatever I discover by its means must be a real quality of extension. I then repeat this idea once, twice, thrice, etc. and find the compound idea of extension, arising from its repetition, always to augment, and become double, triple, quadruple, etc. till at last it swells up to a considerable bulk, greater or smaller, in proportion as I repeat more or less the same idea. When I stop in the addition of parts, the idea of extension ceases to augment; and were I to carry on the addition *in infinitum,* I clearly perceive, that the idea of extension must also become infinite. Upon the whole, I conclude, that the idea of an infinite number of parts is individually the same idea with that of an infinite extension; that no finite extension is capable of containing an infinite number of parts; and consequently that no finite extension is infinitely divisible.

10

KANT · Introduction

Immanuel Kant (1724–1804), although he made no significant original contributions to mathematics proper, was a very competent mathematician and was well acquainted with the most advanced work in this discipline at the middle of the eighteenth century. He taught not only philosophy but also mathematics and physics at the University of Königsberg from 1755 to 1770, when his appointment as Professor of Logic and Metaphysics gave him the opportunity to devote his full attention to philosophy. His discussions of the basic questions of the philosophy of mathematics reflect his familiarity with the basic mathematics texts and his experience in teaching mathematics (an explicit reference to texts appears in 10.7).

Kant's early thinking was particularly influenced by the metaphysics of Leibniz and his followers, and also by the scientific theories (i.e., the natural philosophy) of Newton. He eventually came across some of the writings of Hume, which (as Kant himself described it) woke him from his "dogmatic slumbers." He ultimately formulated his own original theory which he hoped would initiate a revolution in philosophy comparable in scope to the Copernican revolution in astronomy, and which in fact did exert a significant influence on much of the intellectual activity in the Western world during the last two hundred years. Although many elements of his philosophy reflect traditional thinking (e.g., his unquestioning acceptance of the adequacy and certainty of Aristotelean deductive logic and Euclidean geometry), other features of his philosophy were of sufficient originality to justify his hopes of initiating a revolution. Not all of these innovations are of immediate relevance to his philosophy of mathematics and only the most relevant can be discussed here.

Kant believed that all previous attempts to distinguish between necessary and contingent truths were inadequate. He felt that Leibniz and Hume had come the closest to providing adequate analyses of the kinds of knowledge, but he asserted that they had been doomed to failure because they failed to perceive that there are two quite different kinds of distinction which must be considered—the distinction between a priori and a posteriori knowledge (see 10.1–10.2) on the one hand, and between analytic and synthetic judgments on the other (see 10.5). Kant's formulations of these distinctions have been subjected to numerous criticisms by later philosophers, and they are still an important subject of philosophical study and debate today.

There are four logically possible combinations of Kant's two pairs of distinctions: analytic-a priori, analytic-a posteriori, synthetic-a priori, and synthetic-a posteriori. Kant believed that the first and last pairs correspond to Hume's distinction between knowledge of relations of ideas and knowledge of matters of fact, respectively. He denied that there is any analytic-a posteriori knowledge. Mathematics, he believed, provides a prime example of synthetic-a priori knowledge in so far as it is universal in scope, necessary, and non-trivial. He did not attempt to prove that mathematical knowledge is synthetic-a priori; rather, he assumed that this is obvious and then he proceeded to try to ascertain how mathematical knowledge (and synthetic-a priori knowledge in general) is *possible*.

Two features of Kant's distinctions among the kinds of knowledge require comment here. First, as revealed particularly in his definition of "analytic" (see 10.5), he considered as knowledge only judgments which can be expressed in declarative sentences which have a simple subject-predicate form, and he assumed that for a subject or predicate term to be significant or meaningful it must denote or signify an idea or concept (in the broadest sense of these terms). His only break with the tradition here was his quite original account of the nature of the significata of certain terms. Second, his definition of "a priori" in no way involves the notion of knowledge acquired before or prior to sense experience. Kant was attempting to mediate the traditional debate between the philosophers who argued that innate ideas exist (such as Descartes and Leibniz) and those who rejected such ideas (such as Locke and Hume) when he asserted that all knowledge *begins* with experience although it does not all arise out of experience (see 10.1). A full understanding of this apparently self-contradictory statement is dependent on an understanding of one of the most revolutionary features of Kant's philosophy—his theory of perception; but only one element of this theory is of immediate relevance to his philosophy of mathematics, and only this element can be discussed here.

Although some pre-Kantian philosophers had suggested that the mind contributes to perception by processing or adding something to the raw sensory input, the most widely accepted model of perception in the 18th century was still the traditional view that the mind is a *passive* recipient of stimuli generated by external objects; that is, the mind contributes nothing to perception and it only acts on the sensory input after it has been perceived. In contrast, Kant maintained that *all* perception necessarily involves an *interaction* between the perceiver and the object perceived—that perception requires not only an external stimulus and a passive sensory organ, but also a processing of this stimulus and the *addition* of something by the mind. He pointed out that an important corollary of this theory is that it is impossible to have any knowledge whatsoever of the nature of anything beyond what we immediately perceive (the phenomena) and certain inferences we can make concerning the various components of the perceptions. The sources of the stimuli—the "noumena" or "things-in-themselves" which may be said to "underlie" the phenomenal world—are necessarily unknowable. Kant did not believe this to be a philosophy of scepticism; rather, he felt that it cleared the way for assertions of *faith* with regard to issues such as the immortality of the soul, freedom of the will, and the existence of God—that is, the basic questions of ethics and religion.

Kant not only asserted that there is no "pure" perception to which the mind contributes nothing, but also provided a complex and sophisticated account of what the mind contributes and a discussion of how and why it does this. The processing of the raw sensory input (or the "manifold of intuition," in Kant's terms) is accomplished to a great extent by the faculty of the mind which Kant called the "understanding." But prior to this processing (logically prior, at least) the mind adds certain features to the manifold—namely the *forms* of intuition, space and time. In other words, for Kant space and time are not independently existing objective entities as most of his predecessors believed (see for example Newton's account of absolute space in 6.5), nor are they mere relations of ideas (as Leibniz had asserted); rather, they are part of what is put into (or on to) one's experience by the mind. Kant argued that this account provides the only reasonable explanation of how it is possible for mathematical knowledge to be a priori; since the mind *necessarily* applies the forms of space and time to *all* experience, mathematical truths must be universal and necessary—i.e. a priori. However, although this theory of perception clarifies Kant's dictum that all knowledge begins with experience but does not all arise out of experience, it does not explain how it is possible that mathematical knowledge is synthetic. To understand this part of Kant's theory, it is useful to draw some comparisons with Locke's account of the nature of mathematical knowledge.

It will be remembered that Locke also believed that mathematical knowledge is both universally necessary and non-trivial (or in Kant's terminology, synthetic-a priori). His theory differs from Kant's with regard to his explanation of the a prioriness (universal necessity) of the truths of arithmetic and geometry in so far as he believed that the fundamental ideas of unity and of the dimensionless geometrical point could be derived by abstraction only from passively received sensations. But on the question of the synthetic nature of mathematics, Locke's account is much more similar to Kant's in so far as both theories are what is today known as *constructivist* philosophies of mathematics. Geometry for both Locke and Kant is *not* the study of already existing ideas of triangles, spheres, etc.; rather, it involves the construction of these figures by the mind out of simple ideas such as that of the dimensionless point. The theorems of geometry, on this account, are interpreted as being rules for constructing various geometrical figures, and the task of the geometer is essentially to discover the necessary rules for the construction of these figures.

Similarly, both viewed arithmetic as being the search for the rules governing the construction of numbers out of the fundamental concept of unity. They also agreed that these rules must be *discovered:* they are necessary rules and cannot be arbitrarily created or invented by the mind. It is this restriction which led Kant to reject the possibility of non-Euclidean geometries, and it was his failure to anticipate this later (by several decades) development in mathematics which led many philosophers to reject his entire theory of the nature of mathematical knowledge. Some mathematicians began to reconsider his over-all theory at the beginning of the present century, and they formulated a new constructivist philosophy of mathematics based in part on Kant's analysis of time as one of the forms of intuition. This new theory, known as *intuitionism,* has been one of the most influential philosophies of mathematics over the last fifty years, a success which serves as a reminder that one should not be too hasty to discard an entire theory as sophisticated and carefully worked out as Kant's. It is possible that even his philosophy of geometry deserves to be reconsidered to ascertain whether it might be modified in some relatively simple way to accommodate non-Euclidean geometries.

Sources

All selections are from the Norman Kemp Smith translation of Kant's *Critique of Pure Reason.* The numerals refer to the marginal numbering.

10.1	B1–B3.	**10.11**	B146–B147.
10.2	B3–B4.	**10.12**	B297–B299.
10.3	B5–86.	**10.13**	A101–A103.
10.4	B8–B9.	**10.14**	B64–B66.
10.5	B10–B12.	**10.15**	B203–B206.
10.6	B13–B14.	**10.16**	B740–B747.
10.7	B14–B17.	**10.17**	B189–B191.
10.8	B19	**10.18**	B359–B361.
10.9	B20–B21.	**10.19**	B754–B758.
10.10	B33–B35.	**10.20**	B759–B761.

KANT · Selections

A priori and a posteriori knowledge.

10.1 There can be no doubt that all our knowledge begins with experience. For how should our faculty of knowledge be awakened into action did not objects affecting our senses partly of themselves produce representations, partly arouse the activity of our understanding to compare these representations, and, by combining or separating them, work up the raw material of the sensible impressions into that knowledge of objects which is entitled experience? In the order of time, therefore, we have no knowledge antecedent to experience, and with experience all our knowledge begins.

But though all our knowledge begins with experience, it does not follow that it all arises out of experience. For it may well be that even our empirical knowledge is made up of what we receive through impressions and of what our own faculty of knowledge (sensible impressions serving merely as the occasion) supplies from itself. If our faculty of knowledge makes any such addition, it may be that we are not in a position to distinguish it from the raw material, until with long practice of attention we have become skilled in separating it.

This, then, is a question which at least calls for closer examination, and does not allow of any off-hand answer:—whether there is any knowledge that is thus independent of experience and even of all impressions of the senses. Such knowledge is entitled *a priori,* and distinguished from the *empirical,* which has its sources *a posteriori,* that is, in experience.

The expression '*a priori*' does not, however, indicate with sufficient precision the full meaning of our question. For it has been customary to say, even of much knowledge that is derived from empirical sources, that we have it or are capable of having it *a priori,* meaning thereby that we do not derive it immediately from experience, but from a universal rule—a rule which is itself, however, borrowed by us from experience. Thus we would say of a man who undermined the foundations of his house, that he might have known *a priori* that it would fall, that is, that he need not have waited for the experience of its actual falling. But still he could not know this completely *a priori.* For he had first to learn through experience that bodies are heavy, and therefore fall when their supports are withdrawn.

In what follows, therefore, we shall understand by *a priori* knowledge, not knowledge independent of this or that experience, but knowledge absolutely independent of all experience. Opposed to it is empirical knowledge, which is knowledge possible only *a posteriori,* that is, through experience. *A priori* modes of knowledge are entitled pure when there is no admixture of anything empirical. Thus, for instance, the proposition, 'every alteration has its cause,' while an *a priori* proposition, is not a pure proposition, because alteration is a concept which can be derived only from experience.

10.2 What we here require is a criterion by which to distinguish with certainty between pure and empirical knowledge. Experience teaches us that a thing is so and so, but not that it cannot be otherwise. First, then, if we have a proposition which in being thought is thought as *necessary,* it is an *a priori* judgment; and if, besides, it is not derived from any proposition except one which also has the validity of a necessary judgment, it is an absolutely *a priori* judgment. Secondly, experience never confers on its judgments true or strict, but only assumed and comparative *universality,* through induction. We can properly only say, therefore, that, so far as we have hitherto observed, there is no exception to this or that rule. If, then, a judgment is thought with strict universality, that is, in such manner that no exception is allowed as possible, it is not derived from experience, but it is valid absolutely *a priori.* Empirical universality is only an arbitrary extension of a validity holding in most cases to one which holds in all, for instance, in the proposition, 'all bodies are heavy.' When, on the other hand, strict universality is essential to a judgment, this indicates a special source of knowledge, namely, a faculty of *a priori* knowledge. Necessity and strict universality are thus sure criteria of *a priori* knowledge, and are inseparable from one another. But since in the employment of these criteria the contingency of judgments is sometimes more easily shown than their empirical limitation, or, as sometimes also happens, their unlimited universality can be more convincingly proved than their necessity, it is advisable to use the two criteria separately, each by itself being infallible.

10.3 Such *a priori* origin is manifest in certain concepts, no less than in judgments. If we remove from our empirical concept of a body, one by one, every feature in it which is merely empirical, the colour, the hardness or softness, the weight, even the impenetrability, there still remains the space which the body (now entirely vanished) occupied, and this cannot be removed. Again, if we remove from our empirical concept of any object, corporeal or incorporeal, all properties which experience has taught us, we yet cannot take away that property through which the object is thought as substance or as inhering in a substance (although this concept of substance is more determinate than that of an object in general). Owing, therefore, to the necessity with which this concept of substance forces itself upon us, we have no option save to admit that it has its seat in our faculty of *a priori* knowledge.

10.4 Mathematics gives us a shining example of how far, independently of experience, we can progress in *a priori* knowledge. It does, indeed, occupy itself with objects and with knowledge solely in so far as they allow of being exhibited in intuition. But this circumstance is easily overlooked, since this intuition can itself be given *a priori,* and is therefore hardly to be distinguished from a bare and pure concept. Misled by such a proof of the power of reason, the demand for the extension of knowledge recognises no limits. The light dove, cleaving the air in her free flight, and feeling its resistance, might imagine that its flight would be still easier in empty space. It was thus that Plato left the world of the senses, as setting too narrow limits to the understanding, and ventured out beyond it on the wings of the ideas, in the empty space of the pure understanding. He did not observe that with all his efforts he made no advance—meeting no resistance that might, as it were, serve as a support upon which he could take a stand, to which he could apply his powers, and so set his understanding in motion.

Analytic and synthetic knowledge.

10.5 In all judgments in which the relation of a subject to the predicate is thought (I take into consideration affirmative judgments only, the subsequent application to negative judgments being easily made), this relation is possible in two different ways. Either the predicate B belongs to the subject A, as something which is (covertly) contained in this concept A; or B lies outside the concept A, although it does indeed stand in connection with it. In the one case I entitle the judgment analytic, in the other synthetic. Analytic judgments (affirmative) are therefore those in which the connection of the predicate with the subject is thought through identity; those in which this connection is thought without identity should be entitled synthetic. The former, as adding nothing through the predicate to the concept of the subject, but merely breaking it up into those constituent concepts that have all along been thought in it, although confusedly, can also be entitled explicative. The latter, on the other hand, add to the concept of the subject a predicate which has not been in any wise thought in it, and which no analysis could possibly extract from it; and they may therefore be entitled ampliative. If I say, for instance, 'All bodies are extended,' this is an analytic judgment. For I do not require to go beyond the concept which I connect with 'body' in order to find extension as bound up with it. To meet with this predicate, I have merely to analyse the concept, that is, to become conscious to myself of the manifold which I always think in that concept. The judgment is therefore analytic. But when I say, 'All bodies are heavy,' the predicate is something quite different from anything that I think in the mere concept of body in general; and the addition of such a predicate therefore yields a synthetic judgment.

Judgments of experience, as such, are one and all synthetic. For it would be absurd to found an analytic judgment on experience. Since, in framing the judgment, I must not go outside my concept, there is no need to appeal to the testimony of experience in its support. That a body is extended is a proposition that holds *a priori* and is not empirical. For, before appealing to experience, I have already in the concept of body all the conditions required for my judgment. I have only to extract from it, in accordance with the principle of contradiction, the required predicate, and in so doing can at the same time become conscious of the necessity of the judgment—and that is what experience could never have taught me.

10.6 Upon such synthetic, that is, ampliative principles, all our *a priori* speculative knowledge must ultimately rest; analytic judgments are very important, and indeed necessary, but only for obtaining that clearness in the concepts which is requisite for such a sure and wide synthesis as will lead to a genuinely new addition to all previous knowledge.

Mathematical knowledge is synthetic a priori.

10.7 *All mathematical judgments, without exception, are synthetic.* This fact, though incontestably certain and in its consequences very important, has hitherto escaped the notice of those who are engaged in the analysis of human reason, and is, indeed, directly opposed to all their conjectures. For as it was found that all mathematical inferences proceed in accordance with the principle of contradiction (which the nature of all apodeictic certainty requires), it was supposed that the fundamental propositions of the science can themselves be known to be true through that principle. This is an erroneous view. For though a synthetic proposition can indeed be discerned in accordance with the principle of contradiction, this can only be if another synthetic proposition is presupposed, and if it can then be apprehended as following from this other proposition; it can never be so discerned in and by itself.

First of all, it has to be noted that mathematical propositions, strictly so called, are always judgments *a priori*, not empirical; because they carry with them necessity, which cannot be derived from experience. If this be demurred to, I am willing to limit my statement to *pure* mathematics, the very concept of which implies that it does not contain empirical, but only pure *a priori* knowledge.

We might, indeed, at first suppose that the proposition $7 + 5 = 12$ is a merely analytic proposition, and follows by the principle of contradiction from the concept of a sum of 7 and 5. But if we look more closely we find that the concept of the sum of 7 and 5 contains nothing save the union of the two numbers into one, and in this no thought is being taken as to what that single number may be which combines both. The concept of 12 is by no means already thought in merely thinking this union of 7 and 5; and I may analyse my concept of such a possible sum as long as I please, still I shall never find the 12 in it. We have to go outside these concepts, and call in the aid of the intuition which corresponds to one of them, our five fingers, for instance, or, as Segner does in his *Arithmetic*, five points, adding to the concept of 7, unit by unit, the five given in intuition. For starting with the number 7, and for the concept of 5 calling in the aid of the fingers of my hand as intuition, I now add one by one to the number 7 the units which I previously took together to form the number 5, and with the aid of that figure [the hand] see the number 12 come into being. That 5 should be added to 7, I have indeed already thought in the concept of a sum $= 7 + 5$, but not that this sum is equivalent to the number 12. Arithmetical propositions are therefore always synthetic. This is still more evident if we take larger numbers. For it is then obvious that, however we might turn and twist our concepts, we could never, by the mere analysis of them, and without the aid of intuition, discover what [the number is that] is the sum.

Just as little is any fundamental proposition of pure geometry analytic. That the straight line between two points is the shortest, is a synthetic proposition. For my concept of *straight* contains nothing of quantity but only of quality. The concept of the shortest is wholly an addition, and cannot be derived, through any process of analysis, from the concept of the straight line. Intuition, therefore, must here be called in; only by its aid is the synthesis possible. What here causes us commonly to believe that the predicate of such apodeictic judgments is already contained in our concept, and that the judgment is therefore analytic, is merely the ambiguous character of the terms used. We are required to join in thought a certain predicate to a given concept, and this necessity is inherent in the concepts themselves. But the question is not what we *ought* to join in thought to the given concept, but what we *actually* think in it, even if only obscurely; and it is then manifest that, while the predicate is indeed attached necessarily to the concept, it is so in virtue of an intuition which must be added to the concept, not as thought in the concept itself.

Some few fundamental propositions, presupposed by the geometrician, are, indeed, really analytic, and rest on the principle of contradiction. But, as identical propositions, they serve only as links in the chain of method and not as principles; for instance, $a = a$; the whole is equal to itself; or $(a + b) > a$, that is, the whole is greater than its part. And even these propositions, though they are valid according to pure concepts, are only admitted in mathematics because they can be exhibited in intuition.

The general problem of pure reason.

10.8 Much is already gained if we can bring a number of investigations under the formula of a single problem. For we not only lighten our own task, by defining it accurately, but make it easier for others, who would test our results, to judge whether or not we have succeeded in what we set out to do. Now the proper problem of pure reason is contained in the question: How are *a priori* synthetic judgments possible?

10.9 In the solution of the above problem, we are at the same time deciding as to the possibility of the employment of pure reason in establishing and developing all those sciences which contain a theoretical *a priori* knowledge of objects, and have therefore to answer the questions:

How is pure mathematics possible?
How is pure science of nature possible?

Since these sciences actually exist, it is quite proper to ask *how* they are possible; for that they must be possible is proved by the fact that they exist.

The grounds of all knowledge—intuition and concepts.

10.10 In whatever manner and by whatever means a mode of knowledge may relate to objects, *intuition* is that through which it is in immediate relation to them, and to which all thought as a means is directed. But intuition takes place only in so far as the object is given to us. This again is only possible, to man at least, in so far as the mind is affected in a certain way. The capacity (receptivity) for receiving representations *through the mode in which we are* affected by objects, is entitled *sensibility*. Objects are *given* to us by means of sensibility, and it alone yields us *intuitions;* they are *thought* through the understanding, and from the understanding arise *concepts.* But all thought must, directly or indirectly, by way of certain characters, relate ultimately to intuitions, and therefore, with us, to sensibility, because in no other way can an object be given to us.

The effect of an object upon the faculty of representation, so far as we are affected by it, is *sensation.* That intuition which is in relation to the object through sensation, is entitled *empirical.* The undetermined object of an empirical intuition is entitled *appearance.*

That in the appearance which corresponds to sensation I term its *matter;* but that which so determines the manifold of appearance that it allows of being ordered in certain relations, I term the *form* of appearance. That in which alone the sensations can be posited and ordered in a certain form, cannot itself be sensation; and therefore, while the matter of all appearance is given to us *a posteriori* only, its form must lie ready for the sensations *a priori* in the mind, and so must allow of being considered apart from all sensation.

I term all representations *pure* (in the transcendental sense) in which there is nothing that belongs to sensation. The pure form of sensible intuitions in general, in which all the manifold of intuition is intuited in certain relations, must be found in the mind *a priori.* This pure form of sensibility may also itself be called *pure intuition.* Thus, if I take away from the representation of a body that which the understanding thinks in regard to it, substance, force, divisibility, etc., and likewise what belongs to sensation, impenetrability, hardness, colour, etc., something still remains over from this empirical intuition, namely, extension and figure. These belong to pure intuition, which, even without any actual object of the senses or of sensation, exists in the mind *a priori* as a mere form of sensibility.

10.11 To *think* an object and to *know* an object are thus by no means the same thing. Knowledge involves two factors: first, the concept, through which an object in general is thought (the category); and secondly, the intuition, through which it is given. For if no intuition could be given corresponding to the concept, the concept would still indeed be a thought, so far as its form is concerned, but would be without any object, and no knowledge of anything would be possible by means of it. So far as I could know, there would be nothing, and could be nothing, to which my thought could be applied. Now, as the Aesthetic has shown, the only intuition possible to us is sensible; consequently, the thought of an object in general, by means of a pure concept of understanding, can become knowledge for us only in so far as the concept is related to objects of the senses. Sensible intuition is either pure intuition (space and time) or empirical intuition of that which is immediately represented, through sensation, as actual in space and time. Through the determination of pure intuition we can acquire *a priori* knowledge of objects, as in mathematics, but only in regard to their form, as appearances; whether there can be things which must be intuited in this form, is still left undecided. Mathematical concepts are not, therefore, by themselves knowledge, except on the supposition that there are things which allow of being presented to us only in accordance with the form of that pure sensible intuition. Now *things in space and time* are given only in so far as they are perceptions (that is, representations accompanied by sensation)—therefore only through empirical representation. Consequently, the pure concepts of understanding, even when they are applied to *a priori* intuitions, as in mathematics, yield knowledge only in so far as these intuitions—and therefore indirectly by their means the pure concepts also—can be applied to empirical intuitions.

10.12 If the assertion, that the understanding can employ its various principles and its various concepts solely in an empirical and never in a transcendental manner, is a proposition which can be known with certainty, it will yield important consequences. The transcendental employment of a concept in any principle is its application to things *in general and in themselves;* the empirical employment is its application *merely to appearances;* that is, to objects of a possible experience. That the latter application of concepts is alone feasible is evident from the following considerations. We demand in every concept, first, the logical form of a concept (of thought) in general, and secondly, the possibility of giving it an object to which it may be applied. In the absence of such object, it has no meaning and is completely lacking in content, though it may still contain the logical function which is required for making a concept out of any data that may be presented. Now the object cannot be given to a concept otherwise than in intuition; for though a pure intuition can indeed precede the object *a priori,* even this intuition can acquire its object, and therefore objective validity, only through the empirical intuition of which it is the mere form. Therefore all concepts, and with them all principles, even such as are possible *a priori,* relate to empirical intuitions, that is, to the data for a possible experience. Apart from this relation they have no objective validity, and in respect of their representations are a mere play of imagination or of understanding. Take, for instance, the concepts of mathematics, considering them first of all in their pure intuitions. Space has three dimensions; between two points there can be only one straight line, etc. Although all these principles, and the representation of the object with which this science occupies itself, are generated in the mind completely *a priori,* they would mean nothing, were we not always able to present their meaning in appearances, that is, in empirical objects. We therefore demand that a bare concept be *made sensible,* that is, that an object corresponding to it be presented in intuition. Otherwise the concept would, as we say, be without *sense,* that is, without meaning. The mathematician meets this demand by the construction of a figure, which, although produced *a priori,* is an appearance present to the senses. In the same science the concept of magnitude seeks its support and sensible meaning in number, and this in turn in the fingers, in the beads of the abacus, or in strokes and points which can be placed before the eyes. The concept itself is always *a priori* in origin, and so likewise are the synthetic principles or formulas derived from such concepts; but their employment and their relation to their professed objects can in the end be sought nowhere but in experience, of whose possibility they contain the formal conditions.

10.13 For if we can show that even our purest *a priori* intuitions yield no knowledge, save in so far as they contain a combination of the manifold such as renders a thoroughgoing synthesis of reproduction possible, then this synthesis of imagination is likewise grounded, antecedently to all experience, upon *a priori* principles; and we must assume a pure transcendental synthesis of imagination as conditioning the very possibility of all experience. For experience as such necessarily presupposes the reproducibility of appearances. When I seek to draw a line in thought, or to think of the time from one noon to another, or even to represent to myself some particular number, obviously the various manifold representations that are involved must be apprehended by me in thought one after the other. But if I were always to drop out of thought the preceding representations (the first parts of the line, the antecedent parts of the time period, or the units in the order represented), and did not reproduce them while advancing to those that follow, a complete representation would never be obtained: none of the above-mentioned thoughts, not even the purest and most elementary representations of space and time, could arise.

The synthesis of apprehension is thus inseparably bound up with the synthesis of reproduction. And as the former constitutes the transcendental ground of the possibility of all modes of knowledge whatsoever—of those that are pure *a priori* no less than of those that are empirical—the reproductive synthesis of the imagination is to be counted among the transcendental acts of the mind. We shall therefore entitle this faculty the transcendental faculty of imagination.

If we were not conscious that what we think is the same as what we thought a moment before, all reproduction in the series of representations would be useless. For it would in its present state be a new representation which would not in any way belong to the act whereby it was to be gradually generated. The manifold of the representation would never, therefore, form a whole, since it would lack that unity which only consciousness can impart to it. If, in counting, I forget that the units, which now hover before me, have been added to one another in succession, I should never know that a total is being produced through this successive addition of unit to unit, and so would remain ignorant of the number. For the concept of the number is nothing but the consciousness of this unity of synthesis.

Space and time are not objective.

10.14 Let us suppose that space and time are in themselves objective, and are conditions of the possibility of things in themselves. In the first place, it is evident that in regard to both there is a large number of *a priori* apodeictic and synthetic propositions. This is especially true of space, to which our chief attention will therefore be directed in this enquiry. Since the propositions of geometry are synthetic *a priori,* and are known with apodeictic certainty, I raise the question, whence do you obtain such propositions, and upon what does the understanding rely in its endeavour to achieve such absolutely necessary and universally valid truths? There is no other way than through concepts or through intuitions; and these are given either *a priori* or *a posteriori.* In their latter form, namely, as *empirical* concepts, and also as that upon which these are grounded, the *empirical* intuition, neither the concepts nor the intuitions can yield any synthetic proposition except such as is itself also merely empirical (that is, a proposition of experience), and which for that very reason can never possess the necessity and absolute universality which are characteristic of all geometrical propositions. As regards the first and sole means of arriving at such knowledge, namely, in *a priori* fashion through mere concepts or through intuitions, it is evident that from mere concepts only analytic knowledge, not synthetic knowledge, is to be obtained. Take, for instance, the proposition, "Two straight lines cannot enclose a space, and with them alone no figure is possible," and try to derive it from the concept of straight lines and of the number two. Or take the proposition, "Given three straight lines, a figure is possible," and try, in like manner to derive it from the concepts involved. All your labour is vain; and you find that you are constrained to have recourse to intuition, as is always done in geometry. You therefore give yourself an object in intuition. But of what kind is this intuition?

Is it a pure *a priori* intuition or an empirical intuition? Were it the latter, no universally valid proposition could ever arise out of it—still less an apodeictic proposition—for experience can never yield such. You must therefore give yourself an object *a priori* in intuition, and ground upon this your synthetic proposition. If there did not exist in you a power of *a priori* intuition; and if that subjective condition were not also at the same time, as regards its form, the universal *a priori* condition under which alone the object of this outer intuition is itself possible; if the object (the triangle) were something in itself, apart from any relation to you, the subject, how could you say that what necessarily exist in you as subjective conditions for the construction of a triangle, must of necessity belong to the triangle itself? You could not then add anything new (the figure) to your concepts (of three lines) as something which must necessarily be met with in the object, since this object is [on that view] given antecedently to your knowledge, and not by means of it. If, therefore, space (and the same is true of time) were not merely a form of your intuition, containing conditions *a priori*, under which alone things can be outer objects to you, and without which subjective conditions outer objects are in themselves nothing, you could not in regard to outer objects determine anything whatsoever in an *a priori* and synthetic manner. It is, therefore, not merely possible or probable, but indubitably certain, that space and time, as the necessary conditions of all outer and inner experience, are merely subjective conditions of all our intuition, and that in relation to these conditions all objects are therefore mere appearances, and not given us as things in themselves which exist in this manner. For this reason also, while much can be said *a priori* as regards the form of appearances, nothing whatsoever can be asserted of the thing in itself, which may underlie these appearances.

The relation between 'pure' and 'applied' mathematics.

10.15 . . . Appearances are all without exception *magnitudes*, indeed *extensive magnitudes.* As intuitions in space or time, they must be represented through the same synthesis whereby space and time in general are determined.

I entitle a magnitude extensive when the representation of the parts makes possible, and therefore necessarily precedes, the representation of the whole. I cannot represent to myself a line, however small, without drawing it in thought, that is, generating from a point all its parts one after another. Only in this way can the intuition be obtained. Similarly with all times, however small. In these I think to myself only that successive advance from one moment to another, whereby through the parts of time and their addition a determinate time-magnitude is generated. As the element of pure intuition in all appearances is either space or time, every appearance is as intuition an extensive magnitude; only through successive synthesis of part to part in the process of its apprehension can it come to be known. All appearances are consequently intuited as aggregates, as complexes of previously given parts. This is not the case with magnitudes of every kind, but only with those magnitudes which are represented and apprehended by us in this *extensive* fashion.

The mathematics of space (geometry) is based upon this successive synthesis of the productive imagination in the generation of figures. This is the basis of the axioms which formulate the conditions of sensible *a priori* intuition under which alone the schema of a pure concept of outer appearance can arise—for instance, that between two points only one straight line is possible, or that two straight lines cannot enclose a space, etc. These are the axioms which, strictly, relate only to magnitudes (*quanta*) as such.

As regards magnitude (*quantitas*), that is, as regards the answer to be given to the question, 'What is the magnitude of a thing?' there are no axioms in the strict meaning of the term, although there are a number of propositions which are synthetic and immediately certain (*indemonstrabilia*). The propositions, that if equals be added to equals the wholes are equal, and if equals be taken from equals the remainders are equal, are analytic propositions; for I am immediately conscious of the identity of the production of the one magnitude with the production of the other. Consequently, they are not axioms, for these have to be *a priori synthetic* propositions. On the other hand, the evident propositions of numerical relation are indeed synthetic, but are not general like those of geometry, and cannot, therefore, be called axioms but only numerical formulas. The assertion that 7 + 5 is equal to 12 is not an analytic proposition. For neither in the representation of 7, nor in that of 5, nor in the representation of the combination of both, do I think the number 12. (That I must do so in the *addition* of the two numbers is not to the point, since in the analytic proposition the question is only whether I actually think the predicate in the representation of the subject.) But although the proposition is synthetic it is also only singular. So far as we are here attending merely to the synthesis of the homogeneous (of units), that synthesis can take place only in one way, although the *employment* of these numbers is general. If I assert that through three lines, two of which taken together are greater than the third, a triangle can be described, I have expressed merely the function of productive imagination whereby the lines can be drawn greater or smaller, and so can be made to meet at any and every possible angle. The number 7, on the other hand, is possible only in one way. So also is the number 12, as thus generated through the synthesis of 7 with 5. Such propositions must not, therefore, be called axioms (that would involve recognition of an infinite number of axioms), but numerical formulas.

This transcendental principle of the mathematics of appearances greatly enlarges our *a priori* knowledge. For it alone can make pure mathematics, in its complete precision, applicable to objects of experience. Without this principle, such application would not be thus self-evident; and there has indeed been much confusion of thought in regard to it. Appearances are not things in themselves. Empirical intuition is possible only by means of the pure intuition of space and of time. What geometry asserts of pure intuition is therefore undeniably valid of empirical intuition. The idle objections, that objects of the senses may not conform to such rules of construction in space as that of the infinite divisibility of lines or angles, must be given up. For if these objections hold good, we deny the objective validity of space, and consequently of all mathematics, and no longer know why and how far mathematics can be applicable to appearances. The synthesis of spaces and times, being a synthesis of the essential forms of all intuition, is what makes possible the apprehension of appearance, and consequently every outer experience and all knowledge of the objects of such experience. Whatever pure mathematics establishes in regard to the synthesis of the form of apprehension is also necessarily valid of the objects apprehended.

Mathematics as a constructive enterprise.

10.16 Mathematics presents the most splendid example of the successful extension of pure reason, without the help of experience. Examples are contagious, especially as they quite naturally flatter a faculty which has been successful in one field, leading it to expect the same good fortune in other fields. Thus pure reason hopes to be able to extend its domain as successfully and securely in its transcendental as in its mathematical employment, especially when it resorts to the same method as has been of such obvious utility in mathematics. It is therefore highly important for us to know whether the method of attaining apodeictic certainty which is called *mathematical* is identical with the method by which we endeavour to obtain the same certainty in philosophy, and which in that field would have to be called *dogmatic.*

Philosophical knowledge is the *knowledge gained by reason from concepts;* mathematical knowledge is the knowledge gained by reason from the *construction* of concepts. To *construct* a concept means to exhibit *a priori* the intuition which corresponds to the concept. For the construction of a concept we therefore need a *non-empirical* intuition. The latter must, as intuition be a *single* object, and yet none the less, as the construction of a concept (a universal representation), it must in its representation express universal validity for all possible intuitions which fall under the same concept. Thus I construct a triangle by representing the object which corresponds to this concept either by imagination alone, in pure intuition, or in accordance therewith also on paper, in empirical intuition—in both cases completely *a priori,* without having borrowed the pattern from any experience. The single figure which we draw is empirical, and yet it serves to express the concept, without impairing its universality. For in this empirical intuition we consider only the act whereby we construct the concept, and abstract from the many determinations (for instance, the magnitude of the sides and of the angles), which are quite indifferent, as not altering the concept 'triangle.'

Thus philosophical knowledge considers the particular only in the universal, mathematical knowledge the universal in the particular, or even in the single instance, though still always *a priori* and by means of reason. Accordingly, just as this single object is determined by certain universal conditions of construction, so the object of the concept, to which the single object corresponds merely as its schema, must likewise be thought as universally determined.

The essential difference between these two kinds of knowledge through reason consists therefore in this formal difference, and does not depend on difference of their material or objects. Those who propose to distinguish philosophy from mathematics by saying that the former has its object *quality* only and the latter *quantity* only, have mistaken the effect for the cause. The form of mathematical knowledge is the cause why it is limited exclusively to quantities. For it is the concept of quantities only that allows of being constructed, that is, exhibited *a priori* in intuition; whereas qualities cannot be presented in any intuition that is not empirical. Consequently reason can obtain a knowledge of qualities only through concepts. No one can obtain an intuition corresponding to the concept of reality otherwise than from experience; we can never come into possession of it *a priori* out of our own resources, and prior to the empirical consciousness of reality. The shape of a cone we can form for ourselves in intuition, unassisted by any experience, according to its concept alone, but the colour of this cone must be previously given in some experience or other. I cannot represent in intuition the concept of a cause in general except in an example supplied by experience; and similarly with other concepts. Philosophy, as well as mathematics, does indeed treat of quantities, for instance, of totality, infinity, etc. Mathematics also concerns itself with qualities, for instance, the difference between lines and surfaces, as spaces of different quality, and with the continuity of extension as one of its qualities. But although in such cases they have a common object, the mode in which reason handles that object is wholly different in philosophy and in mathematics. Philosophy confines itself to universal concepts; mathematics can achieve nothing by concepts alone but hastens at once to intuition, in which it considers the concept *in concreto,* though not empirically, but only in an intuition which it presents *a priori,* that is, which it has constructed, and in which whatever follows from the universal conditions of the construction must be universally valid of the object of the concept thus constructed.

Suppose a philosopher be given the concept of a triangle and he be left to find out, in his own way, what relation the sum of its angles bears to a right angle. He has nothing but the concept of a figure enclosed by three straight lines, and possessing three angles. However long he meditates on this concept, he will never produce anything new. He can analyse and clarify the concept of a straight line or of an angle or of the number three, but he can never arrive at any properties not already contained in these concepts. Now let the geometrician take up these questions. He at once begins by constructing a triangle. Since he knows that the sum of two right angles is exactly equal to the sum of all the adjacent angles which can be constructed from a single point on a straight line, he prolongs one side of his triangle and obtains two adjacent angles, which together are equal to two right angles. He then divides the external angle by drawing a line parallel to the opposite side of the triangle, and observes that he has thus obtained an external adjacent angle which is equal to an internal angle—and so on. In this fashion, through a chain of inferences guided throughout by intuition, he arrives at a fully evident and universally valid solution of the problem.

But mathematics does not only construct magnitudes (*quanta*) as in geometry; it also constructs magnitude as such (*quantitas*), as in algebra. In this it abstracts completely from the properties of the object that is to be thought in terms of such a concept of magnitude. It then chooses a certain notation for all constructions of magnitude as such (numbers), that is, for addition, subtraction, extraction of roots, etc. Once it has adopted a notation for the general concept of magnitudes so far as their different relations are concerned, it exhibits in intuition, in accordance with certain universal rules, all the various operations through which the magnitudes are produced and modified. When, for instance, one magnitude is to be divided by another, their symbols are placed together, in accordance with the sign for division, and similarly in the other processes; and thus in algebra by means of a symbolic construction, just as in geometry by means of an ostensive construction (the geometrical construction of the objects themselves), we succeed in arriving at results which discursive knowledge could never have reached by means of mere concepts.

Now what can be the reason of this radical difference in the fortunes of the philosopher and the mathematician, both of whom practise the art of reason, the one making his way by means of concepts, the other by means of intuitions which he exhibits *a priori* in accordance with concepts? The cause is evident from what has been said above, in our exposition of the fundamental transcendental doctrines. We are not here concerned with analytic propositions, which can be produced by mere analysis of concepts (in this the philosopher would certainly have the advantage over his rival), but with synthetic propositions, and indeed with just those synthetic propositions that can be known *a priori*. For I must not restrict my attention to what I am actually thinking in my concept of a triangle (this is nothing more than the mere definition); I must pass beyond it to properties which are not contained in this concept, but yet belong to it. Now this is impossible unless I determine my object in accordance with the conditions either of empirical or of pure intuition. The former would only give us an empirical proposition (based on the measurement of the angles), which would not have universality, still less necessity; and so would not at all serve our purpose. The second method of procedure is the mathematical one, and in this case is the method of geometrical construction, by means of which I combine in a pure intuition (just as I do in empirical intuition) the manifold which belongs to the schema of a triangle in general, and therefore to its concept. It is by this method that universal synthetic propositions must be constructed.

It would therefore be quite futile for me to philosophise upon the triangle, that is, to think about it discursively. I should not be able to advance a single step beyond the mere definition, which was what I had to begin with. There is indeed a transcendental synthesis framed from concepts alone, a synthesis with which the philosopher is alone competent to deal; but it relates only to a thing in general, as defining the conditions under which the perception of it can belong to possible experience. But in mathematical problems there is no question of this, nor indeed of existence at all, but only of the properties of the objects in themselves, that is to say, solely in so far as these properties are connected with the concept of the objects.

Mathematical reasoning.

10.17 The universal, though merely negative, condition of all our judgments in general, whatever be the content of our knowledge, and however it may relate to the object, is that they be not self-contradictory; for if self-contradictory, these judgments are in themselves, even without reference to the object, null and void. But even if our judgment contains no contradiction, it may connect concepts in a manner not borne out by the object, or else in a manner for which no ground is given, either *a priori* or *a posteriori,* sufficient to justify such judgment, and so may still, in spite of being free from all inner contradiction, be either false or groundless.

The proposition that no predicate contradictory of a thing can belong to it, is entitled the principle of contradiction, and is a universal, though merely negative, criterion of all truth. For this reason it belongs only to logic. It holds of knowledge, merely as knowledge in general, irrespective of content; and asserts that the contradiction completely cancels and invalidates it.

But it also allows of a positive employment, not merely, that is, to dispel falsehood and error (so far as they rest on contradiction), but also for the knowing of truth. For, *if the judgment is analytic,* whether negative or affirmative, its truth can always be adequately known in accordance with the principle of contradiction. The reverse of that which as concept is contained and is thought in the knowledge of the object, is always rightly denied. But since the opposite of the concept would contradict the object, the concept itself must necessarily be affirmed of it.

The principle of contradiction must therefore be recognised as being the universal and completely sufficient *principle of all analytic knowledge;* but beyond the sphere of analytic knowledge it has, as a *sufficient* criterion of truth, no authority or no field of application. The fact that no knowledge can be contrary to it without self-nullification, makes this principle a *conditio sine qua non,* but not a determining ground, of the truth of our non-analytic knowledge. Now in our critical enquiry it is only with the synthetic portion of our knowledge that we are concerned; and in regard to the truth of this kind of knowledge we can never look to the above principle for any positive information, though, of course, since it is inviolable, we must always be careful to conform to it.

10.18 A distinction is commonly made between what is immediately known and what is merely inferred. That in a figure which is bounded by three straight lines there are three angles, is known immediately; but that the sum of these angles is equal to two right angles, is merely inferred. Since we have constantly to make use of inference, and so end by becoming completely accustomed to it, we no longer take notice of this distinction, and frequently, as in the so-called deceptions of the senses, treat as being immediately perceived what has really only been inferred. In every process of reasoning there is a fundamental proposition, and another, namely the conclusion, which is drawn from it, and finally, the inference (logical sequence) by which the truth of the latter is inseparably connected with the truth of the former. If the inferred judgment is already so contained in the earlier judgment that it may be derived from it without the mediation of a third representation, the inference is called immediate (*consequentia immediata*)—I should prefer to entitle it inference of the understanding. But if besides the knowledge contained in the primary proposition still another judgment is needed to yield the conclusion, it is to be entitled an inference of the reason. In the proposition: "All men are mortal," there are already contained the propositions: "some men are mortal," "some mortal beings are men," "nothing that is not mortal is a man"; and these are therefore immediate conclusions from it. On the other hand, the proposition: "All learned beings are mortal," is not contained in the fundamental judgment (for the concept of learned beings does not occur in it at all), and it can only be inferred from it by means of a mediating judgment.

In every syllogism I first think a *rule* (the major premiss) through the *understanding.* Secondly, I *subsume* something known under the condition of the rule by means of *judgment* (the minor premiss). Finally, what is thereby known I *determine* through the predicate of the rule, and so *a priori* through *reason* (the conclusion). The relation, therefore, which the major premiss, as the rule, represents between what is known and its condition is the ground of the different kinds of syllogism. Consequently, syllogisms, like judgments, are of three kinds, according to the different ways in which, in the understanding, they express the relation of what is known; they are either *categorical, hypothetical,* or *disjunctive.*

If, as generally happens, the judgment that forms the conclusion is set as a problem—to see whether it does not follow from judgments already given, and through which a quite different object is thought—I look in the understanding for the assertion of this conclusion, to discover whether it is not there found to stand under certain conditions according to a universal rule. If I find such a condition, and if the object of the conclusion can be subsumed under the given condition, then the conclusion is deduced from the rule, *which is also valid for other objects of knowledge.* From this we see that in inference reason endeavours to reduce the varied and manifold knowledge obtained through the understanding to the smallest number of principles (universal conditions) and thereby to achieve in it the highest possible unity.

10.19 The exactness of mathematics rests upon definitions, axioms and demonstrations. I shall content myself with showing that none of these, in the sense in which they are understood by the mathematician, can be achieved or imitated by the philosopher. I shall show that in philosophy the geometrician can by his method build only so many houses of cards, just as in mathematics the employment of a philosophical method results only in mere talk. Indeed it is precisely in knowing its limits that philosophy consists; and even the mathematician, unless his talent is of such a specialised character that it naturally confines itself to its proper field, cannot afford to ignore the warnings of philosophy, or to behave as if he were superior to them.

1. *Definitions.*—To *define* as the word itself indicates, really only means to present the complete, original concept of a thing within the limits of its concept. If this be our standard, an *empirical* concept cannot be defined at all, but only *made explicit.* . . . And indeed what useful purpose could be served by defining an empirical concept, such, for instance, as that of water? When we speak of water and its properties, we do not stop short at what is thought in the word, water, but proceed to experiments. The word, with the few characteristics which we attach to it, is more properly to be regarded as merely a designation than as a concept of the thing; the so-called definition is nothing more than a determining of the word. In the second place, it is also true that no concept given *a priori,* such as substance, cause, right, equity, etc., can, strictly speaking, be defined. For I can never be certain that the clear representation of a given concept, which as given may still be confused, has been completely effected, unless I know that it is adequate to its object. . . . Since, then, neither empirical concepts nor concepts given *a priori* allow of definition, the only remaining kind of concepts, upon which this mental operation can be tried, are arbitrarily invented concepts. A concept which I have invented I can always define; for since it is not given to me either by the nature of understanding or by experience, but is such as I have myself

deliberately made it to be, I must know what I have intended to think in using it. I cannot, however, say that I have thereby defined a true object. For if the concept depends on empirical conditions, as e.g. the concept of a ship's clock, this arbitrary concept of mine does not assure me of the existence or of the possibility of its object. I do not even know from it whether it has an object at all, and my explanation may better be described as a declaration of my project than as a definition of an object. There remain, therefore, no concepts which allow of definition, except only those which contain an arbitrary synthesis that admits of *a priori* construction. Consequently, mathematics is the only science that has definitions. For the object which it thinks it exhibits *a priori* in intuition, and this object certainly cannot contain either more or less than the concept, since it is through the definition that the concept of the object is given—and given originally, that is, without its being necessary to derive the definition from any other source.

10.20 In short, the definition in all its precision and clarity ought, in philosophy, to come rather at the end than at the beginning of our enquiries. In mathematics, on the other hand, we have no concept whatsoever prior to the definition, through which the concept itself is first given. For this reason mathematical science must always begin, and it can always begin, with the definition.

. . . Mathematical definitions can never be in error. For since the concept is first given through the definition, it includes nothing except precisely what the definition intends should be understood by it. But although nothing incorrect can be introduced into its content, there may sometimes, though rarely, be a defect in the form in which it is clothed, namely as regards precision. Thus the common explanation of the circle that it is a *curved* line every point in which is equidistant from one and the same point (the center), has the defect that the determination, curved, is introduced unnecessarily. For there must be a particular theorem, deduced from the definition and easily capable of proof, namely, that if all points in a line are equidistant from one and the same point, the line is curved (no part of it straight). Analytic definitions, on the other hand, may err in many ways, either through introducing characteristics which do not really belong to the concept, or by lacking that completeness which is the essential feature of a definition. The latter defect is due to the fact that we can never be quite certain of the completeness of the analysis. For these reasons the mathematical method of definition does not admit of imitation in philosophy.

2. *Axioms.*—These, in so far as they are immediately certain, are synthetic *a priori* principles. Now one concept cannot be combined with another synthetically and also at the same time immediately, since, to be able to pass beyond either concept, a third something is required to mediate our knowledge. Accordingly, since philosophy is simply what reason knows by

means of concepts, no principle deserving the name of an axiom is to be found in it. Mathematics, on the other hand, can have axioms, since by means of the construction of concepts in the intuition of the object it can combine the predicates of the object both *a priori* and immediately, as, for instance, in the proposition that three points always lie in a plane. But a synthetic principle derived from concepts alone can never be immediately certain, for instance, the proposition that everything which happens has a cause. . . .

3. *Demonstrations.*—An apodeictic proof can be called a demonstration, only in so far as it is intuitive. Experience teaches us what is, but does not teach us that it could not be other than what it is. Consequently, no empirical grounds of proof can ever amount to apodeictic proof. Even from *a priori* concepts, as employed in discursive knowledge, there can never arise intuitive certainty, that is, demonstrative evidence, however apodeictically certain the judgment may otherwise be. Mathematics alone, therefore, contains demonstrations, since it derives its knowledge not from concepts but from the construction of them, that is, from intuition, which can be given *a priori* in accordance with the concepts. Even the method of algebra with its equations, from which the correct answer, together with its proof, is deduced by reduction, is not indeed geometrical in nature, but is still constructive in a way characteristic of the science. The concepts attached to the symbols, especially concerning the relations of magnitudes, are presented in intuition; and this method, in addition to its heuristic advantages, secures all inferences against error by setting each one before our eyes. While philosophical knowledge must do without this advantage, inasmuch as it has always to consider the universal *in abstracto* (by means of concepts), mathematics can consider the universal *in concreto* (in the single intuition) and yet at the same time through pure *a priori* representation, whereby all errors are at once made evident.

11

MILL · Introduction

John Stuart Mill (1806–1873) was educated under the strict tutelage of his father James, a well known philosopher and historian. He began the study of Greek at the age of three, and of Latin at the age of eight. His education also included rigorous training in mathematics and logic, although it probably did not extend to the frontiers of mathematics at that time, in contrast to the educations of most of the other philosophers represented in this book. In addition, there is no indication in Mill's later discussions of the philosophy of mathematics that he was familiar with many of the significant developments which occurred during his lifetime, such as the invention of non-Euclidean geometries, the "rigorization" of analysis and number theory, and the advances made in logic by Boole, De Morgan, and others. It was the problems generated by such developments that led to the increased concentration on the foundations of mathematics at the end of the 19th century, on the part of both mathematicians and philosophers. It was probably as much Mill's failure to discuss these developments (and his apparent ignorance of them) as it was the sharp criticisms of his theories by mathematicians such as Gottlob Frege which led to the general dismissal of his philosophy of mathematics by the turn of the century. Although his work in this area was not taken very seriously during the first half of the present century, in recent years philosophers who are also quite competent mathematicians have begun to reconsider some of Mill's arguments and conclusions, while others have more or less independently formulated new philosophies of mathematics which resemble Mill's theory in significant ways.

Mill was influenced by the French positivists Comte and Saint-Simon, as well as by other philosophers and scientists, but the strongest influence on his general outlook and on his philosophy of mathematics in particular came from the English philosophers dating back to Francis Bacon and including Hobbes, Locke, Berkeley, and Hume. Many of the following selections reflect ideas first discussed by others, especially Hobbes and Berkeley, but Mill added several features which were new and original, and he ultimately rejected every one of the general theories of the nature of mathematics formulated by his predecessors.

One of the most important features of Mill's philosophy of mathematics—a feature which is more anticipatory of future work in philosophy than reminiscent of the past—is indicated in the title of the work from which all of the following passages have been drawn, *A System of Logic*. Indeed, this work contains discussions of many of the traditional problems of ontology and epistemology, but the main emphasis is on problems of logic (both inductive and deductive) and of the philosophy of language (both semantics and syntax). This focus on the problems of logic and language provided a model which was to be emulated during the next century by many philosophers—particularly by those writing in the English language—although the similarities in method did not produce corresponding similarities in the conclusions reached.

Mill's intensive analysis of logical and linguistic matters produced mixed results, and the evaluations of his work have varied considerably over the years. His formulation of a set of rules of induction has been widely credited as a major contribution to the philosophy of the natural sciences, but most philosophers have rejected Mill's claim that these rules are also applicable to mathematical practice. He contributed nothing to the development of the *techniques* of deductive logic, and although his *philosophy* of deductive logic was thought for many years to be fundamentally mistaken, recent work in this area suggests that Mill might have been ahead of his time. Also, through various elements of his philosophy of language Mill anticipated or influenced later work in this area. In particular, his distinction between the connotative and denotative meanings of terms (see 11.8) provided a new tool for dealing with some traditional philosophical problems, such as that of the meanings of abstract general terms. (It must be noted that Mill's definitions of "connotation" and "denotation" are quite different from the meanings which are ordinarily associated with these terms in nonphilosophic discourse today.)

Mill's concentration on logical and linguistic matters also had some negative consequences, especially in so far as it led him to ignore or give inadequate analyses of other important problems. For example, in his discussion of the meaning of propositions (see 11.4), he concluded that for a proposition to be meaningful at least two terms in the proposition must signify *things* or properties of things. But nowhere in the *System of Logic* did Mill provide any explanation of exactly what a "thing" is. In another work, *An Examination of Sir William Hamilton's Philosophy*, he rejected the traditional theories of mental and physical substances, and also the totally negative theories (such as Hume's); but all that he said in a positive vein was that things in some way require a "Permanent Possibility of Experience," and few philosophers have considered this to be adequate.

Mill's handling of the term "thing" also reflects his working assumption that if a word functions well in ordinary discourse then there is little need for philosophers to worry further about it. Even if this assumption is sound, Mill did not always apply it correctly; for example, the term "thing" is used in a variety of different ways in everyday discourse, and problems are generated by the resulting ambiguities which the philosopher must clarify. An even more serious set of problems arises concerning the term "certainty," problems which are compounded rather than resolved by Mill's discussion of the kinds of knowledge and his own use(s) of the term in these discussions. For example, in 11.14 Mill asserts that "what is commonly called mathematical certainty . . . which comprises the twofold conception of unconditional truth and perfect accuracy" belongs only to those mathematical truths "which relate to pure number, as distinguished from quantity in the more enlarged sense." In other passages (e.g., 11.11) Mill asserts that *all* mathematical truths are known by induction from experience, and in still other passages he explicitly denies that *any* mathematical truths can be absolutely certain (see 11.15).

These discussions of mathematics and certainty have provided one of the prime targets for Mill's critics, and justifiably so. However, it should be noted that the real problem is not necessarily one of simple inconsistency on Mill's part, but rather it may be more a case of careless use of certain terms. And it is possible that the confusions in the use of crucial terms are the product of Mill's propensity to use ordinary terminology as much as possible in his philosophical discourse. If he believed that people do in fact acquire certainty about mathematical truths by means of inductions from experience, and if this is what they normally mean by the terms "mathematical certainty" and "unconditional truth," then Mill is not necessarily guilty of being inconsistent. In any case, Mill was at least careless in his use of many crucial terms, and although his reliance on common terms rather than technical jargon makes his writings appear easier to read than those of Kant, his use of these common terms also produces (and frequently conceals) serious philosophical problems.

Sources

All selections are from John Stuart Mill's *A System of Logic*, New York, Harper & Brothers Publishers, 1869. The numerals refer to the book, chapter, and section, respectively.

11.1 Introduction, section 4, pp. 3–4
11.2 Introduction, section 4, p. 5
11.3 I, i, 1, pp. 11–12
11.4 I, i, 2, pp. 12–13
11.5 I, ii, 3, pp. 17–18
11.6 I, ii, 4, p. 18
11.7 I, ii, 4, p. 19
11.8 I, ii, 5, pp. 20–21
11.9 I, iii, 7, pp. 51–52
11.10 I, iii, 5, pp. 48–49
11.11 III, xxiv, 4, p. 364
11.12 III, xxiv, 5, pp. 364–367
11.13 III, xxiv, 6, pp. 367–368
11.14 II, vi, 3, pp. 168–169
11.15 II, v, 1, pp. 148–150
11.16 II, v, 2, p. 151
11.17 II, v, 1, p. 150
11.18 II, v, 3, 4, pp. 151–152
11.19 III, xxiv, 7, pp. 369–371
11.20 III, xxiv, 8, p. 371
11.21 II, v, 5, pp. 154–155
11.22 II, vi, 2, pp. 164–168
11.23 II, v, 6, pp. 156–158

MILL · Selections

The kinds of truths—intuitive and inferred.

11.1 Truths are known to us in two ways: some are known directly, and of themselves; some through the medium of other truths. The former are the subject of intuition or consciousness; the latter, of inference. The truths known by intuition are the original premises from which all others are inferred. Our assent to the conclusion being grounded on the truth of the premises, we never could arrive at any knowledge by reasoning, unless something could be known antecedently to all reasoning.

Examples of truths known to us by immediate consciousness, are our own bodily sensations and mental feelings. I know directly, and of my own knowledge, that I was vexed yesterday, or that I am hungry today. Examples of truths which we know only by way of inference, are occurrences which took place while we were absent, the events recorded in history, or the theorems of mathematics. The two former we infer from the testimony adduced, or from the traces of those past occurrences which still exist; the latter, from the premises laid down in books of geometry under the title of definitions and axioms. Whatever we are capable of knowing must belong to the one class or to the other; must be in the number of the primitive data, or of the conclusions which can be drawn therefrom.

With the original data, or ultimate premises of our knowledge; with their number or nature, the mode in which they are obtained, or the tests by which they may be distinguished; logic, in a direct way at least, has, in the sense in which I conceive the science, nothing to do. These questions are partly not a subject of science at all, partly that of a very different science.

Whatever is known to us by consciousness is known beyond possibility of question. What one sees, or feels, whether bodily or mentally, one cannot but be sure that one sees or feels. No science is required for the purpose of establishing such truths; no rules of art can render our knowledge of them more certain than it is in itself. There is no logic for this portion of our knowledge.

11.2 The province of logic must be restricted to that portion of our knowledge which consists of inferences from truths previously known; whether those antecedent data be general propositions, or particular observations and perceptions. Logic is not the science of Belief, but the science of Proof, or Evidence. So far forth as belief professes to be founded upon proof, the office of logic is to supply a test for ascertaining whether or not the belief is well grounded. With the claims which any proposition has to belief on its own intrinsic evidence, that is, without evidence in the proper sense of the word, logic has nothing to do.

An adequate theory of meaning is necessary for logic and philosophy.

11.3 Since reasoning, or inference, the principal subject of logic, is an operation which usually takes place by means of words, and, in complicated cases, can take place in no other way, those who have not a thorough insight into the signification and purposes of words will be under almost a necessity of reasoning or inferring incorrectly. And logicians have generally felt that unless, in the very first stage, they removed this fertile source of error, unless they taught their pupil to put away the glasses which distort the object, and to use those which are adapted to his purpose in such a manner as to assist, not perplex, his vision; he would not be in a condition to practice the remaining part of their discipline with any prospect of advantage. Therefore it is that an inquiry into language, so far as is needful to guard against the errors to which it gives rise, has at all times been deemed a necessary preliminary to the science of logic.

But there is another reason, of a still more fundamental nature, why the import of words should be the earliest subject of the logician's consideration: because without it he cannot examine into the import of propositions. Now this is a subject which stands on the very threshold of the science of logic.

The object of logic, as defined in the introductory chapter, is to ascertain how we come by that portion of our knowledge (much the greatest portion) which is not intuitive; and by what criterion we can, in matters not self-evident, distinguish between things proved and things not proved, between what is worthy and what is unworthy of belief. Of the various questions which the universe presents to our inquiring faculties, some are soluble by direct consciousness; others only by means of evidence. Logic is concerned with these last. . . . But before inquiring into the mode of resolving questions, it is necessary to inquire, what are the questions which present themselves? what questions are conceivable? what inquiries are there, to which men have either obtained or been able to imagine it possible that they should obtain, and answer? This point is best ascertained by a survey and analysis of propositions.

11.4 The answer to every question which it is possible to frame, is contained in a proposition, or assertion. Whatever can be an object of belief, or even of disbelief, must, when put into words, assume the form of a proposition. All truth and all error lie in propositions. What, by a convenient misapplication of an abstract term, we call a truth, is simply a true proposition; and errors are false propositions. To know the import of all possible propositions would be to know all questions which can be raised, all matters which are susceptible of being either believed or disbelieved. How many kinds of inquiries can be propounded; how many kinds of judgments can be made; and how many kinds of propositions is it possible to frame with a meaning, are but different forms of one and the same question. Since, then the objects of all belief and of all inquiry express themselves in propositions; a sufficient scrutiny of propositions and of their varieties will apprise us what questions mankind have actually asked themselves, and what, in the nature of answers to those questions, they have actually thought they had grounds to believe.

Now the first glance at a proposition shows that it is formed by putting together two names. A proposition, according to the common simple definition, which is sufficient for our purpose, is *discourse, in which some thing is affirmed or denied of something*. . . .

Every proposition consists of three parts: the subject, the predicate, and the copula. The predicate is the name denoting that which is affirmed or denied. The subject is the name denoting the person or thing which something is affirmed or denied of. The copula is the sign denoting that there is an affirmation or denial; and thereby enabling the hearer or reader to distinguish a proposition from any other kind of discourse. . . .

Dismissing, for the present, the copula, of which more will be said hereafter, every proposition, then, consists of at least two names; brings together two names, in a particular manner. This is already a first step towards what we are in quest of. It appears from this, that for an act of belief, *one* object is not sufficient; the simplest act of belief, supposes, and has something to do with, *two* objects: two names, to say the least; and (since the names must be names of something) two *nameable things.*

The different classifications of words.

11.5 The distinction . . . between *general* names, and *individual* or *singular* names, is fundamental; and may be considered as the first grand division of names.

A general name is familiarly defined, a name which is capable of being truly affirmed, in the same sense, of each of an indefinite number of things. An individual or singular name is a name which is only capable of being truly affirmed, in the same sense, of one thing.

Thus, *man* is capable of being truly affirmed of John, George, Mary, and other persons without assignable limits: and it is affirmed of all of them in the same sense; for the word "man" expresses certain qualities, and when we predicate it of those persons, we assert that they all possess those qualities. But *John* is only capable of being truly affirmed of one single person, at least in the same sense. For although there are many persons who bear that name, it is not conferred upon them to indicate any qualities, or any thing which belongs to them in common; and cannot be said to be affirmed of them in any *sense* at all, consequently not in the same sense. "The king who succeeded William the Conqueror" is also an individual name. For, that there cannot be more than one person at a time of whom it can be truly affirmed, is implied in the meaning of the words. Even "the king," when the occasion or the context defines the individual of whom it is to be understood, may justly be regarded as an individual name.

It is not unusual, by way of explaining what is meant by a general name, to say that it is the name of a *class.* But this, though a convenient mode of expression for some purposes, is objectionable as a definition, since it explains the clearer of two things by the more obscure. It would be more logical to reverse the proposition, and turn it into a definition of the word *class:* "A class is the indefinite multitude of individuals denoted by a general name."

It is necessary to distinguish *general* from *collective* names. A general name is one which can be predicated of *each* individual of a multitude; a collective name cannot be predicated of each separately, but only of all taken together. "The seventy-sixth regiment of foot in the British army," which is a collective name, is not a general but an individual name; for although it can be predicated of a multitude of individual soldiers taken jointly, it cannot be predicated of them severally. We may say, "Jones is a soldier, and Thompson is a soldier, and Smith is a soldier," but we cannot say, "Jones is the seventy-sixth regiment, and Thompson is the seventy-sixth regiment, and Smith is the seventy-sixth regiment." We can only say, "Jones, and Thompson, and Smith, and Brown, and so forth (enumerating all the soldiers) are the seventy-sixth regiment."

11.6 The second general division of names is into *concrete* and *abstract.* A concrete name is a name which stands for a thing; an abstract name is a name which stands for an attribute of a thing. Thus, *John, the sea, this table* are names of things. *White,* also, is a name of a thing, or rather of things. Whiteness, again, is the name of a quality or attribute of those things. Man is a name of many things; humanity is a name of an attribute of those things. *Old* is a name of things; *old age* is a name of one of their attributes.

11.7 It may be objected to our definition of an abstract name, that not only the names which we have called abstract, but adjectives, which we have placed in the concrete class, are names of attributes; that *white,* for example, is as much the name of the color, as *whiteness* is. But (as before remarked) a word ought to be considered as the name of that which we intend to be understood by it when we put it to its principal use, that is, when we employ it in predication. When we say, "snow is white," "milk is white," "linen is white," we do not mean it to be understood that snow, or linen, or milk, is a color. We mean that they are things having the color. The reverse is the case with the word whiteness; what we affirm to *be* whiteness is not snow but the color of snow. Whiteness, therefore, is the name of the color exclusively, white is a name of all things whatever having the color; a name not of the quality whiteness, but of every white object. It is true, this name was given to all those various objects on account of the quality; and we may therefore say, without impropriety, that the quality forms part of its signification; but a name can only be said to stand for, or to be a name of, the things of which it can be predicated.

11.8 This leads us to the consideration of the third great division of names, into *connotative* and *non-connotative,* the latter sometimes, but improperly, called *absolute.* This is one of the most important distinctions which we shall have occasion to point out, and one of those which go deepest into the nature of language.

A non-connotative term is one which signifies a subject only, or an attribute only. A connotative term is one which denotes a subject and implies an attribute. By a subject is here meant anything which possesses attributes. Thus John, or London, or England, are names which signify a subject only. Whiteness, length, virtue, signify an attribute only. None of these names, therefore, are connotative. But *white, long, virtuous,* are connotative. The word *white,* denotes all white things, as snow, paper, the foam of the sea, etc., and implies, or as it was termed by the schoolmen, *connotes,* the attribute *whiteness.* The word *white* is not predicated of the attribute, but of the subjects, snow, etc.; but when we predicate it of them we imply, or connote, that the attribute whiteness belongs to them. . . .

All concrete general names are connotative. The word *man,* for example, denotes Peter, Jane, John, and an indefinite number of other individuals, of whom, taken as a class, it is the name. But it is applied to them, because they possess, and to signify that they possess, certain attributes. These seem to be corporeity, animal life, rationality, and a certain external form, which for distinction, we call the human. Every existing thing, which possessed all these attributes, would be called a man; and anything which possessed none of them, or only one, or two, or even three of them without the fourth, would not be so called. For example, if in the interior of Africa there were to be discovered a race of animals possessing reason equal to that of human beings, but with the form of an elephant, they would not be called men. Swift's Houyhnhnms would not be so called. Or if such newly-discovered beings possessed the form of man without any vestige of reason, it is probable that some other name than that of man would be found for them. How it happens that there can be any doubt about the matter, will appear hereafter. The word *man,* therefore, signifies all these attributes, and all subjects which possess these attributes. But it can be predicated only of the subjects. What we call men, are the subjects, the individual Stiles and Nokes; not the qualities by which their humanity is constituted. The name, therefore, is said to signify the subjects *directly,* the attributes *indirectly;* it *denotes* the subjects, and implies, or involves, or indicates, or as we shall say henceforth, *connotes,* the attributes. It is a connotative name.

11.9 As the result, therefore, of our analysis, we obtain the following as an enumeration and classification of all nameable things:—

1st. Feelings, or states of consciousness.

2nd. The minds which experience those feelings.

3rd. The bodies, or external objects which excite certain of those feelings, together with the powers or properties whereby they excite them; these last being included rather in compliance with common opinion, and because their existence is taken for granted in the common language from which I cannot prudently deviate, than because the recognition of such powers or properties as real existences appears to be warranted by a sound philosophy.

4th, and last. The successions and coexistences, the likenesses and unlikenesses, between feelings or states of consciousness. Those relations, when considered as subsisting between other things, exist in reality only between the states of consciousness which those things, if bodies, excite, if minds, either excite or experience.

This, until a better can be suggested, must serve as a substitute for . . . the Categories of Aristotle. The practical application of it will appear when we commence the inquiry into the Import of Propositions, in other words, when we inquire what it is which the mind actually believes, when it gives what is called its assent to a proposition.

These four classes comprising, if the classification be correct, all namable things, these or some of them must of course compose the signification of all names; and of these or some of them is made up whatever we call a fact.

Number-words connote quantitative attributes of things.

11.10 Let us imagine two things, between which there is no difference (that is, no dissimilarity), except in quantity alone: for instance, a gallon of water, and more than a gallon of water. A gallon of water, like any other external object, makes its presence known to us by a set of sensations which it excites. Ten gallons of water are also an external object, making its presence known to us in a similar manner; and as we do not mistake ten gallons of water for a gallon of water, it is plain that the set of sensations is more or less different in the two cases. In like manner, a gallon of water, and a gallon of Madeira are two external objects, making their presence known by two sets of sensations, which sensations are different from each other. In the first case, however, we say that the difference is in quantity; in the last there is a difference in quality, while the quantity of the water and of the Madeira is the same. What is the real distinction between the two cases? It is not the province of logic to analyze it; nor to decide whether it is susceptible of analysis or not. For us the following considerations are sufficient. It is evident that the sensations I receive from the gallon of water and those I receive from the gallon of Madeira are not the same, that is, not precisely alike; neither are they altogether unlike: they are partly similar, partly dissimilar; and that in which they resemble is precisely that in which alone the gallon of water and the ten gallons do not resemble. That in which the gallon of water and the gallon of wine are like each other, and in which the gallon and the ten gallons of water are unlike each other, is called their quantity. This likeness and unlikeness I do not pretend to explain, no more than any other kind of likeness or unlikeness. But my object is to show that when we say of two things that they differ in quantity, just as when we say that they differ in quality, the assertion is always grounded upon a difference in the sensations which they excite. Nobody, I presume, will say, that to see, or to lift, or to drink, ten gallons of water, does not include in itself a different set of sensations from those of seeing, lifting, or drinking one gallon; or that to see or handle a foot-rule, and to see or handle a yard-measure made exactly like it, are the same sensations. I do not undertake to say what the difference in the sensations is. Everybody knows, and nobody can tell; no more than anyone could tell what white is, to a person who had never had the sensation. But the difference, so far as cognizable by our faculties, lies in the sensations. Whatever difference we say there is in the things themselves, is, in this as in all other cases, grounded, and grounded exclusively, on a difference in the sensations excited by them.

The truths of arithmetic are inductions from experience.

11.11 That things equal to the same thing are equal to another, or that two straight lines which have once intersected with one another continue to diverge, are inductive truths; resting indeed, like the law of universal causation, only upon induction *per enumerationem simplicem;* upon the fact that they have been perpetually found true and never once false. But as we have seen in a recent chapter that this evidence, in the case of a law so completely universal as the law of causation, amounts to the fullest proof attainable by the human faculties, so is this even more evidently true of the general propositions to which we are now adverting; because, as a perception of their truth in any individual case whatever, requires only the simple act of looking at the objects in a proper position, there never could have been in their case (what, for a long period, in the case of the law of causation, there were) instances which were apparently, though not really, exceptions to them. Their infallible truth was recognized from the very dawn of speculation; and as their extreme familiarity made it impossible for the mind to conceive the objects under any other law, they were, and still are, generally considered as truths recognized by their own evidence, or by instinct.

11.12 There is something which seems to require explanation, in the fact that the immense multitude of truths (a multitude still as far from being exhausted as ever) comprised in the mathematical sciences, can be elicited from so small a number of elementary laws. One sees not, at first, how it is that there can be room for such an infinite variety of true propositions, on subjects apparently so limited.

To begin with the science of number. The elementary or ultimate truths of this science are the common axioms concerning equality, namely, "Things which are equal to the same thing are equal to one another," and "Equals added to equals make equal sums," (no other axioms are necessary,) together with the definitions of the various numbers. Like other so-called definitions, these are composed of two things, the explanation of a name and the assertion of a fact: of which the latter alone can form a first principle or premiss of a science. The fact asserted in the definition of a number is a physical fact. Each of the numbers two, three, four, etc., denotes physical phenomena, and connotes a physical property of those phenomena. Two, for instance, denotes all pairs of things, and twelve all dozens of things, connoting what makes them pairs, or dozens; and that which makes them so is something physical; since it cannot be denied that two apples are physically distinguishable from three apples, two horses from one horse, and so forth; that they are a different visible and tangible phenomenon. I am not undertaking to say what the difference is; it is enough that there is a difference of which the senses can take cognizance. And although an hundred and two horses are not so easily distinguished from an hundred and three, as two horses are from three—though in most positions the senses do not perceive any difference—yet they *may* be so placed that a difference will be perceptible, or else we should never have distinguished them, and given them different names. Weight is confessedly a physical property of things; yet small differences between great weights are as imperceptible to the senses in most situations, as small differences between great numbers; and are only put in evidence by placing the two objects in a peculiar position, namely, in the opposite scales of a delicate balance.

What, then, is that which is connoted by a name of number? Of course some property belonging to the agglomeration of things which we call by the name; and that property is, the characteristic manner in which the agglomeration is made up of, and may be separated into, parts. We will endeavor to make this more intelligible by a few explanations.

When we call a collection of objects *two, three,* or *four,* they are not two, three, or four in the abstract; they are two, three, or four things of some particular kind; pebbles, horses, inches, pounds weight. What the name of number connotes is, the manner in which single objects of the given kind must be put together, in order to produce that particular aggregate. If the aggregate be of pebbles, and we call it *two,* the name implies that to compose the aggregate, one pebble must be joined to one pebble. If we call it *three,* we mean that one and one and one pebble must be brought together to produce it, or else that one pebble must be joined to an aggregate of the kind called *two,* already existing. The aggregate which we call *four* has a still greater number of characteristic modes of formation. One and one and one and one pebble may be brought together; or two aggregates of the kind called *two* may be united; or one pebble may be added to an aggregate of the kind called *three.* Every succeeding number in the ascending series, may be formed by the junction of smaller numbers in a progressively greater variety of ways. Even limiting the parts to two, the number may be formed, and consequently may be divided, in as many different ways as there are numbers smaller than itself; and, if we admit of threes, fours, etc., in a still greater variety. Other modes of arriving at the same aggregate present themselves, not by the union of smaller, but by the dismemberment of larger aggregates. Thus, *three pebbles* may be formed by taking away one pebble from an aggregate of four; *two pebbles,* by an equal division of a similar aggregate; and so on.

Every arithmetical proposition; every statement of the result of an arithmetical operation; is a statement of one of the modes of the formation of a given number. It affirms that a certain aggregate might have been formed by putting together certain other aggregates, by withdrawing certain portions of some aggregate; and that, by consequence, we might reproduce those aggregates from it, by reversing the process.

Thus, when we say that the cube of 12 is 1728, what we affirm is this: That if, having a sufficient number of pebbles or of any other objects, we put them together in the particular sort of parcels or aggregates called twelve; and put together these twelves again into similar collections; and, finally, make up twelve of these largest parcels; the aggregate thus formed will be such a one as we call 1728; namely, that which (to take the most familiar of its modes of formation) may be made by joining the parcel called a thousand pebbles, the parcel called seven hundred pebbles, the parcel called twenty pebbles, and the parcel called eight pebbles. The converse proposition, that the cube root of 1728 is 12, asserts that this large aggregate may again be decomposed into the twelve twelves of twelves of pebbles which it consists of.

The modes of formation of any number are innumerable; but when we know one mode of formation of each, all the rest may be determined deductively. If we know that a is formed from b and c, b from d and e, c from d and f, and so forth, until we have included all the numbers of any scale we choose to select, (taking care that for each number the mode of formation is really a distinct one, not bringing us round again to the former numbers, but introducing a new number,) we have a set of propositions from which we may reason to all the other modes of formation of those numbers from one another. Having established a chain of inductive truths connecting together all the numbers of the scale, we can ascertain the formation of any one of those numbers from any other by merely travelling from the one to the other along the chain. Suppose that we know only the following modes of formation: $6 = 4 + 2$, $4 = 7 - 3$, $7 = 5 + 2$, $5 = 9 - 4$. We could determine how 6 may be formed from 9. For $6 = 4 + 2 = 7 - 3 + 2 = 5 + 2 - 3 + 2 = 9 - 4 + 2 - 3 + 2$. It may therefore be formed by taking away 4 and 3, and adding 2 and 2. If we know besides that $2 + 2 = 4$, we obtain 6 from 9 in a simpler mode, by merely taking away 3.

It is sufficient, therefore, to select one of the various modes of formation of each number, as a means of ascertaining all the rest. And since things which are uniform, and therefore simple, are most easily received and retained by the understanding, there is an obvious advantage in selecting a mode of formation which shall be alike for all; in fixing the connotation of names of number on one uniform principle. The mode in which our existing numerical nomenclature is contrived possesses this advantage, with the additional one, that it happily conveys to the mind two of the modes of formation of every number. Each number is considered as formed by the addition of an unit to the number next below it in magnitude, and this mode of formation is conveyed by the place which it occupies in the series. And each is also considered as formed by the addition of a number of units less than ten, and a number of aggregates each equal to one of the successive powers of ten: and this mode of its formation is expressed by its spoken name, and by its numerical character.

What renders arithmetic a deductive science, is the fortunate applicability to it of a law so comprehensive as "The sums of equals are equals:" or (to express the same principle in less familiar but more characteristic language), whatever is made up of parts is made up of the parts of those parts. This truth, obvious to the senses in all cases which can be fairly referred to their decision, and so general as to be coextensive with nature itself, being true of all sorts of phenomena (for all admit of being numbered), must be considered an inductive truth, or law of nature, of the highest order. And every arithmetical operation is an application of this law, or of other laws capable of being deduced from it. This is our warrant for all calculations. We believe that five and two are equal to seven, on the evidence of this inductive law, combined with the definitions of those numbers. We arrive at that conclusion (as all know who remember how they first learned it) by adding a single unit at a time: $5 + 1 = 6$, therefore $5 + 1 + 1 = 6 + 1 = 7$: and again $2 = 1 + 1$, therefore $5 + 2 = 5 + 1 + 1 = 7$.

11.13 Innumerable as are the true propositions which can be formed concerning particular numbers, no adequate conception could be gained, from these alone, of the extent of the truths composing the science of number. Such propositions as we have spoken of are the least general of all numerical truths. It is true that even these are coextensive with all nature; the properties of the number four are true of all objects that are divisible into four equal parts, and all objects are either actually or ideally so divisible. But the propositions which compose the science of algebra are true, not of a particular number, but of all numbers; not of all things under the condition of being divided in a particular way, but of all things under the condition of being divided in *any* way—of being designated by a number at all.

Since it is impossible for different numbers to have any of their modes of formation completely in common, it looks like a paradox to say, that all propositions which can be made concerning numbers relate to their modes of formation from other numbers, and yet that there are propositions which are true of all numbers. But this very paradox leads to the real principle of generalization concerning the properties of numbers. Two different numbers cannot be formed in the same manner from the same numbers; but they may be formed in the same manner from different numbers; as nine is formed from three by multiplying it into itself, and sixteen is formed from four by the same process. Thus there arises a classification of modes of formation, or, in the language commonly used by mathematicians, a classification of Functions. Any number, considered as formed from any other number, is called a function of it; and there are as many kinds of functions as there are modes of formation. The simple functions are by no means numerous, most functions being formed by the combination of several of the operations which form simple functions, or by successive repetitions of some one of those operations. The simple functions of any number x are all reducible to the following forms: $x + a$, $x - a$, $a\,x$, $\frac{x}{a}$, x^a, $\sqrt[a]{x}$, log. x (to the base a), and the same expressions varied by putting x for a and a for x, wherever that substitution would alter the value: to which perhaps we ought to add (with M. Comte) sin x, and arc (sin $= x$). All other functions of x are formed by putting some one or more of the simple functions in the place of x or a, and subjecting them to the same elementary operations.

One assumption is necessary to make mathematical truths certain.

11.14 The inductions of arithmetic are of two sorts: first, those which we have just expounded, such as "one and one are two," "two and one are three," etc., which may be called the definitions of the various numbers, in the improper or geometrical sense of the word definition; and secondly, the two following axioms: "The sums of equals are equal," "The differences of equals are equal." These two are sufficient; for the corresponding propositions respecting unequals may be proved from these, by . . . a *reductio ad absurdum.*

These axioms, and likewise the so-called definitions, are, as already shown, results of induction; true of all objects whatever, and, as it may seem, exactly true, without any hypothetical assumption of unqualified truth where an approximation to it is all that exists. The conclusions, therefore, it will naturally be inferred, are exactly true, and the science of number is an exception to other demonstrative sciences in this, that the absolute certainty which is predicable of its demonstrations is independent of all hypothesis.

On more accurate investigation, however, it will be found that, even in this case, there is one hypothetical element in the ratiocination. In all propositions concerning numbers, a condition is implied, without which none of them would be true; and that condition is an assumption which may be false. The condition is, that $1 = 1$; that all the numbers are numbers of the same or of equal units. Let this be doubtful, and not one of the propositions of arithmetic will hold true. How can we know that one pound and one pound make two pounds, if one of the pounds may be troy and the other avoirdupois? They may not make two pounds of either, or of any weight. How can we know that a forty-horse power is always equal to itself, unless we assume that all horses are of equal strength? It is certain that 1 is always equal in *number* to 1; and, where the mere number of objects, or of the parts of an object, without supposing them to be equivalent in any other respect, is all that is material, the conclusions of arithmetic, so far as they go to that alone, are true without mixture of hypothesis. There are a few such cases; as, for instance, an inquiry into the amount of the population of any country. It is indifferent to that inquiry whether they are grown people or children, strong or weak, tall or short; the only thing we want to ascertain is their number. But whenever, from equality or inequality of number, equality or inequality in any other respect is to be inferred, arithmetic carried into such inquiries becomes as hypothetical a science as geometry. All units must be assumed to be equal in that other respect; and this is never precisely true, for one pound weight is not exactly equal to another, nor one mile's length to another; a nicer balance, or more accurate measuring instruments, would always detect some difference.

What is commonly called mathematical certainty, therefore, which comprises the twofold conception of unconditional truth and perfect accuracy, is not an attribute of all mathematical truths, but of those only which relate to pure number, as distinguished from quantity in the more enlarged sense; and only so long as we abstain from supposing that the numbers are a precise index to actual quantities. The certainty usually ascribed to the conclusions of geometry, and even to those of mechanics, is nothing whatever but certainty of inference. We can have full assurance of particular results under particular suppositions, but we cannot have the same assurance that these suppositions are accurately true, nor that they include all the data which may exercise an influence over the result in any given instance.

Geometrical truths contain an hypothetical element.

11.15 If . . . the foundation of all sciences, even deductive or demonstrative sciences, is induction; if every step in the ratiocinations even of geometry is an act of induction; and if a train of reasoning is but bringing many inductions to bear upon the same subject of inquiry, and drawing a case within one induction by means of another; wherein lies the peculiar certainty always ascribed to the sciences which are entirely, or almost entirely, deductive? Why are they called the exact sciences? Why are mathematical certainty, and the evidence of demonstration, common phrases to express the very highest degree of assurance attainable by reason? Why are mathematics by almost all philosophers, and (by many) even those branches of natural philosophy which, through the medium of mathematics, have been converted into deductive sciences, considered to be independent of the evidence of experience and observation, and characterized as systems of necessary truth?

The answer I conceive to be, that this character of necessity, ascribed to the truths of mathematics, and (even with some reservations to be hereafter made) the peculiar certainty attributed to them, is an illusion; in order to sustain which, it is necessary to suppose that those truths relate to, and express the properties of, purely imaginary objects. It is acknowledged that the conclusions of geometry are deduced, partly at least, from the so-called definitions and that those definitions are assumed to be correct representations, as far as they go, of the objects with which geometry is conversant. Now we have pointed out that, from a definition as such, no proposition, unless it be one concerning the meaning of a word, can ever follow; and that what apparently follows from a definition, follows in reality from an implied assumption that there exists a real thing conformable thereto. This assumption, in the case of the definitions of geometry, is not strictly true; there exist no real things exactly conformable to the definitions. There exist no points without magnitude; no lines without breadth, nor perfectly straight; no circles with all their radii exactly equal, nor squares with all their angles perfectly right. It will perhaps be said that the assumption does not extend to the actual, but only to the possible, existence of such things. I answer that, according to any test we have of possibility, they are not even possible. Their existence, so far as we can form any judgment, would seem to be inconsistent with the physical constitution of our planet at least, if not of the universe. To get rid of this difficulty, and at the same time to save the credit of the supposed systems of necessary truth, it is customary to say that the points, lines, circles, and squares which are the subject of geometry, exist in our conceptions merely, and are part of our minds, which minds, by working on their own materials, construct an *a priori* science, the evidence of which is purely mental and has nothing whatever to do with outward experience. By howsoever high authorities this doctrine may have been sanctioned, it appears to me psychologically incorrect. The points, lines, circles, and squares, which anyone has in his mind are (I apprehend) simply copies of the points, lines, circles, and squares which he has known in his experience. Our idea of a point I apprehend to be simply our idea of the *minimum visible,* the smallest portion of surface which we can see. A line, as defined by geometers, is wholly inconceivable. We can reason about a line as if it had no breadth, because we have a power, which is the foundation of all the control we can exercise over the operations of our minds; the power, when a perception is present to our senses, or a conception to our intellects, of *attending* to a part only of that perception or conception instead of the whole. But we cannot *conceive* a line without breadth; we can form no mental picture of such a line: all the lines which we have in our minds are lines possessing breadth. If anyone doubts this, we may refer him to his own experience. I much question if anyone who fancies that he can conceive what is called a mathematical line, thinks so from the evidence of his consciousness: I suspect it is rather because he supposes that unless such a conception were possible, mathematics could not exist as a science: a supposition which there will be no difficulty in showing to be entirely groundless.

Since, then, neither in nature, nor in the human mind, do there exist any objects exactly corresponding to the definitions of geometry, while yet that science cannot be supposed to be conversant about nonentities; nothing remains but to consider geometry as conversant with such lines, angles, and figures as really exist; and the definitions, as they are called, must be regarded as some of our first and most obvious generalizations concerning those natural objects. The correctness of those generalizations, *as* generalizations, is without a flaw: the equality of all the radii of a circle is true of all circles, so far as it is true of any one: but it is not exactly true of any circle: it is only nearly true: so nearly that no error of any importance in practice will be incurred by feigning it to be exactly true. When we have occasion to extend these inductions, or their consequences, to cases in which the error would be appreciable—to lines of perceptible breadth or thickness, parallels which deviate sensibly from equidistance, and the like—we correct our conclusions, by combining with them a fresh set of propositions relating to the aberration; just as we also take in propositions relating to the physical or chemical properties of the material, if those properties happen to introduce any modification into the result, which they easily may, even with respect to figure and magnitude, as in the case, for instance, of expansion by heat. So long, however, as there exists no practical necessity for attending to any of the properites of the object except its geometrical properties, or to any of the natural irregularities in those, it is convenient to neglect the consideration of the other properties and of the irregularities, and to reason as if these did not exist: accordingly, we formally announce, in the definitions, that we intend to proceed on this plan. But it is an error to suppose, because we resolve to confine our attention to a certain number of the properties of an object, that we therefore conceive, or have an idea of, the object, denuded of its other properties. We are thinking, all the time, of precisely such objects as we have seen and touched, and with all the properties which naturally belong to them; but, for scientific convenience, we feign them to be divested of all properties, except those which are material to our purpose in regard to which we design to consider them.

The peculiar accuracy, supposed to be characteristic of the first principles of geometry thus appears to be fictitious. The assertions on which the reasonings of the science are founded, do not, any more than in other sciences, exactly correspond with the fact; but we *suppose* that they do so, for the sake of tracing the consequences which follow from the supposition. The opinion of Dugald Stewart respecting the foundations of geometry is, I conceive, substantially correct; that it is built on hypotheses; that it owes to this alone the peculiar certainty supposed to distinguish it; and that in any science whatever, by reasoning from a set of hypotheses, we may obtain a body of conclusions as certain as those of geometry, that is, as strictly in accordance with the hypotheses, and as irresistibly compelling assent, *on condition* that those hypotheses are true.

11.16 Those who say that the premisses of geometry are hypotheses, are not bound to maintain them to be hypotheses which have no relation whatever to fact. Since an hypothesis framed for the purpose of scientific inquiry must relate to something which has real existence (for there can be no science respecting non-entities), it follows that any hypothesis we make respecting an object, to facilitate our study of it, must not involve anything which is distinctly false, and repugnant to its real nature: we must not ascribe to the thing any property which it has not; our liberty extends only to suppressing some of those which it has, under the indispensable obligation of restoring them whenever, and in as far as, their presence or absence would make any material difference in the truth of our conclusions. Of this nature, accordingly, are the first principles involved in the definitions of geometry. In their positive part they are observed facts; it is only in their negative part that they are hypothetical. That the hypotheses should be of this particular character, is, however, no further necessary, than inasmuch as no others could enable us to deduce conclusions which, with due corrections, would be true of real objects: and in fact, when our aim is only to illustrate truths and not to investigate them, we are not under any such restriction. We might suppose an imaginary animal, and work out by deduction, from the known laws of physiology, its natural history; or an imaginary commonwealth, and from the elements composing it, might argue what would be its fate. And the conclusions which we might thus draw from purely arbitrary hypotheses, might form a highly useful intellectual exercise: but as they could only teach us what *would* be the properties of objects which do not really exist, they would not constitute any addition to our knowledge: while on the contrary, if the hypothesis merely divests a real object of some portion of its properties, without clothing it in false ones, the conclusions will always express, under known liability to correction, actual truth.

11.17 When, therefore, it is affirmed that the conclusions of geometry are necessary truths, the necessity consists in reality only in this, that they correctly follow from the suppositions from which they are deduced. Those suppositions are so far from being necessary, that they are not even true; they purposely depart, more or less widely, from the truth. The only sense in which necessity can be ascribed to the conclusions of any scientific investigation, is that of legitimately following from some assumption, which, by the conditions of the inquiry, is not to be questioned. In this relation, of course, the derivative truths of every deductive science must stand to the inductions, or assumptions, on which the science is founded, and which, whether true or untrue, certain or doubtful in themselves, are always supposed certain for the purposes of the particular science.

11.18 Some of the axioms of Euclid might, no doubt, be exhibited in the form of definitions, or might be deduced, by reasoning, from propositions similar to what are so called. Thus, if instead of the axiom, "magnitudes which can be made to coincide are equal," we introduce a definition, "equal magnitudes are those which may be so applied to one another as to coincide;" the three axioms which follow (Magnitudes which are equal to the same are equal to one another—If equals are added to equals, the sums are equal—If equals are taken from equals, the remainders are equal) may be proved by an imaginary superposition, resembling that by which the fourth proposition of the first book of Euclid is demonstrated. But though these and several others may be struck out of the list of first principles, because, though not requiring demonstration, they are susceptible of it; there will be found in the list of axioms two or three fundamental truths, not capable of being demonstrated: among which . . . [is] the proposition that two straight lines cannot inclose a space (or its equivalent, straight lines which coincide in two points coincide altogether) and some property of parallel lines, other than that which constitutes their definition: the most suitable, being that selected by Professor Playfair: "Two straight lines which intersect each other cannot both of them be parallel to a third straight line."

The axioms, as well those which are indemonstrable as those which admit of being demonstrated, differ from that other class of fundamental principles which are involved in the definitions, in this, that they are true without any mixture of hypothesis. That things which are equal to the same thing are equal to one another, is as true of the lines and figures in nature, as it would be of the imaginary ones assumed in the definitions. In this respect, however, mathematics is only on a par with most other sciences. In almost all sciences there are some general propositions which are exactly true, while the greater part are only more or less distant approximations to the truth. Thus, in mechanics, the first law of motion, (the continuance of a movement once impressed, until stopped or slackened by some resisting force) is true without qualification or error. . . . The rotation of the earth in twenty-four hours, of the same length as in our time, has gone on since the first accurate observations, without the increase or diminution of one second in all that period. These are inductions which require no fiction to make them be received as accurately true: but along with them there are others, as for instance the propositions respecting the figure of the earth, which are but approximations to the truth; and in order to use them for the further advancement of our knowledge, we must feign that they are exactly true, though they really want something of being so.

It remains to inquire, what is the ground of our belief in axioms—what is the evidence on which they rest? I answer, they are experimental truths; generalizations from observation. The proposition, "Two straight lines cannot inclose a space"—or, in other words, "Two straight lines which have once met, do not meet again, but continue to diverge"—is an induction from the evidence of our senses.

Geometry is a strictly physical science.

11.19 If the extreme generality and remoteness, not so much from sense as from the visual and tactual imagination, of the laws of number, render it a somewhat difficult effort of abstraction to conceive those laws as being in reality physical truths obtained by observation; the same difficulty does not exist with regard to the laws of extension. The facts of which those laws are expressions, are of a kind peculiarly accessible to the sense, and suggesting eminently distinct images to the fancy. That geometry is a strictly physical science would doubtless have been recognized in all ages, had it not been for the illusions produced by two causes. One of these is the characteristic property, already noticed, of the facts of geometry, that they may be collected from our ideas or mental pictures of objects as effectually as from the objects themselves. The other is, the demonstrative character of geometrical truths; which was at one time supposed to constitute a radical distinction between them and physical truths, the latter, as resting on merely *probable* evidence, being deemed essentially uncertain and unprecise. The advance of knowledge has, however, made it manifest that physical science, in its better understood branches, is quite as demonstrative as geometry: the task of deducing its details from a few comparatively simple principles being found to be anything but the impossibility it was once supposed to be; and the notion of the superior certainty of geometry being an illusion arising from the ancient prejudice which in that science mistakes the ideal data from which we reason, for a peculiar class of realities while the corresponding ideal data of any deductive physical science are recognized as what they really are, mere hypotheses.

Every theorem in geometry is a law of external nature, and might have been ascertained by generalizing from observation and experiment, which in this case resolve themselves into comparison and measurement. But it was found practicable, and being practicable, was desirable, to deduce these truths by ratiocination from a small number of general laws of nature, the certainty and universality of which was obvious to the most careless observer, and which compose the first principles and ultimate premisses of the science. Among these general laws must be included the same two which we have noticed as ultimate principles of the Science of Number also, and which are applicable to every description of quantity: viz., the sums of equals are equal, and things which are equal to the same thing are equal to one another; the latter of which may be expressed in a manner more suggestive of the inexhaustible multitude of its consequences by the following terms: Whatever is equal to any one of a number of equal magnitudes, is equal to any other of them. To these two must be added, in geometry, a third law of equality, namely, that lines, surfaces, or solid spaces, which can be so applied to one another as to coincide, are equal. Some writers have asserted that this law of nature is a mere verbal definition: that the expression "equal magnitudes" *means* nothing but magnitudes which can be so applied to one another as to coincide. But in this opinion I cannot agree. The quality of two geometrical magnitudes cannot differ fundamentally in its nature from the equality of two weights, two degrees of heat, or two portions of duration, to none of which would this pretended definition of equality be suitable. None of these things can be so applied to one another as to coincide, yet we perfectly understand what we mean when we call them equal. Things are equal in magnitude, as things are equal in weight, when they are felt to be exactly similar in respect of the attribute in which we compare them: and the application of the objects to each other in the one case, like the balancing them with a pair of scales in the other, is but a mode of bringing them into a position in which our senses can recognize deficiencies of exact resemblance that would otherwise escape our notice.

Along with these three general principles or axioms, the remainder of the premisses of geometry consist of the so-called definitions, that is to say, propositions asserting the real existence of the various objects therein designated, together with some one property of each. In some cases more than one property is commonly assumed, but in no case is more than one necessary. It is assumed that there are such things in nature as straight lines, and that any two of them setting out from the same point, diverge more and more without limit. This assumption, (which includes and goes beyond Euclid's axiom that two straight lines cannot inclose a space,) is as indispensable in geometry, and as evident, resting upon as simple, familiar, and universal observation, as any of the other axioms. It is also assumed that straight lines diverge from one another in different degrees; in other words, that there are such things as angles, and that they are capable of being equal or unequal. It is assumed that there is such a thing as a circle, and that all its radii are equal; such things as ellipses, and that the sums of the focal distances are equal for every point in an ellipse; such things as parallel lines, and that those lines are everywhere equally distant.

11.20 It is a matter of something more than curiosity to consider to what peculiarity of the physical truths which are the subject of geometry, it is owing that they can all be deduced from so small a number of original premisses; why it is that we can set out from only one characteristic property of each kind of phenomenon, and with that and two or three general truths relating to equality, can travel from mark to mark until we obtain a vast body of derivative truths, to all appearance extremely unlike those elementary ones.

The explanation of this remarkable fact seems to lie in the following circumstances. In the first place, all questions of position and figure may be resolved into questions of magnitude. The position and figure of any object is determined, by determining the position of a sufficient number of points in it; and the position of any point may be determined by the magnitude of three rectangular coordinates, that is, of the perpendiculars drawn from the point to three axes at right angles to one another, arbitrarily selected. By this transformation of all questions of quality into questions only of quantity, geometry is reduced to the single problem of the measurement of magnitudes, that is, the ascertainment of the equalities which exist between them. Now when we consider that by one of the general axioms, any equality, when ascertained, is proof of as many other equalities as there are other things equal to either of the two equals; and that by another of those axioms, any ascertained equality is proof of the equality of as many pairs of magnitudes as can be formed by the numerous operations which resolve themselves into the addition of the equals to themselves or to other equals: we cease to wonder that in proportion as a science is conversant about equality, it should afford a more copious supply of marks; and that the sciences of number and extension, which are conversant with little else than equality, should be the most deductive of all sciences.

11.21 The foundations of geometry would . . . be laid in direct experience, even if the experiments (which in this case consist merely in attentive contemplation) were practiced solely upon what we call our ideas, that is, upon the diagrams in our minds, and not upon outward objects. For in all systems of experimentation we take some objects to serve as representatives of all which resemble them, and in the present case the conditions which qualify a real object to be the representative of its class, are completely fulfilled by an object existing only in our fancy. Without denying, therefore, the possibility of satisfying ourselves that two straight lines cannot inclose a space, by merely thinking of straight lines without actually looking at them; I contend, that we do not believe this truth on the ground of the imaginary intuition simply, but because we know that the imaginary lines exactly resemble real ones, and that we may conclude from them to real ones with quite as much certainty as we could conclude from one real line to another. The conclusion, therefore, is still an induction from observation. And we should not be authorized to substitute observation of the image in our mind for observation of the reality, if we had not learned by long continued experience that all the properties of the reality are faithfully represented in the image; just as we should be scientifically warranted in describing . . . an animal which we have never seen from a picture made of it with a daguerreotype, but not until we had learned by ample experience that observation of such a picture is precisely equivalent to observation of the original.

These considerations also remove the objection arising from the impossibility of ocularly following the lines in their prolongation to infinity. For though, in order actually to see that two given lines never meet, it would be necessary to follow them to infinity: yet without doing so we may know that if they ever do meet, or if, after diverging from one another, they begin again to approach, this must take place not at an infinite, but at a finite distance. Supposing, therefore, such to be the case, we can transport ourselves thither in imagination, and can frame a mental image of the appearance which one or both of the lines must present at that point, which we may rely on as being precisely similar to the reality. Now, whether we fix our contemplation upon this imaginary picture or call to mind the generalizations we have had occasion to make from former ocular observation, we learn by the evidence of experience that a line which, after diverging from another straight line, begins to approach to it, produces the impression on our senses which we describe by the expression, "a bent line," not by the expression, "a straight line."

An argument against formalism.

11.22 This theory [formalism] attempts to solve the difficulty apparently inherent in the case, by representing the propositions of the science of numbers as merely verbal, and its processes as simple transformations of language, substitutions of one expression for another. The proposition, "Two and one is equal to three," according to these writers, is not a truth, is not the assertion of a really existing fact, but a definition of the word three; a statement that mankind have agreed to use the name three as a sign exactly equivalent to two and one; to call by the former name whatever is called by the other more clumsy phrase. According to this doctrine, the longest process in algebra is but a succession of changes in terminology, by which equivalent expressions are substituted one for another; a series of translations of the same fact from one into another language: though how, after such a series of translations, the fact itself comes out changed (as when we demonstrate a new geometrical theorem by algebra) they have not explained; and it is a difficulty which is fatal to their theory.

It must be acknowledged that there are peculiarities in the processes of arithmetic and algebra which render the theory in question very plausible, and have not unnaturally made those sciences the stronghold of Nominalism. The doctrine that we can discover facts, detect the hidden processes of nature, by an artful manipulation of language, is so contrary to common sense, that a person must have made some advances in philosophy to believe it; men fly to so paradoxical a belief to avoid, as they think, some even greater difficulty which the vulgar do not see. What has led many to believe that reasoning is a mere verbal process, is that no other theory seemed reconcilable with the nature of the science of numbers. For we do not carry any ideas along with us when we use the symbols of arithmetic or of algebra. In a geometrical demonstration we have a mental diagram, if not one on paper; AB, AC, are present to our imagination as lines, intersecting other lines, forming an angle with one another, and the like; but not so *a* and *b*. These may represent lines or any other magnitudes, but those magnitudes are never thought of; nothing is realized in our imagination but *a* and *b*. The ideas which, on the particular occasion, they happen to represent, are banished from the mind during every intermediate part of the process between the beginning when the premises are translated from things into signs, and the end, when the conclusion is translated back from signs into things. Nothing, then, being in the reasoner's mind but the symbols, what can seem more inadmissible than to contend that the reasoning process has to do with anything more? We seem to have come to one of Bacon's prerogative instances; an *experimentum crucis* on the nature of reasoning itself.

Nevertheless, it will appear on consideration that this apparently so decisive instance is no instance at all; that there is in every step of an arithmetical or algebraical calculation a real induction, a real inference of facts from facts; and that what disguises the induction is simply its comprehensive nature and the consequent extreme generality of the language. All numbers must be numbers of something: there are no such things as numbers in the abstract. *Ten* must mean ten bodies, or ten sounds, or ten beatings of the pulse. But though numbers must be numbers of something, they may be numbers of anything. Propositions, therefore, concerning numbers, have the remarkable peculiarity that they are propositions concerning all things whatever; all objects, all existences of every kind, known to our experience. All things possess quantity, consist of parts which can be numbered; and in that character possess all the properties which are called properties of numbers. That half of four is two must be true whatever the word four represents, whether four men, four miles, or four pounds weight. We need only conceive a thing divided into four equal parts (and all things may be conceived as so divided) to be able to predicate of it every property of the number four, that is, every arithmetical proposition in which the number four stands on one side of the equation. Algebra extends the generalization still farther, every number represents that particular number of all things without distinction, but every algebraical symbol does more; it represents all numbers without distinction. As soon as we conceive a thing divided into equal parts, without knowing into what number of parts, we may call it a or x, and apply to it, without danger of error, every algebraical formula in the books. The proposition, $2(a + b) = 2a + 2b$, is a truth coextensive with all nature. Since, then, algebraical truths are true of all things whatever, and not, like those of geometry, true of lines only or of angles only, it is no wonder that the symbols should not excite in our minds ideas of any things in particular. When we demonstrate the forty-seventh proposition of Euclid, it is not necessary that the words should raise in us an image of all right-angled triangles, but only of some one right-angled triangle: so in algebra we need not, under the symbol a, picture to ourselves all things whatever, but only some one thing; why not, then, the letter itself? The mere written characters, a, b, x, y, z, serve as well for representatives of things in general, as any more complex and apparently more concrete conception. That we are conscious of them, however, in their character of things, and not of mere signs, is evident from the fact that our whole process of reasoning is carried on by predicating of them the properties of things. In resolving an algebraic equation, by what rules do we proceed? By applying at each step to a, b, and x the proposition that equals added to equals make equals; that equals taken from equals leave equals; and other propositions founded on these two. These are not properties of language or of signs as such, but of magnitudes, which is as much as to say, of all things. The inferences, therefore, which are successively drawn, are inferences concerning things, not symbols; though as any things whatever will serve the turn, there is no necessity for keeping the idea of the thing at all distinct, and consequently the process of thought may, in this case, be allowed without danger to do what all processes of thought, when they have been performed often, will do if permitted, namely, to become entirely mechanical. Hence the general language of albegra comes to be used familiarly without exciting ideas, as all other general language is prone to do from mere habit, though in no other case than this can it be done with complete safety. But when we look back to see from whence the probative force of the process is derived, we find that at every single step, unless we suppose ourselves to be thinking and talking of the things, and not the mere symbols, the evidence fails.

There is another circumstance which, still more than that which we have now mentioned, gives plausibility to the notion that the propositions of arithmetic and algebra are merely verbal. This is, that when considered as propositions respecting things, they all have the appearance of being identical propositions. The assertion, "Two and one are equal to three," considered as an assertion respecting objects, as for instance, "Two pebbles and one pebble are equal to three pebbles," does not affirm equality between two collections of pebbles, but absolute identity. It affirms that if we put one pebble to two pebbles, those very pebbles are three. The objects, therefore, being the very same, and the mere assertion that "objects are themselves" being insignificant, it seems but natural to consider the proposition, "Two and one are equal to three," as asserting mere identity of signification between the two names.

This, however, though it looks so plausible, will not stand examination. The expression "two pebbles and one pebble" and the expression, "three pebbles" stand, indeed, for the same aggregation of objects, but they by no means stand for the same physical fact. They are names of the same objects, but of those objects in two different states: though they *de*note the same things, their *con*notation is different. Three pebbles in two separate parcels, and three pebbles in one parcel, do not make the same impression on our senses; and the assertion that the very same pebbles may by an alteration of place and by arrangement be made to produce either the one set of sensations or the other, though it is a very familiar proposition, is not an identical one. It is a truth known to us by early and constant experience: an inductive truth: and such truths are the foundation of the science of number. The fundamental truths of that science all rest on the evidence of sense; they are proved by showing to our eyes and our fingers that any given number of objects—ten balls, for example—may by separation and rearrangement exhibit to our senses all the different sets of numbers the sum of which is equal to ten. All the improved methods of teaching arithmetic to children proceed on knowledge of this fact. All who wish to carry the child's *mind* along with them in learning arithmetic; all who . . . wish to teach numbers, and not mere ciphers—now teach it through the evidence of the senses, in the manner we have described.

We may, if we please, call the proposition, "Three is two and one," a definition of the number three, and assert that arithmetic, as it has been asserted that geometry, is a science founded on definitions. But they are definitions in the geometrical sense, not the logical; asserting not the meaning of a term only, but along with it an observed matter of fact. The proposition, "A circle is a figure bounded by a line which has all its points equally distant from a point within it," is called the definition of a circle; but the proposition from which so many consequences follow, and which is really a first principle in geometry is that figures answering to this description exist. And thus we may call, "three is two and one" a definition of three; but the calculations which depend on that proposition . . . follow from the . . . arithmetical theorem presupposed in it, namely, that collections of objects exist, which while they impress the senses thus, ०°० , may be separated into two parts, thus

OO O. This proposition being granted, we term all such parcels threes, after which the enunciation of the above-mentioned physical fact will serve also for a definition of the word *three.*

The science of number is thus no exception to the conclusions we previously arrived at, that the processes even of deductive sciences are altogether inductive, and that their first principles are generalizations from experience.

Hume's definition of mathematical necessity is unacceptable.

11.23 This . . . is the principle asserted: that propositions, the negation of which is inconceivable, or in other words, which we cannot figure to ourselves as being false, must rest on evidence of a higher and more cogent description than any which experience can afford. . . .

Now I cannot but wonder that so much stress should be laid on the circumstance of inconceivableness, when there is such ample experience to show that our capacity or incapacity of conceiving a thing has very little to do with the possibility of the thing in itself; but is, in truth, very much an affair of accident, and depends on the past history and habits of our own minds. There is no more generally acknowledged fact in human nature, than the extreme difficulty at first felt in conceiving anything as possible, which is in contradiction to long established and familiar experience; or even to old and familiar habits of thought. And this difficulty is a necessary result of the fundamental laws of the human mind. When we have often seen and thought of two things together, and have never in any one instance either seen or thought of them separately, there is by the primary laws of association an increasing difficulty, which may in the end become insuperable, of conceiving the two things apart. This is most of all conspicuous in uneducated persons, who are in general utterly unable to separate any two ideas which have once become firmly associated in their minds; and if persons of cultivated intellect have any advantage on the point, it is only because, having seen and heard and read more, and being more accustomed to exercise their imagination, they have experienced their sensations and thoughts in more varied combinations, and have been prevented from forming many of these inseparable associations. But this advantage has necessarily its limits. The man of most practiced intellect is not exempt from the universal laws of our conceptive faculty. If daily habit presents to anyone for a long period two facts in combination, and if he is not led during that period either by accident or intention to think of them apart, he will probably in time become incapable of doing so even by the strongest effort; and the supposition that the two facts can be separated in nature will at last present itself to his mind with all the characters of an inconceivable phenomenon. There are remarkable instances of this in the history of science: instances in which the wisest men rejected as impossible, because inconceivable, things which their posterity, by earlier practice and longer perserverance in the attempt, found it quite easy to conceive, and which everybody now knows to be true. There was a time when men of the most cultivated intellects, and the most emancipated from the dominion of early prejudice could not credit the existence of antipodes; were unable to conceive, in opposition to old association, the force of gravity acting upward instead of downward. . . .

Now, in the case of a geometrical axiom, such, for example, as that two straight lines cannot inclose a space—a truth which is testified to us by our very earliest impressions of the external world—how is it possible (whether those external impressions be or be not the ground of our belief) that the reverse of the proposition *could* be otherwise than inconceivable to us? What analogy have we, what similar order of facts in any other branch of our experience, to facilitate to us the conception of two straight lines inclosing a space? Nor is even this all. I have already called attention to the peculiar property of our impressions of form, that the ideas or mental images exactly resemble their prototypes, and adequately represent them for the purpose of scientific observation. From this, and from the intuitive character of the observation, which in this case reduces itself to simple inspection, we cannot so much as call up on our imagination two straight lines, in order to attempt to conceive them inclosing a space, without by that very act repeating the scientific experiment which establishes the contrary. Will it really be contended that the inconceivableness of the thing, in such circumstances, proves anything against the experimental origin of the conviction? Is it not clear that in whichever mode our belief in the proposition may have originated, the impossibility of our conceiving the negative of it must, on either hypothesis, be the same?

12

FREGE · Introduction

Gottlob Frege (1848–1925) was a mathematician who influenced work on the philosophy of mathematics and on mathematical logic as much as any other individual, or more. His writings, particularly his contributions to symbolic logic, were so completely original that neither mathematicians nor philosophers were at first able to comprehend or appreciate them. Since Bertrand Russell and others "discovered" and publicized his work around the turn of the century, so much discussion and elaboration of Frege's original ideas have taken place that it is impossible even to begin to summarize it all here. Suffice it to say that he not only initiated the movement which led to the establishment of mathematical logic and the philosophy of mathematics as quasi-independent disciplines, but he influenced work in other areas—such as philosophy of language and general epistemology.

Although Frege is sometimes spoken of as being the first philosopher of mathematics, the previous sections provide sufficient evidence to demonstrate that he was at most the initiator of the recent period of intensive concentration on this area by specialists using the tools of mathematical logic. Frege considered himself to be working entirely within the tradition of Plato, Descartes, Leibniz, etc. with regard to his work on the philosophy of mathematics. Indeed, his basic work on this topic, *The Foundations of Arithmetic*, contains detailed analyses of the various theories which had been promulgated by his predecessors and contemporaries, including Leibniz, Kant, and Mill. Only after providing what he considered fatal criticisms of all of the other theories did he assert his conclusion that the Leibnizian ideal that all mathematics is reducible to the laws of logic is correct. He went far beyond Leibniz, however, in that he provided the logical tools necessary for actually carrying out this derivation of mathematics from logic. The project was continued on by Bertrand Russell and Alfred North Whitehead in their famous *Principia Mathematica*. That the project was never satisfactorily completed and has now been abandoned in no way reflects poorly on Frege's work or even his basic goal. The failure of the attempt to derive mathematics from the laws of logic resulted partly from unforeseeable technical problems, the discovery of which was important and had numerous ramifications for a variety of other kinds of philosophical investigations. But another kind of problem also provided an obstacle to the project, namely the problem of ascertaining exactly what are and what are not the "laws of logic," a problem which still occupies an important position in philosophical debates.

The philosophy of mathematics espoused by Leibniz, Frege, and Russell has been given the title of "logicism," but this separate name does not necessarily imply that this theory incorporates totally new and unique ontological and epistemological concepts. In point of fact, the logicist must still answer the traditional questions of ontology and epistemology with regard to the nature and objects of *logical* knowledge in order to answer the questions of the philosophy of mathematics. Frege and Russell both adopted a Platonic view that the truths of logic are about objects which cannot be perceived through the senses or abstracted from sense experience, but which exist independently of every particular mind. Full justice cannot be given to the arguments which they gave in support of this theory; the interested reader is referred to the relevant portions of Frege's *Foundations* and Russell's *Problems of Philosophy.*

The following essay is included in this volume not as a complete statement of Frege's philosophy of mathematics, but rather as an example of the kind of criticism which he was able to direct at competing theories. It is a complete translation of Frege's article "Le Nombre Entier" ("The Whole Number") which originally appeared in the *Révue de Métaphysique et de Morale,* volume iii (1895). The English translation by V. H. Dudman first appeared in *Mind,* volume LXXIX, no. 316 (1970). Although the arguments are directed at a specific mathematician, one M. Ballue, the points being criticized will be recognized as similar if not identical to concepts and arguments propounded by various philosophers represented in the preceding sections of this volume. And although Frege's arguments are essentially critical, it is possible to see how they can be used to support his own general position on the basic ontological and epistemological questions of the philosophy of mathematics.

FREGE · Selections

I have noticed that this Journal is trying to reconcile mathematics and philosophy, and to me this seems very valuable. Indeed, these sciences cannot but profit by exchanging ideas. It is this that prompts me to enter the discussion. The views put forward by M. Ballue in the May number[1] are doubtless shared by most mathematicians. But they embody logical difficulties which to me seem serious enough to be worth exposing—all the more because they might obscure the issues, and make philosophers stop bothering about the principles of arithmetic. To begin with, it seems worth pointing out that a frequent fault of mathematicians is their mistaking symbols for the objects of their investigations. In fact, symbols are only the means—albeit very useful: indispensable, even—of investigation, not its objects. These latter are represented by the symbols. The *shapes* of the signs, and their physical and chemical properties, can be more or less appropriate, but they are not essential. There is no symbol that cannot be replaced by another of different shape and qualities, the connection between things and symbols being purely conventional. This goes for any system of signs and for any language. Language is a powerful instrument of the human intelligence, no doubt: but one language can be as useful as another. So it is necessary not to overrate words and symbols, either by ascribing to them quasi-magical powers over things, or by mistaking them for the actual things of which they are at most the (more or less accurate) representations. It hardly seems worth insisting on the point, but M. Ballue's article is perhaps not immune from the error in question. His topic is whole numbers. What are they? M. Ballue says: "Pluralities are represented by symbols called whole numbers." According to him, then, whole numbers are symbols, and it is of these symbols that he wants to speak. But symbols are not, and cannot be the foundation of mathematical analysis. When I write down 1 + 2 = 3 I am putting forward a proposition about the numbers 1, 2 and 3, but it is not those symbols that I am talking about. I could substitute A, B and Γ for them; I could write *p* instead of + and *e'* instead of =. By writing A*p*B*e'*Γ I should then express the same thought as before—but by means of different symbols. The theorems of arithmetic are never about symbols, but about the things they represent. True, these objects are neither palpable, visible, nor even real, if what is called real is what can exert or suffer an influence. Numbers do not undergo change, for the theorems of arithmetic embody eternal truths. We can say, therefore, that these objects are outside time; and from this it follows that they are not subjective percepts or ideas, because these are continually changing in conformity with psychological laws. Arithmetical laws form no part of psychology. It is not as if every man had a number of his own, called *one*, forming part of his mind or his consciousness: there is just one number of that name, the same for everybody, and objective. Numbers are therefore very curious objects, uniting in themselves the apparently contrary qualities of being objective and of being unreal. But it emerges from a more serious consideration that there is no contradiction here. Negative numbers, fractions, etc. are of the same nature; and perhaps that is why in arithmetic too much store is often set by the symbols. Because of the difficulty of identifying objects which are neither discernible to the sense nor psychological, visible objects have been substituted for them. But this is to forget that these symbols are not what we want to study. And so numbers have a double nature conferred upon them: they are called symbols; and yet they are themselves represented—they are given names. M. Ballue writes: "Like all symbols, the whole number admits of a double representation: the sound which it produces upon the ear, the impression which its written name produces upon the sight . . . Besides this, the whole number possesses its own particular written representation, requiring the use of special characters called *figures*. The aim of *numeration* is to study the means of representing all whole numbers with a small number of words and figures." What is it then that the figure 2 designates? A number; which is to say, a symbol according to M. Ballue. Is it the word *deux*? If that were the case, we Germans should have numbers which were different from those of Frenchmen, and our arithmetic would be a different science from theirs, having different objects of investigation. Perhaps M. Ballue's opinion is that the word *deux* represents the same number as the figure 2. But whatever this number might be, it represents a plurality, and is itself represented by the figure 2. What then do we want with this somewhat mysterious intermediary? Why will it not do to designate the plurality directly by the figure?

It might be thought that this is only a verbal slip on M. Ballue's part, which could easily be corrected by substituting *plurality* for *whole number* in the title of the article. For it is pluralities of which whole numbers are the symbolic representatives, according to M. Ballue. But this will not save us from all the difficulties. What is a plurality? M. Ballue replies: "The assemblage of several distinct objects, considered as distinct, without attention to the nature or shape of these objects, is called a *plurality*. It will be seen that a plurality is an assemblage of units."

This definition is not as clear as the author seems to think. The sense of the word *plurality* could be found contained in the word *several* and in the plural form, but M. Ballue adds some restrictions, saying "distinct objects, considered as distinct, without attention to the nature or shape of these objects." What he is calling *distinct* here he has previously called *isolated*, saying: "An isolated object, considered as isolated, in abstraction from its nature or shape, is given the name of unit." It will perhaps be objected that if the objects were absolutely isolated there would be no assemblage. And besides, it is to be doubted whether an absolutely isolated object exists, every material particle being related to every other by gravitation. So the precise degree of isolation required would have to be specified. I shall not labour this point, but I do want to enquire more closely into what M. Ballue intends by the words "considered as distinct, without attention to the nature or shape of these objects," and by the words "considered as isolated, in abstraction from its nature or shape." What strikes me here is that the way of considering an object, and the abstractions performed in the mind of a subject, seem to be being taken for qualities of the object. I ask: after the object has been considered as isolated, is it the same object as before?—or has one created a new object by considering it? In the former case, nothing essential would be changed. And surely, if I consider the planet Jupiter as distinct or isolated, its gravitational ties to the other heavenly bodies do not become any the weaker;

and if I abstract from its mass and its spheroidal shape, Jupiter loses neither its mass nor its shape. So what is the point of performing this abstraction? There would be a psychological difficulty over this as well. While I am considering an object, I can be sure that it is being considered. But in conducting a proof I have to fix my attention upon other objects successively, for I am incapable of considering each of even a hundred objects at the same time. It is all the more difficult in that, without getting the objects mixed up, I am not to attend to their nature or shape. [?Tr.[2]] I should thus lose the assurance that these objects were all in fact units. Certainly they would not be units with respect to me: perhaps with respect to other people they would but I should probably know nothing of this. And even if I did know, it would be useless from the point of view of my proof, for I could infer nothing thence.

Orion is an assemblage of stars. If it is possible in general to consider objects as distinct, without attention to the nature or shape of these objects, then it will be possible in this case. After performing this consideration we shall say, if we take M. Ballue at his word, that the constellation is a plurality. And since the name *Orion* is a symbol for that plurality, we shall regard this word as a number. Admittedly he does not actually say that the stars are considered as distinct, etc. But that is neither here not there: granted that the constellation is a plurality, the name of the constellation is a symbol for a plurality.

Let us examine the alternative conception, that the considered object is different from the original object. The sun for instance, as a material, luminous body, having a shape and occupying a position, would be different from the sun considered as distinct, in abstraction from its nature or shape. It might be said that the latter is created by the act of considering it and that, since an external object cannot be created in this way, it would have to be a subjective idea or something of the sort in the mind of the person performing that consideration and that abstraction. By thus considering the sun, everyone would form such an idea of his own, distinct from anyone else's. Pluralities would then be subjective too. And that would be at variance with the fact that naturalists are giving objective information when they specify the precise number of pistils in a flower.

What could be the effect of abstracting from the nature or shape of an object? Does it lose its nature and shape? This is apparently the effect M. Ballue is after. But it is obvious that an external object cannot be changed in this way. As for the idea of an object that someone forms for himself, there is no need for abstraction in order that *it* lack the qualities of the object itself. An idea of the sun is not a material, luminous body. But still, such an idea is in general qualitatively different from the same person's idea of the moon. M. Ballue's abstraction can efface the difference between these ideas.—But then what remains of the plurality?

Closely bound up with this is another difficulty. M. Ballue says: "The simplest plurality is formed by adding one unit to another unit." But if there exist more than two units there will be several pluralities formed by adding one unit to another unit, and the definite article in the singular that M. Ballue uses is therefore inaccurate. It should be: The simplest pluralities are formed, etc. But the number *two* is neither a particular plurality of that sort nor the symbol for such a plurality. It is perhaps nearer the truth to say that it is the species or class of pluralities formed by adding one unit to another unit. But then, to meet the demands of precision, we want good definitions of *unit* and *plurality*. Readers of this Journal will easily verify that the first of these terms is not used with any uniformity by authors. Comparing M. Ballue's thesis ("The simplest plurality is formed by adding one unit to another unit") with what MM. Le Roy and Vincent say (in their article, published in the September number, page 519: "The possibility of the mind's forming whole numbers indefinitely by adding unity to itself") we see that the latter use this term <*unité*> as a proper name, whereas M. Ballue uses it as a general term by assuming the existence of several units. At the same time we see that the words *assemblage* and *adding* used by M. Ballue, require explanation. MM. Le Roy and Vincent employ the verb *to add*: apparently this is an action that takes place in people's minds. But it is difficult to conceive how a thing could be added to itself. What sort of relations are they that yield these assemblages? Are they physical, historical, geometrical or psychological?—or are they purely logical?

Readers will perhaps be dissatisfied, because I have done no more than enter protests and pose questions. But since positive solutions to the problems are set forth in my works cited below,[3] I can confine myself at present to showing that some pretty thorny issues are involved, and that the matter is more complicated than it appears at first sight.

Notes

[1] *Le Nombre entier considéré comme fondement de l'analyse mathématique.*

[2] *C'est d'autant plus difficile que je ne dois pas me préoccuper de la nature ou de la forme des objects sans les confondre.*

[3] *Die Grundlagen der Arithmetik*, Breslau, Wilhelm Köbner, 1884; *Grundgesteze der Arithmetik* I, Jena, Hermann Pohle, 1893.

13

THREE CONTEMPORARY VIEWS · Introduction

The three essays included in this section are offered as examples of recent work in the philosophy of mathematics. They have been selected for a variety of reasons, not the least of which is their intrinsic interest and their exemplification of good philosophical reasoning. None of the three represents anything that could be called a dominant or consensus view of the nature of mathematics, since the situation in this area is today more unsettled than it has been at any time during the last one hundred years—so unsettled that there is presently no consensus. Thus the three essays might be considered representative in so far as they reflect the diversity of views now being debated; it should be recognized that there are many other theories currently being espoused by philosophers which bear little resemblance to any of the three included here. (Many of these other theories are discussed in books and essays listed in the Bibliography; *all* of them resemble in various ways the 'traditional' theories presented in this volume.)

The three following essays are not representative of all the recent literature of the philosophy of mathematics, particularly in so far as none of them is primarily concerned with technical problems in symbolic logic or metamathematics. The work on the technical problems is certainly important, but it has often tended to obscure rather than resolve many of the traditional problems in ways such as those discussed in the opening sections of Max Black's essay. It seems quite doubtful that the questions of the nature of mathematical knowledge and the ontological status of mathematical entities (if there are any) could ever be answered solely by work on the technical problems of logic and metamathematics.

The three essays in this section are similar in so far as they are all concerned with the fundamental philosophical questions discussed in the previous sections. Each of the three authors is addressing a distinct problem from his own special perspective. Erik Stenius takes what is perhaps closest to a traditionalistic approach to the essentially *epistemological* and *semantical* problem of the nature of mathematical truth. Sidney Axinn analyzes the *methodology* of mathematics from his perspective as an experimentalist. Max Black uses the techniques of linguistic analysis in an attempt to solve (or perhaps dissolve) the problem of the *ontological* status of mathematical sets or classes.

These contemporary pieces are related to the work of the earlier philosophers in ways other than the fact that they are concerned with the same problems. Some of these ties are readily apparent; for example, Professor Stenius' debt to Kant is clearly reflected even in the title of his essay. Other ties are less obvious, but are present nonetheless. Professor Axinn's essay might appear at first glance to be in the tradition of Mill and Hobbes, but the author has indicated that he has been most strongly influenced by the writings of Kant. Professor Black's approach, to take a final example, may be viewed by some readers as very similar to that of Aristotle with its sensitivity for the nuances of meanings of basic terms. But despite the variety of ties with and similarities to the traditional philosophies of mathematics, each of these essays also contains significant new ideas, and thus together they provide evidence that the tradition is not ended, but that it is continuing and hopefully will continue into the future with the formulation of new and better theories to deal with the fundamental questions concerning the nature of mathematics.

STENIUS · Introduction

Erik Stenius is presently Professor of Philosophy at the University of Helsinki, Finland. He received his undergraduate degree in mathematics and taught mathematics and physics for several years before completing his doctorate in philosophy. His original interest in the epistemological foundations of mathematics led him to work in the philosophy of language and in the philosophies of Kant and Wittgenstein. He is the author of several books, including *Das Interpretationsproblem der formalisierten Zahlentheorie und ihre formale Widerspruchsfreiheit, Wittgenstein's Tractatus—A Critical Exposition of Its Main Lines of Thought,* and *Critical Essays.* He has also authored many papers in addition to the one reprinted here, which first appeared in *The Philosophical Review,* vol. lxxiv, no. 3 (July, 1965), pp. 357–372, under the title "Are True Numerical Statements Analytic or Synthetic?"

STENIUS · Selection

SO FAR as I know, the view that mathematical truths, like logical truths, have nothing to do with empirical observation is almost universally accepted among analytic philosophers. In spite of this unanimity, I think the problem of the semantical and epistemological status in this respect of numerical truths in particular is still worthy of a thorough examination. This paper is presented in the hope that it will incite a discussion of this problem on what I believe to be a somewhat new basis.[1]

I

Let us consider the controversy between logical empiricism and the Kantian view on the subject.

Kant held that there are synthetic propositions which are known to be true a priori. Accordingly, he held that there are general synthetic propositions of the truth of which we have certain knowledge, and that, in particular, all mathematical statements are of this kind. The logical empiricists opposed these tenets, laying down the three well-known theses:

LE 1. *No synthetic propositions can be (known to be true) a priori.*

LE 2. *All general synthetic propositions are hypothetical.*

LE 3. *All true mathematical propositions are analytic.*

For my part, I have certain doubts concerning the validity of thesis LE 3 in respect of numerical statements—that is, in respect of what could be called singular arithmetical statements, such as $7 + 5 = 12$, or $2 \times 5 = 10$, or $2^{10} = 1024$. General arithmetical propositions, such as $a + b = b + a$, I shall not discuss in this context, nor shall I discuss propositions of geometry. The thesis I am going to discuss is, accordingly, the one we get when we replace the word "mathematical" with the word "numerical" in LE 3, or, as it could also be formulated:

LE 3a. *All numerical truths are analytic.*

My doubts concerning LE 3a belong to an old controversy of mine with logical empiricism. But to make clear the character of this controversy, I must add the following declaration. It has become common to think that almost all of what logical empiricism emphasized was wrong or misleading. So, for instance, there are philosophers who say that thesis LE 1 is unjustified—for different reasons. I am not one of these philosophers. I still think LE 1 is a very important epistemological principle, the acceptance of which gives anyone who analyzes the foundations of scientific thought a fruitful task. This is the task of showing about any principle which seems to form a synthetic a priori proposition that it is either not a priori or not synthetic. So I accept LE 1 and LE 2. Moreover, I do not embrace Mill's conception of numerical statements as inductive generalizations.[2] But, nevertheless, I question LE 3a. I will try in the sequel to explain of what kind my doubts are.

II

First, some observations concerning the fundamental concepts occurring in our problem—that is, the two distinctions, on the one hand between "analytic" and "synthetic," and on the other hand between "a priori" and "a posteriori."

In their eagerness to criticize logical empiricism, many contemporary philosophers have introduced great confusion about these concepts—so great as to make the issue between the logical empiricists and Kantian philosophy more or less unintelligible. So I must recall how these concepts were understood in the context of this issue.

What did Kant mean by an "analytic" proposition? I think that Kant's main characterization is found in his statement that analytic propositions "add nothing to our knowledge" since they are only "explicative" (*Prolegomena*, §2). This means that analytic propositions are empty so far as "matters of fact" are concerned; that we, accordingly, can state their truth by a mere analysis of the concepts occurring in them. The same features were stressed by the logical empiricists. Of course, Kant's definition of "analyticity" was not clear, and the logical empiricists tried to amend it—without final success. It is, however, not my intention to discuss here the question of an adequate definition of analyticity. In this context, I want only to stress some essential features of the Kantian and empiricist distinction between "analytic" and "synthetic," as contrasted with their distinction between "a priori" and "a posteriori."

I think some clarity is gained by pointing out that the distinction between "analytic" and "synthetic" could be called a *semantical* distinction, whereas the distinction between "a priori" and "a posteriori" is an *epistemological* distinction.

An *analytic* statement is characterized by the fact that a semantical analysis of the terms involved shows it to have a vacuous content and, thus, to be uninformative about matters of empirical observation.

A *synthetic* statement is, accordingly, characterized by the fact that a semantical analysis of the terms involved shows it to have a nonvacuous content and, thus, to be really informative about matters of empirical observation.

An *a priori* statement is characterized by the fact that we can *know* it to be true without consulting experience; our knowledge of its truth is not *founded* on empirical observation.

An *a posteriori* statement is characterized by the fact that we *cannot* know it to be true without consulting experience: our knowledge of its truth *is* founded on empirical observation.

It follows from these definitions that an analytic statement is a priori: since it does not inform us about matters of empirical observation, we need not consult experience for its verification. It is true independently of what is the case. And this is the only kind of a priori truth admitted by the logical empiricists. The main tenet of Kant's theoretical philosophy is, however, that this semantical kind of a priori is not the only kind. There are statements which are a priori in spite of being synthetic, and all mathematical statements are of this kind.

Thus Kant's contention that mathematical statements are synthetic statements a priori contains two partial contentions. True mathematical statements are: (1) synthetic: they are not vacuous but really informative about matters of empirical observation; (2) a priori: our knowledge of their truth is not founded on empirical observation.

Principle LE 1 implies that no statement can fulfill these conditions simultaneously. From this observation the logical empiricists inferred that mathematical statements must be analytic. The argument for this conclusion is what I want to examine in respect of numerical statements.

III

I shall found my criticism of these arguments on their exposition in A. J. Ayer's discussion of the subject in Chapter IV of *Language, Truth and Logic.*[3]

Although the title of this chapter reads "The *A Priori*," Ayer uses, for the most part, the expression "necessary truths about matters of fact" rather than the main Kantian term "synthetic propositions a priori." Now the introduction of the concept of "necessity" into this context is historically correct, but since it is the source of additional difficulties, I shall state Ayer's view in terms of the "a priori."

With regard to the Kantian contentions (1) and (2) of the preceding section, Ayer sees two alternatives for an empiricist. Of numerical statements he must say either: (a) that they are after all *not* a priori (in this case he must account for the universal conviction that they are); or (b) that they are *not* synthetic (in this case he must explain how a proposition which is empty of all factual content can be true, useful, and surprising).

IV

I have formulated Ayer's alternatives with regard only to numerical statements. Ayer's own formulation refers to mathematical statements in general and, in addition, to logical statements. This means that he puts the question of the semantical and epistemological status of mathematical statements on the same level as the corresponding question about truths of logic. For my part, I want to make a sharp distinction between these two questions. To be sure, the semantical status of truths of logic has also been discussed recently, but in this paper I shall consider it as decided in the traditional way. If by "truths of logic" is meant what we ordinarily call "logical truths" such as, for instance, "either it is raining or it is not raining" or "all men are men," I regard their semantical nature as unquestionable, because I take the concept of a "logical truth" as being defined in such a way that a statement which is logically true is always analytic. To say that the statement "either it is raining or it is not raining" or "all men are men" is logically true is to say that for semantical reasons it is void of factual content and accordingly is not synthetic but analytic.

V

Ayer first treats alternative (a), which he identifies with Mill's opinion that all numerical propositions are inductive generalizations from experience.[4] Since I agree in the main with Ayer's argument against Mill, I shall not dwell upon it, but shall only state that it leads to a rejection of Mill's view.

Next Ayer reasons like this:

Numerical statements are not inductive generalizations from experience. Thus alternative (a) is excluded, and there remains only alternative (b).

Because this is an important point I shall give the argument the form of a syllogism, which I call Ayer's syllogism:

LE 1. *No synthetic Propositions can be a priori.*
LE 4. *Numerical truths are a priori.*

LE 3a. *Numerical truths are analytic.*

Ayer adds that Kant made the mistake of confusing a logical argument with a psychological one, when he arrived at the conclusion that the statement "$7 + 5 = 12$" is synthetic. I think Ayer's and Kant's arguments are logical and psychological to the same degree. Both thinkers refer to their intuitions concerning the meaning of the statement "$7 + 5 = 12$," but since these intuitions are different, they arrive at different results concerning its semantical status.

VI

What are the arguments *against* the analyticity of a numerical statement like "$7 + 5 = 12$"?

I think three types of arguments are of interest:

(1) It is difficult to believe that "7 + 5 = 12" is a mere tautology.

For my part, I think it is intuitively satisfactory to say that the proposition "either it is raining or it is not raining" does not inform one about the weather and that it is devoid of factual content at least in this sense. But I think it is not in accordance with one's intuitive feeling to say that the same is true of "7 + 5 = 12." It seems in some way absurd to say that "7 + 5 = 12" is devoid of factual content.

(2) Against this objection, adherents of the analyticity of numerical statements often argue that, although the statement "either it is raining or it is not raining" does not inform us about the weather, it does inform us about certain linguistic conventions. The same might be true of "7 + 5 = 12," and this accounts for our feeling that this statement is not empty.

But does the statement "7 + 5 = 12" give us information about a linguistic convention? I think this contention, too, is counterintuitive. (By the way, I think it is also misleading to say that the statement "either it is raining or it is not raining" informs us about certain linguistic conventions. But it would take us too far to go into this question here.)

(3) The second argument is connected with the following objection against the analyticity of numerical statements. We do not originally learn to know what statements of the form "a + b = c" are correct simply by learning them by heart. In fact, we learn a certain *procedure* by means of which we determine the number which is the sum of two other numbers. This procedure is called "addition," and is one of the four simple rules of arithmetic. Thus we could say that we verify or falsify a statement of the form "a + b = c" by performing the addition a + b and looking for the number we have got: if this number is c, then the statement is verified; if it is not c, the statement is falsified. At least prima facie, this procedure has the character of an experiment; but if the statement "7 + 5 = 12" is verified by empirical observation of the result of an experiment it is a posteriori and synthetic.

This seems to me to be the most important objection against the analyticity of numerical statements.

VII

What is Ayer's reply to an argument of this kind? His reply is actually found in explicit form in his exposition (pp. 75 f.), though in this context he discusses not the statement "7 + 5 = 12" but the statement "2 × 5 = 10." Ayer remarks that the best way to substantiate the assertion that the truths of pure mathematics are a priori is to examine cases in which they might seem to be confuted:

> It might easily happen, for example, that when I came to count what I had taken to be five pairs of objects,[5] I found that they amounted only to

> nine. And if I wished to mislead people I might say that on this occasion twice five was not ten. But in that case I should not be using the complex sign "$2 \times 5 = 10$" in the way in which it is ordinarily used. I should be taking it not as the expression of a purely mathematical proposition, but as the expression of an empirical generalization, to the effect that whenever I counted what appeared to me to be five pairs of objects I discovered that they were ten in number. The generalization may very well be false. But if it proved false in a given case, one would not say that the mathematical proposition "$2 \times 5 = 10$" had been confuted. One would say that I was wrong in supposing that there were five pairs of objects to start with, or that one of the objects had been taken away while I was counting, or that two of them had coalesced, or that I had counted wrongly. One would adopt as an explanation whatever empirical hypothesis fitted best with the accredited facts. The one explanation which would in no circumstance be adopted is that ten is not always the product of two and five.

Suppose that the objects we have counted are matchsticks. Let us then assume that we are counting the number of sticks in five pairs of sticks and arrive at the result that they are nine. According to Ayer's account, we could explain this result by assuming that we had dropped one of the sticks to the floor without observing it, that two sticks had coalesced or that we had counted wrongly. Of these alternatives, the last is by far the most probable. In fact, we may classify all the different possibilities as so many different ways of *counting wrongly*. For in any case we have not correctly counted how many objects there are in five pairs of objects.

We may therefore sum up Ayer's argument as follows:

(A) *If in counting five pairs of objects we arrive at the result that they are nine in number we infer that in one way or another we have counted wrongly. Therefore no observation can refute the statement "$2 \times 5 = 10$."*

According to this argument, then, numerical statements are a priori—that is, LE 4 is valid.

VIII

There is, however, a weak point in this argument. This is that it is founded on the assumption that we *know in advance* that $2 \times 5 = 10$. A child who does not know this in advance cannot realize that he has counted wrongly on the mere ground that he arrives at the result nine. In short, an essential condition for the validity of (A) is that we *know* that $2 \times 5 = 10$. In order to make (A) precise we must add this condition. Then we arrive at the following thesis:

(B) *If we know that 2 × 5 = 10 we also know that we have made a calculation error if on some occasion we arrive at the reslut 2 × 5 = 9.*

But this formulation does not imply the apriority of a numerical statement. It seems rather to be a sheer logical circle; and if this is so, must we not reject the premise LE 4 of Ayer's syllogism as unproved and, accordingly, also the conclusion LE 3a?

Now we should not accept this objection too hastily. Thesis (B) is not so pure a logical circle as it might seem to be. In fact, Ayer's argument speaks in favor of the correctness of one important observation: 2 × 5 must by logical necessity equal 10 *once for all*, or it must once for all not equal 10.

Independently of our knowing whether 2 × 5 = 10 or not, we know that 2 × 5 cannot one day be 10 and another day 9. This observation could be given the following formulation:

(C) *If two calculations of the product 2 × 5 give different results, at least one of them must be a miscalculation.*

Now as a result of Ayer's argument, I shall assume that proposition (C) is certain and a priori and, hence, analytic. I therefore *accept* the view that (C) is analytic.

IX

Thesis (C) could also be formulated in this way:

(D) *If we once have calculated the product 2 × 5 correctly, then every time we perform this calculation correctly we shall arrive at the same result.*

Thus Ayer's argument shows that (D) is a priori and analytic, but the same is not true of the statement "2 × 5 = 10" itself.

The word "correctly" should be stressed in the formulation of (D). In the discussion of my paper at Harvard, Professor R. Firth expressed, referring to Mill, his doubts concerning the analyticity even of thesis (D). If I understood him rightly, he interpreted Mill as contending, not that our knowledge of the truth of a particular numerical statement—such as, for instance, "2 × 5 = 10"—is as such a product of inductive generalization, but rather that our knowledge of the general *uniqueness* of the result of all elementary arithmetical operations is a product of inductive generalization. I have not been able to form a definite opinion as to how Mill ought to be interpreted: but whether or not it can be regarded as a view of Mill's, I think there are arguments in favor of taking the knowledge of this uniqueness as founded on experience.

Even though I admit this, however, I think thesis (D), as it has been formulated above, must be regarded as a priori and analytic, since it forms a *standard* by which the correctness of a numerical calculation is tested. What might be regarded as a product of inductive generalization is our knowledge that this sort of standard can be upheld; that is, that we are, with sufficient care, able to perform a simple numerical calculation again and again, every time arriving at the same result, and that we can therefore eliminate miscalculations as products of some kind of carelessness. That this is so is a prerequisite for our ability to form the concept of, for instance, the sum of two numbers which is "timelessly" the same.

Thus I see no objection to accepting (D) as analytic, so I shall do so in the sequel. My contention in this respect could be generalized in the following way:

(E) *It is analytically true that the result of a correct performance of a simple numerical calculation like the calculation of a sum or product of two numbers is unique.*

This is so, because the uniqueness of the result is a standard for the correctness of the calculation.

X

It follows from the analyticity of (D) that if one wants to adhere to the view (proposed above in Section VI) that "$2 \times 5 = 10$" is a synthetic statement a posteriori, one must infer that it is a statement which can be verified by *one single correct calculation*, that is, by one single *correct empirical observation*. This means that the statement "$2 \times 5 = 10$," considered as a synthetic statement, is singular. We could express this conclusion as follows:

(F) *The statement "$2 \times 5 = 10$" is logically equivalent to a singular synthetic statement.*

This singular statement could be of the form: "The person A has at time t performed the calculation of the product 2×5 correctly and observed the result to be 10." What person and what time are referred to in this statement are irrelevant because of the analyticity of (D).

I take the following as a standard method for the calculation of the product 2×5. We form the series

(1) one, two, one, two, one, two, one, two, one, two,

checking that it consists of five pairs of the type "one, two." The series (1) we put into one-one correspondence to an initial segment of the series of numerals (we must know the order of the numerals in order to know what "number" every individual numeral "denotes") like this:

	one,	two,	one,	two,	one,	two,	one,	two,	one,	two
(2)	↕	↕	↕	↕	↕	↕	↕	↕	↕	↕
	one,	two,	three,	four,	five,	six,	seven,	eight,	nine,	ten

The result of the multiplication is observed by noting what numeral—in this instance "ten"—occurs as the last one in the initial segment corresponding to (1).

A calculation of this product by means of the counting of matches is just one way of establishing with what initial segment the series (1) is in one-one correspondence.

The corresponding standard method for the calculation of 7 + 5 can be written

$$
\begin{array}{lcccccccccccc}
 & 1, & 2, & 3, & 4, & 5, & 6, & 7, & 1, & 2, & 3, & 4, & 5. \\
(3) & \updownarrow & \updownarrow & \updownarrow & \updownarrow & \updownarrow & \updownarrow & \updownarrow & \updownarrow & \updownarrow & \updownarrow & \updownarrow & \updownarrow \\
 & 1, & 2, & 3, & 4, & 5, & 6, & 7, & 8, & 9, & 10, & 11, & 12.
\end{array}
$$

This is the establishment of a one-one correspondence of the "sum" of the initial segments corresponding to seven and five and the initial segment corresponding to twelve. Since the initial segment corresponding to seven is trivially in one-one correspondence to itself, what is essential in this procedure is the establishment of the equivalence between the series of numerals from one to five and the series of numerals from eight to twelve.

This is, by the way, the method of verification of "7 + 5 = 12" described by Kant (*Prolegomena* §2c; *Critique of Pure Reason*, B 15 f.).

XI

Our result accordingly, is this: Ayer's argument does not show that the certainty of a numerical statement is a priori. It shows at most that a numerical statement is logically equivalent to a singular statement. But if numerical statements are singular they can, of course, be certain, even if they are a posteriori; they can be verified by a single correct observation. Thus numerical statements can without contradiction be regarded as *synthetic* propositions which can be verified a posteriori.

XII

This argument reveals an ambiguity in the formulation of LE 2. What is meant by a "general synthetic proposition"? Consider the proposition:

(4) All men are men and the earth is round.

In a sense this proposition is general, since it treats of all men. In a sense it is also *synthetic*, since it implies the synthetic proposition "the earth is round." Nevertheless, it would be misleading to call it a "general synthetic proposition," since by this expression we usually understand a proposition which entails an unlimited number of independent synthetic instances. One could perhaps clarify this point by adopting the following terminology:

If conceived of as *general*, proposition (4) is *quasi-analytic*.
If conceived of as *synthetic*, proposition (4) is *singular*.
If we adopt this terminology LE 2 ought, in order really to be a consequence of LE 1, to be formulated like this:

LE 2′ *All propositions are hypothetical which, if conceived of as synthetic, are general.*

The result of our analysis can, therefore, be given the following formulation.

Ayer's argument shows that if a numerical statement is conceived of as a *general* statement it is not genuinely synthetic. But this does not mean that it must be conceived of as an analytic statement; it can also naturally be taken as a quasi-analytic statement—that is, as a *singular synthetic statement*, which is verified a posteriori.

XIII

To the last point of my argument one might object, from Ayer's point of view, that there *are* no statements which can, strictly speaking, be regarded as singular, since on closer scrutiny, all statements turn out to be general.

As applied to our reasoning, this objection takes the form of a request for a criterion of the correctness of an arithmetical calculation. A necessary condition for regarding the statement "$2 \times 5 = 10$" as verified a posteriori is that we can say with certainty of at least one of our calculations of the product 2×5 that it is correct. By what criterion is such a certainty claimed?

In practice we rely on the uniqueness of the result. If we have once arrived at a certain result, and then by carefully checking it arrive at the same result again, we regard this result as confirmed. But nevertheless we must ask ourselves the question whether we can leave out of consideration the possibility that we have always made a calculation error when we have arrived at the result 10 in performing the calculation 2×5. Could our belief that $2 \times 5 = 10$ be founded on a perpetual miscalculation? The question might be worthy of a thorough examination; and I think it cannot be answered by an unambiguous "yes" or "no." A tentative answer is this: that our elementary numerical calculations are correct is (almost) "as certain as that two and two is four." And this kind of answer means as much as that we must always take some things for granted before we can even start to discuss whether other things are "certain." So in our present discussion I think one is justified in being dogmatic and claiming that the statement "$2 \times 5 = 10$" is verified with certainty a posteriori.

XIV

I have tried to show that numerical statements *can* naturally be regarded as synthetic in a way that does not make them counterinstances to LE 1 or LE 2. From this it does not follow that they *must* be so regarded. In fact there is a very simple reason for regarding numerical statements as analytic. This is the fact that they can be logically deduced from analytic premises, as we do in formalized number theory. The recursive definitions of sum and product can be regarded as conventions about the use of the signs "$+$" and "$\times$," according to which certain formulas containing these signs are stipulated as vacuous and thus made analytically true. From these formulas we logically derive further formulas which then must also be regarded as analytic. Among these formulas occur "5 + 7 = 12" and "2 × 5 = 10," which, therefore, must be regarded as analytic statements. In this way all numerical statements can be considered analytic.

But one thing must be stressed in this context. The *synthetic* numerical statement "2 × 5 = 10" is not the same statement as the *analytic* statement "2 × 5 = 10"; the synthetic and the analytic statements are two different interpretations of the same formula, which is thus used to formulate two different statements.

What is the relation between these two interpretations? I cannot go into any detail concerning this question here, but must restrict myself to a hint. Suppose that the formula "2 × 5 = 10 is interpreted as expressing an *analytic* statement, and consider the statement

(5) The statement "2 × 5 = 10" is analytic.

According to some logicians, statement (5) is again an analytic statement. This, I think, is wrong. In order to show that (5) is true, we have to check that the *formal* manipulations leading from the recursive definitions to the theorem "2 × 5 = 10" are in accordance with our formal logical rules. But this means that we have to observe, by means of our optical faculty, whether certain formal relations hold between premises and conclusions occurring in the proof. Thus to check that a certain string of formulas really constitutes a formal proof of the formula "2 × 5 = 10" is to perform an empirical observation. Since, therefore, we need empirical observation in order to establish the truth of (5), it is a *synthetic* proposition (and moreover a singular one). Now I think that the establishment of the fact that a certain string of formulas is a proof of "2 × 5 = 10" is a procedure which, among other things, amounts to a verification of the *synthetic* proposition "2 × 5 = 10" (or something very near to it).[6]

Professor Paul Bernays is reported to have remarked that "syntax" is but a branch of number theory.[7] I think the number theory of which Bernays is speaking here must be one in which numerical statements are regarded as synthetic.

I do not think that the relation between analytic and synthetic numerical statements is analogous to the relation between "pure" and "applied" geometry. Statements of applied geometry are definitely *not* mathematical statements whereas, according to my view, synthetic numerical statements are.

XV

To sum up: we certainly can interpret numerical formulas in such a way that they become analytic statements. But it is natural—and, as I think, more natural—to interpret them in such a way that they become synthetic. If we do so, and if we still accept LE 1 as a fundamental principle, we have to accept Ayer's first alternative and say that true numerical statements are not a priori but a posteriori. But we must not make the mistake of believing that this implies that numerical statements are inductive generalizations. As I have shown, Ayer's first alternative can be accepted in a way that allows us to fulfill Ayer's requirement of giving an "account for the general conviction" that these statements are a priori. For what leads people to think that statements like "$2 \times 5 = 10$" are a priori is the analyticity of thesis (D) or the truth of thesis (E). That this is so seems to me striking in the case of Kant.

Of course, such an interpretation of numerical statements gives a simple solution of Ayer's problem as to how they can be "true, useful and surprising." I think our interpretation also sheds light on other epistemological problems of mathematics, and in particular on many of the problems discussed by Wittgenstein in the *Remarks on the Foundations of Mathematics*.

Notes

[1] This paper is based on a paper read in slightly different versions at Lund, Sweden, in the spring of 1961 and at several universities and colleges in the United States during the spring term of 1962.

[2] Following Ayer and many others I here take this as being Mill's conception. Cf., however, below, Sec. IX.

[3] (2nd ed.; London, 1946), pp. 71 ff.

[4] Cf. above, n. 2.

[5] Ayer seems to take the counting of five pairs of objects as the natural procedure for a calculation of the product 2×5. I should rather consider it the natural procedure for a calculation of the products 5×2. But for the sake of convenience, I here follow Ayer in this respect.

[6] Cf. Wittgenstein's formulation: "Thus Russell teaches us a new calculus for reaching 5 from 2 and 3; and that is true even if we say that a logical calculus is only—frills tacked on the arithmetical calculus," in *Remarks on the Foundations of Mathematics*, p. 67.

[7] See Hao Wang, "Eighty Years of Foundational Studies," *Dialectica*, 12 (1958), 470.

AXINN · Introduction

Sidney Axinn is presently Professor of Philosophy at Temple University. A student of E. A. Singer, Jr., C. West Churchman and Nelson Goodman, he describes his philosophical position as *experimentalism,* and he views every question of fact as being a question of experiment. He has also been influenced by traditional philosophers, particularly Kant, and he is now engaged in combining the scientific consequences of Kant's work with more recent developments. He is the author of numerous papers in addition to the one reprinted here from *Philosophia Mathematica,* vol. 5, no. 1-2 (June–December, 1968), pp. 1–10, under the title "Mathematics As An Experimental Science."

AXINN · Selection

This paper holds that experiments take place and are just as necessary in mathematics as in the physical sciences. Examples of experiments in both areas will be studied, and it will be argued that theory and observation play just the same role in each. A mathematical proof turns out to satisfy each of the requirements of a proper experiment.

Galileo's Experiments

Before generalizing on the nature of an experiment, let's consider some examples of what may safely be called classics. We will start with one of the pendulum experiments that Galileo describes in his *Dialogues Concerning Two New Sciences.*[1] In the section in which Galileo is discussing the question of whether two bodies differing greatly in weight will fall with the same speed, the resistance due to air causes him a number of complications, and he turns to a special case. ". . . it occurred to me to allow the bodies to fall along a plane slightly inclined to the horizontal. For in such a plane, just as well as in a vertical plane, one may discover how bodies of different weight behave; and besides this, I also wished to rid myself of the resistance which might arise from contact of the moving body with the aforesaid inclined plane. Accordingly I took two balls, one of lead and one of cork, the former more than a hundred times heavier than the later, and suspended them by means of two equally fine threads, each 4 or 5 cubits long. Pulling each ball aside from the perpendicular, I let them go at the same instant, and they, falling along the circumferences of circles each having these equal strings for semi-diameters, past beyond the perpendicular and returned along the same path. This free vibration repeated 100 times showed clearly that the heavy body maintains so nearly the period of the light body that neither in a hundred swings nor even in a thousand will the former anticipate the later by as much as a single moment, so perfectly do they keep step. We can also observe the effect of the medium . . ." [2]

You may recall that this experiment is preceded by a considerable theoretical discussion, and it is followed by a detailed theoretical discussion. This experiment is part of an argument designed to prove that heavy and light objects maintain equal speeds in equal arcs. There are several sentences that explain to the audience what are the assumptions and what is the expected conclusion: then the audience is asked to observe that the two balls keep step perfectly, although they are of considerably different weight. The observer is asked to take particular note of this result: substituting the heavy lead ball for the light cork ball will not alter the frequency of vibration. Of course, this experiment, this argument, has a considerable number of assumptions. The two balls are suspended by what are called equally fine threads and equally long. They are pulled back from the perpendicular and "let go in the same instant." It must then be noticed that "they keep step." In each of these matters the observer must decide where to station himself, and how to check on himself to be sure that the premises of this argument have been satisfied. There is the clearly implied invitation to the audience to carry out the same operations themselves, to make one-to-one substitution for each of the essential elements, and see if their results are not identical.

Let us turn to another of Galileo's classic experiments, this time the inclined plane experiment. From the same source book we find, in the course of a discussion on the subject of accelerated motion, the following experiment. A piece of wooden molding about 12 cubits long, half a cubit wide and 3 finger-breadths thick was chosen; on its edge was cut a channel a little more than one finger in breadth. Having made this groove very straight, smooth and polished and having lined it with parchment also as smooth and polished as possible, "we rolled along it a hard, smooth, and very round bronze ball. Having placed this board in a sloping position, by lifting one end from one or two cubits above the other, we rolled the ball . . . along the channel, noting, in a manner presently to be described, the time required to make the descent. We repeated this experiment more than once in order to measure the time with an accuracy such that the deviation between two observations never exceeded 1/10 of a pulse beat. Having performed this operation and having assured ourselves of its reliability, we now rolled the ball only one quarter the length of the channel: and having measured the time of descent, we found it precisely one half of the former. Next we tried distances, comparing the time for the whole length with that for the half, with two thirds, with three fourths,

. . . In such experiments, repeated a full 100 times, we always found that the spaces traversed were to each other as the squares of the time, and this was true for all inclinations of the plane, i.e. of the channel along which we rolled the ball." [3]

Because we are told that the deviation between two observations was never more than 1/10 of a pulse beat, we might assume that pulse beats were used to measure the time. How about the excitement that a successful experiment might cause, and a possible resultant increase in the speed of the pulse beat? Galileo has met our question by a beautiful and impressive device. (Perhaps it should be called the laundry-tub technique.) This was a stage in the history of technology when balance scales were available, but accurate chronometers were not. Turning again to Galileo's text we read: "for the measurement of time, we employed a large vessel of water placed in an elevated position; to the bottom of this vessel was soldered a pipe of small diameter giving a thin jet of water, which we collected in a small glass during the time of each descent, whether for the whole length of the channel or for part of its length; the water thus collected was weighed, after each descent, on a very accurate balance; the differences of these weights gave us the differences and ratios of the times, and this with such accuracy that although the operation was repeated many, many times, there was no appreciable discrepancy in the results." [4]

This inclined plane experiment was an argument that established in his terms "that the spaces traversed were to each other as the squares of the time." Distance was proportional to the time squared, provided that the experiment was satisfactory; if the conclusion followed logically from the assumptions and the rules of procedure, and if anyone with similar equipment could repeat the operations and observe the same results. In Galileo's *Dialogue* immediately after this inclined plane experiment he has Simplicio say, "I would like to have been present at these experiments; but feeling confidence in the care in which you performed them, and in the fidelity in which you relate them, I am satisfied and accept them as true and valid." [5] Whether this use of the word "valid" in this translation should really have the meaning that I want to give it I must leave to the linguists. Before leaving Galileo, let me generalize a bit on what an experiment involves.

In these experiments we can isolate the following four aspects. *Number one*: a clear statement of *purpose*. Galileo leaves us in no doubt about the conclusion for which he is arguing. *Number two*: a body of *theory*, the assumptions. I hope that the short excerpt reminded you of the rich body of theory with which Galileo surrounds each of these passages. The idea of a vacuum, of the resistance of a medium, of the meaning of speed and ways of measuring it, of momentum, of specific gravity, and of gravity, are just a few of his concepts. The balls of lead and of cork must have their weights established before the experiment starts and this takes some external operations and attendant theory. The inclined plane must have marks for one-third, two-thirds, one-quarter, one-half, three-quarters, etc. This takes at least a moderate amount of geometry. Of course, the laundry tub technique for measuring time takes not only a laundry tub and water, but also quite a bit of theory to satisfy us that comparing the weights of the water that escapes for each passage will be a satisfactory indication of the lapse of times. *Number three*: Some *apparatus* whose operations will be pertinent to the question. *Number four*: A set of *operations* that provide us with data.

Are these four elements a complete list of the essentials in an experiment? Where, after all, is the expected term *observation*? Observations are required for *each* of them. A statement of purpose must be seen or heard to be communicated or touched, as in Braille. The theory must again be expressed in symbols or words that can be observed. The theory must be developed enough to include some notion of tolerable error, of course. We must know from our theory when the operations have come close enough to expectations to satisfy the purpose—to let us assert the hypothesis—and when they fail. In contemporary terms, we want to know when to reject the null hypothesis. The apparatus, of course, must be observed to be located; and the operations, clearly enough, must be observed to be recorded and then utilized.

There is no claim here that just these four classifications are sacred in the characterization of experiments. One could easily break them into more than four or combine them into less. I discover that somewhat to my own surprise, I have characterized Galileo's experiments in just about the way that his chief opponent, Aristotle, analyzed the notion of a complete explanation. In Aristotle's so-called "four causes," or modes of explanation, there were four categories: formal, material, efficient, and final. Final cause is purpose; formal cause is theory, definitions; material cause, of course, is the apparatus; and lastly the efficient cause, in my version, is the operation or operations. Although these terms are barely enough to characterize an experiment, let's assume that we are satisfied temporarily and see whether mathematics contains anything to meet even this thin notion of an experiment.

Euclid's Experiments

For my purpose in this next section I must ask you to consider only what a mathematical proof *is* and ignore for the moment the question of what a proof *means*. To look at it, a proof seems to be a series of statements or symbolic strings. The first one or more such statements are set forth as premises; and the claim is made, or the argument is offered, that *if* the audience will kindly permit these to be asserted, by using them and certain agreed upon rules (including substitution rules) a series of additional statements can be produced. The last of these new statements will be the desired goal or conclusion. Again, please ignore what the mathematician says or thinks about while he carries out these operations; consider just what he actually does. There are two very separate aspects to constructing a mathematical proof. First, the proof must be *discovered*, and then it must be *checked*. It has already been established that only a few areas in mathematics and logic can be explored mechanically: outside of these, the actual discovery of a proof is a matter of trial and error, of following out hunches to see where they lead and what can be found. As Professor Quine puts it in his well-known *Mathematical Logic*, "a proof once discovered can be mechanically checked, but the actual discovery of the proof is a hit and miss matter." [6] Once again, "the mathematician hits upon his proof by unregimented insight and good fortune, but afterwards other mathematicians can check his proof." [7] The elegance and power of a well made proof come later. Quine talks about a proof as a discovery, and the students in our first year logic courses and mathematics courses know what he means. "I understand the problem when someone else puts it on the board," they say, "but whatever made him think of it?" (Of course, there are a variety of so-called helpful hints that one gives students; there are after all just a finite number of operations that are allowable and the symbols, the apparatus upon which these operations may be performed, is, again, finite.) We find proofs even in those areas in which we know that there are no mechanical procedures available. But, the point has to be recognized: proofs must be *found*, must be *discovered*, and discovered just as much as experiments, arguments, are discovered in physics.

As an example of a mathematical experiment or proof let's take Proposition 47 of Book 1 of Euclid's *Elements*. This is the familiar Pythagorean theorem that the square on the hypotenuse equals the sum of the squares on the legs of a right triangle. The purpose of this experiment for Euclid was to prove this assertion for any right triangle. The theory involved includes all the earlier material in Book 1: the 23 definitions, 5 postulates, 5 "common notions," and the 46 propositions already established at that point. Of course, at our point in the history of mathematics we now talk about a number of additional principles that Euclid assumed, tacitly. Occasionally a principle of continuity is needed to establish that circles will intersect, and so on. I've mentioned the purpose and the Theory available to Euclid for experiment 47; now for the apparatus and the operations. Turning to the translation by Sir Thomas Heath, the apparatus is described so: "let ABC be a right-angled triangle having the angle BAC right; let there be described on BC the square BDEC, and on BA, AC, the squares GB, HC; through A let AL be drawn parallel to either BD or CE, and let AD, FC be joined." [8] This diagram is the apparatus to which Euclid directs us. (Later on he asks that we join AE, and BK.) The operations consist of producing a series of additional lines of the argument, each justified by reference to some already admitted part of the theoretical system. Reading just the right hand side of the page we find one line justified by Proposition 14, then Common Notion number two, then

Proposition four, then Proposition 41, again 41, again Common Notion two; and I have skipped a number of definitions employed on the way. Now, we are to make *the operation of substituting* the new statements that Euclid makes for the statements that he refers us to in the previous propositions and common notions. This operation of checking on a one-to-one substitution must be looked at separately.

Substitution is a very critical activity. What can and what cannot be changed in a legitimate substitution instance? To soundly claim that a particular string of symbols is a one-to-one substitution in an already permitted rule, we must point out several matters to the observer. First, the rule has some variables and some fixed relationships between them. We can substitute for the variables only, not the fixed relationships. What may be substituted for these variables? Only new entities that have the same essential classifications. What stands for a line in the rule, can be replaced by a symbol for a line, but presumably not by another kind of entity. What represents a point can be replaced by a representation for another point, but not by another sort of entity. Of course, there are rules of duality, and we know from these that we can in some situations exchange a point for a line and a line for a point, etc. But they are separate experiments.

To stay on the question of substitution rules for another moment, consider the familiar situation where we are looking for examples of substitutions for logical tautologies. Take a law of transitivity such as: for any three statements P, Q, and R, if it is true that if P holds then Q holds, and if Q holds then R holds, it will be true if P holds then R holds. What can, and what can't, we substitute for P, Q, and R in producing examples or instances of this law? Two typical substitution rules are these: First, the same letter *may* be substituted for different letters in a tautology. We can substitute the letter A for P and also for Q, if we choose. However, a second rule states that different letters *may not* be substituted for the same letter in a tautology. We cannot substitute A for P in one occurrence, but something else for P in its second occurrence. Now, how do we know that a particular example offered to us is in fact an example of this transitive law. We know by making observations in which we note each instance of P in the tautology and see if whatever is substituted for one of these occurrences is also substituted for the second occurrence. We then check the variable Q and the variable R and we cry "foul," if either of these two rules is broken. Of course, not merely single letters, but compounds of any complexity may be substituted for the variables in a tautology. The more complex the substitution pattern, the slower and the more cautious is our observation or inspection procedure. However, it takes an observation to check on the operation of a substitution rule, even in the simplest kind of case, the case in which we merely have alphabetic variance.

Returning to Euclid's "experiment" 47, Book 1, we find that the apparatus involved presents us with nine triangles, most of them overlapping the others in at least one section; three squares; and two parallelograms. In my edition there are twelve separate steps or operations concluding with, "the square on the side BC is equal to the squares on the side BA and AC. Therefore, etc. QED." Actually, this version with twelve steps has a number of telescoped steps: it would be longer without these condensations. These operations are each justified by substitutions in several previous propositions, to repeat, numbers 46, 14, 4, and 41, and several "common notions." Of course, there are a number of other ways of establishing the Pythagorean theorem and much simpler ways of establishing special cases, such as the isosceles right triangle, etc. But the question before us is this: Does Proposition 47 satisfy the requirements of an experiment?

Some Possible Objections

Let us consider some possible objections.

1. An experiment must be repeatable. This particular experiment has been repeated for about 2500 years.

2. There is the objection that in this case the particular diagram or the particular symbols in the lines of proof, what I am calling the apparatus, is irrelevant. It doesn't matter if we draw a particular diagram with a somewhat shaky hand, it doesn't matter that the particular square that we erect on one leg does not have perfectly straight lines or obviously right right angles. Of course the same remark holds for Galileo's experiments: no one particular set of data is completely critical. The point of Galileo's argument is clear even if the particular inclined plane that we use in our lab some afternoon has a few bumps in it, or the "very round bronze ball" is not perfectly round or completely bronze.

3. Euclid's proof gives us an absolutely exact result, but Galileo's experiment merely gives an approximation. But Galileo's claim is for an exact *result:* The distance is exactly proportional to the square of the time. Let us not confuse what Galileo called the "discrepancy" in some actual operation with the theoretical assertion that he takes the experiment to support. To walk up to the blackboard after a demonstration of Proposition 47, measure the areas of the three squares involved, and point out that the theorem does not quite hold, would be a case of ignorance of the point at issue. The point is this: we make experiments in the phenomenal world in order to gain confidence for beliefs about both noumenal and phenomenal worlds.

4. Another version of the third. This is the assertion that the theory of error requires that in a particular repetition of Galileo's work we do not get exactly what our theory predicts. However, a repetition of Euclid's work always gives us just and only what our theory predicts. Not so. There are many different ways of reaching Euclid's last line: a student who just memorizes his proof is no more persuasive to himself than one who memorizes

Galileo's data. Also, we do not get exactly Galileo's data when we repeat his work; but we do get exactly his conclusion.

This objection amounts to saying that the lucky mathematician can work happily without any fear of making an error. Now, this does seem to be an exaggeration. Perhaps there is something rather different and special about the kinds of errors the mathematicians make. However, the history of mathematics is the history of human behavior. A mathematics examination is a human predicament, and humans are capable of beautiful errors. Some mistakes are more interesting and exciting than others, of course. One of the most facinating of all the errors in the history of science is certainly the remarkable work of Gerolamo Saccheri and the conclusions that he himself drew about his own work. Saccheri tried to defend Euclid's fifth postulate and his system by showing the contradictions that would result from a non-Euclidian foundation. This man, apparently the first one in 2000 years to systematically consider other alternatives beyond Euclid's and to work out with great force and precision many consequences of these non-Euclidian hypotheses, died without realizing what he had actually done. A century before the founders of non-Euclidian geometry this man developed it (non-Euclidian geometry), in order to prove that it could not be developed. Occasionally in the history of science we make clear the purpose and the theory, but are unable to manage the apparatus and the operation that it would take to carry out an experiment. Occasionally, as Saccheri has shown, one can have the apparatus and even carry out many of the operations but never imagine the theory that would adequately support the experiment. Returning to our list of objections, how do we evaluate the position that says that either there are no errors to be made in a mathematical proof, or that the errors are of a totally different nature? In any area, there are two kinds of errors that can be committed: The Type One error consists in denying a statement that is actually true and the Type Two error in asserting a statement that is false. Mathematics has obviously a rich potential for both of these errors. One is tempted to say that in mathematics there is no degree of either one of these errors; one either commits or doesn't commit a Type One or a Type Two. However, I don't think this holds up. One can make general assertions that actually hold only for a special case, and be partly in error. My main rejoinder to this objection consists in emphasizing the fact that in experimental science one must separate the theory from the degree to which the operations with some particular apparatus confirm the theory. Math has this separation. How confident should one be before publishing a proof?

Objection No. 5 consists in pointing out that it takes but a single glance to see whether X and Y have been substituted for P and Q in a small string of symbols. However, the observations required are much more complicated in any physical experiment. Recall Galileo's experiment in which a heavy and a light ball are suspended as pendula. After a certain number of swings, it took only a single glance to let him note that they were still both "in step." The glance that finds that the bronze ball and the cork still march together need be no more piercing than the glance that compares X with P and Y with Q.

Objection No. 6. My examples of a mathematical experiment are restricted to geometry. In algebra there is no "apparatus" to observe. Can we admit that there are algebraic *operations,* but hold that they are not carried out on anything at all? (Perhaps this should be named the null operation.) What, in fact, is *done* when an algebraic operation is "carried out"? The string of symbols is manipulated according to whatever directions the operation describes. We might rearrange the members of the string, make substitution for some or for all members of the string, add or drop members of the string, etc. The operations are carried *out on the symbols.* I hope that the reader is already convinced that it is not only possible but sometimes necessary that one observe an operation carried out on or with symbols. It's entirely too dangerous to leave these symbols unwatched.

This paper has not committed itself on the question of the separation of logic and mathematics. Presumably the distinction can be established, at least for some parts of mathematics. For the limited purposes here, we might define mathematics ostensively as the material acceptable in certain journals.

Suppose these views were accepted, what would change, if anything? We would have *one* theory of knowledge for everything knowable. Instead of an epistemological separation between deductive and inductive sciences, it would be seen that experimental patterns hold everywhere. Theory and data are correlative, but *knowledge* of either requires experiment.

Can the distinction between deduction and induction be maintained? My suggestion is that it can be, but not on the basis that induction requires hard experimental work, while deductions just magically arrive here and there and need only a casual glance to convince us of their eternal truth. Experimental work is essential in both patterns: to assert the conclusion in either one requires the confidence that comes only from a carefully carried out experiment. However, the particular apparatus differs. For deductions, the apparatus itself consists of symbols. Substitution, replacement of symbols by other symbols, is the basic operation in formal science.[9]

Notes

[1] Translation by Henry Crew and Alfonso de Salvio, reprinted in part in Russell Kahl, editor; *Studies in Explanation,* Prentice Hall, Englewood Cliffs, New Jersey, 1963.

[2] *Ibid.,* pp. 43–44.

[3] *Ibid.,* pp. 57–58.

[4] *Ibid.,* p. 58.

[5] *Ibid.,* p. 58.

[6] Willard Van Orman Quine, *Mathematical Logic,* Revised Edition, Harvard University Press, Cambridge, 1951, p. 6.

[7] *Ibid.,* p. 87.

[8] *The Thirteen Books of Euclid's Elements,* trans. by Sir Thomas L. Heath, Second Edition, Dover Pubs., New York, 1956, volume 1, p. 349.

[9] The position of substitution is not diminished by pointing out that some topics, e.g. the propositional calculus, can be developed without a substitution rule. Where a *modus ponens* rule is used, the explanation and defense of this in the metalanguage does require a substitution rule.

BLACK · Introduction

Max Black is presently Susan Linn Sage Professor of Philosophy and a Senior Member of the Program on Science, Technology and Society at Cornell University. He received his B.A. with honors in mathematics from Cambridge University, and his Ph.D. in mathematical logic from the University of London. He is the author of many books, including *The Nature of Mathematics, Critical Thinking, Models and Metaphors, A Companion to Wittgenstein's Tactatus,* and *The Labyrinth of Language,* as well as of numerous journal articles. He was an editor of the *Journal of Symbolic Logic,* and is presently editor of *The Philosophical Review.* He is past president of the American Philosophical Association and a Fellow of the American Academy of Arts and Sciences. The following essay originally appeared in *The Review of Metaphysics,* vol. xxiv, no. 4(1971), pp. 614–636, under the title "The Elusiveness of Sets."

BLACK · Selection

> If we remove the veil and look underneath, if, laying aside the expressions, we set ourselves attentively to consider the things themselves which are supposed to be expressed or marked thereby, we shall discover much emptiness, darkness and confusion; nay, if I mistake not, direct impossibilities and contradictions.
>
> (Berkeley, *The Analyst,* sec. 8)

Whether there can be science of the conclusion where there is not science of the principles? And whether a man can have science of the principles without understanding them? And, therefore, whether the

mathematicians of the present age act like men of science, in taking so much more pains to apply their principles than to understand them?
(Berkeley, *op. cit.,* sec. 50, query 36)

The most important and most basic term to be found in modern mathematics and logic is that of *set* or class. . . . The modern mathematical theory of sets is one of the most remarkable creations of the human mind. Because of the unusual boldness of some of the ideas found in its study, and because of some of the singular methods of proof to which it has given rise, the theory of sets is indescribably fascinating. But above this, the theory has assumed tremendous importance for almost the whole of mathematics. It has enormously enriched, clarified, extended and generalized many domains of mathematics, and its influence on the study of the foundations of mathematics has been profound.
(H. Eves and C. V. Newsom, *An Introduction to the Foundations and Fundamental Concepts of Mathematics.* New York, 1958, p. 226)

To bring clearly before the mind what is meant by *class,* and to distinguish this notion from all the notions to which it is allied, is one of the most difficult and important problems of mathematical philosophy.
(Russell, *Principles of Mathematics,* p. 66)

Nowadays, even schoolchildren babble about "null sets" and "singletons" and "one-one correspondences," as if they knew what they were talking about. But if they understand even less than their teachers, which seems likely, they must be using the technical jargon with only an illusion of understanding. Beginners are taught that a set having three members is a single thing, wholly constituted by its members but distinct from them. After this, the theological doctrine of the Trinity as "three in one" should be child's play.

Bourbaki once said, "As every one knows, all mathematical theories can be considered as extensions of the general theory of sets." Paul Cohen said, "By analyzing mathematical arguments, logicians became convinced that the notion of 'set' is the most fundamental concept of mathematics." [1] One might therefore expect mathematicians and logicians to possess a firm concept of 'set.' But then they owe laymen and beginners—and philosophers, too—full explanation of a concept so fundamental and so important.

Can the notion of 'set' be too basic to permit elucidation and too familiar to need it? So some pundits claim. A report on the teaching of mathematics, commissioned by the Office for European Economic Coordination, urges that "From the beginning . . . the teacher should see to it that students acquire, by their own effort, an understanding of the concept of 'set,' building largely upon *examples* that they have encountered in their social life, their experiences at school and in the world about them." And it adds, "In this manner, students will be taken into the confidence of the teacher." [2] It is reassuring to know that the meaning of 'set' is not to be kept confidential.

We can learn what familiar examples to expect from the opening of Professor Paul R. Halmos' book, *Naive Set Theory* (New York, 1960): "A pack of wolves, a bunch of grapes, or a flock of pigeons are all examples of sets of things" (p. 1). It ought then to make sense, at least sometimes, to speak of being pursued by a set, or eating a set, or putting a set to flight. But perhaps such ways of speaking reflect what Halmos calls "erroneous understanding." For, on the same page, he says that the "point of view adopted assumes that the reader has the ordinary, human, intuitive (and *frequently erroneous*) understanding of what sets are; the purpose of the exposition is to delineate some of the many things that one can correctly do with them."

How erroneous is the "intuitive" notion? Frege once took Ernst Schroeder to task, with characteristic vehemence, for regarding sets as "collections of individuals." Frege said, "I regard as futile the attempt to make it [a set] rest, not on the concept, but on single things. . . . The extension of a concept does not consist of objects falling under the concept, in the way, e.g., that a wood consists of trees." [3] Frege thought that logic and mathematics have nothing to do with packs and bunches and flocks. And many other experts, from Hermann Weyl to Adolf Fraenkel and Willard Quine, agree in insisting that sets are "abstract objects" (about which more later). But no abstract object ever ran across the steppes, or hung from a bush; so perhaps the "intuitive" notion is radically wrong? As wrong, perhaps, as the idea that numbers are made of chalk and ink—which some advanced thinkers are still ready to maintain.

The confusions and evasions that disfigure contemporary text-books on set theory can be attributed to two main reasons. One is a persistent and unresolved dispute between theorists who profess a highbrow view of sets as "abstract entities" and others who endorse some refinement of the low-brow view of sets as collections (aggregates, groups, multitudes). It would be optimistic to expect any definitive resolution of this conceptual hassle. A more tractable source of expository muddle is an outmoded and inadequate conception of the purposes of definition.

Writers on set theory often seem to assume that the only admissible form of definition is *per genus et differentiam.* Where that is inapplicable, as must often be the case, they seem to think an author is excused from explaining how he uses his basic ("undefined") terms. Now the only available synonyms for 'set'—and approximate ones at that—may well be such equally problematic words as 'class,' or 'group,' or 'collection.' But this by no means exempts a writer from explaining how he uses 'set,' especially if he intends to deviate from lay senses of the word.

A word is, among other things, an instrument for expression, communication, and reference, and synonymous words can be viewed as verbal instruments having the same specific uses. But the use of a tool is not usually explained by producing an equivalent tool: nobody would be foolish enough to insist on having the use of a hammer explained by some hammer-substitute; then why not be as liberal in our demands on explanations of meaning? Perhaps 'set' cannot be defined, upon some restrictive interpretation of definition; but its employment can surely be elucidated. The point is not to "define" the word, but to delineate its functions—and that, too, deserves to be called "definition."

Professional mathematicians are often content to treat 'set' as primitive or "indefinable," in the sense in which 'point,' 'between,' and other primitive terms of an axiomatic geometry are indefinable. Well, we do have elegant axiomatic set theories—but they are used, "applied," outside those theories. Otherwise, 'set,' for all that mathematicians cared, could mean any objects we pleased—footballs or walking sticks—that satisfied the set-theoretical axioms. (And no axiomatic theory can preclude such deviant interpretations.) But when we have occasion to talk about "the set of integers," we intend to mean something definite. A set must be something better than "the ghost of a departed quantity" (to borrow Berkeley's splendid phrase). So back to the rigors of "honest toil."

If we are to be satisfied with something other than a formal definition of 'set,' then what are we to demand? What should be our criterion for a sufficiently clear concept? I can conceive of no general answer to this very general demand. But the notion of '(natural) number' provides a satisfactory exemplar: Any reasonable man should be content, for a start, to understand 'set' as well as he understands 'number.'

To have mastered the primary uses of 'number' means *inter alia:* to know when and how to count and how to get the answer right; and to know how to calculate (to do sums) and how to get the answers right. In short: to use numbers, according to accepted rules and procedures, outside as well as inside arithmetic. The correspondingly modest requirements upon the concept of set are: to know when and how to exhibit specific sets; and to know how to calculate with sets—how to pass from premises about them to warranted conclusions. The first demand might be called *quasi-ostentation;* it is flunked by almost all expositors, however distinguished.

Cantor's Explanation

Cantor's famous formula (1895) does sound like a recipe for quasi-ostentation: "By a 'set' we understand any assembly (*Zusammenfassung*) into a whole M of definite and well-distinguished objects m of our perception (*Anschauung*) or thought." [4] Elsewhere, Cantor remarks: "Every set of well-distinguished things can be regarded *as a unified thing in itself* (*für sich*), in which those things are components or constitutive elements." [5]

A hundred text-book writers have thought Cantor's formula sufficiently illuminating to be echoed, sometime with a side-remark about its not counting as a genuine definition. The following version is characteristic: "As a description of the idea [of a set] it is enough to say: A set is a collection of well-defined objects thought of as a whole." [6] Is it enough?

Let us strip Cantor's formula of its inessentials. The reference to "objects of our *perception* or *thought*" is otiose. (To translate *"Anschauung"* as "intuition" rather than as "perception" will not help.) Suppose some "objects" really exist; then that suffices to make them eligible elements of a set. Whether we perceive them, think about them, name them, or describe them, has nothing to do with their capacity to generate authentic sets. Sets are not thought or designated into existence. (Cantor was a "realist," with reservations: he thought that mathematical entities have "transeunt" reality, although they can be introduced into mathematical inquiry only through adequate symbolization.) So we can simply expunge the references to perception and thought in Cantor's formula.

The reference to "definite" and "well-distinguished" objects does have a serious point, however: Cantor wants to recognize as elements of a set only sharply demarcated objects (numbers, but not clouds), subject to sharp criteria of identity and difference (men but not electrons). Allen's epithet "well-defined," in the passage quoted above, sufficiently captures Cantor's intentions.

So Cantor's formula, stripped to essentials, runs quite simply: "A set is an assembly into a whole of (well-defined) objects." Here, the phrase, "assembly into a whole," certainly suggests that something is *to be done* to the elements, in order for the "whole" or "the unified thing," which *is* the set to result. But *what* is to be done, if not merely thinking about, the set? And what difference can thought make to distinct objects?

What kind of unification is in point? Tell a child to take three pennies and make a "whole" out of them; what is he supposed to do, and how shall we tell him what to do?

The supposed assembly or collection prerequisite to the emergence of a set cannot involve any physical manipulation, if only because the elements may be abstract entities, such as numbers. A typical dictionary definition of 'collection' is a "group of objects or an amount of material accumulated in one location, especially for some purpose or as a result of some process" (*Random House Dictionary*). Consider, then, the set composed of the Hapsburgs and the resurrection that would be needed to bring *them* into "one location."

Quine (who is far from being a Laputan) says: "Sets are classes. The notion of class is so fundamental to thought that we cannot hope to define it in more fundamental terms. We can say that a class is any aggregate, any collection, any combination of objects of any sort; if this helps, well and good. But even this will be less help than hindrance when we keep clearly in mind that the aggregating or collecting or combining here is to connote no actual displacement of the objects . . ." [7]

That is to say, a set may be regarded, "if this helps," as a collection of things that are not collected—or an aggregate of things that are not aggregated, a combination of things that are not combined. I do not think this "helps." This way of talking is no better, and no more intelligible, than defining a Pickwickian omelet as what results from breaking and cooking eggs, even though no eggs are broken or cooked. In fact, sets are in a worse case, since it would be a conceptual blunder to think that the physical proximity of objects had *anything at all* to do with their constituting a set.

If talk about "assembling" or "unifying" is irredeemably figurative, what literal sense, if any, lurks behind it?

Some philosophers,—Husserl, for one—have thought that the problematic assembly and unification of objects into a set could be accomplished by some peculiar act of the mind. And text-book writers, for want of anything better, sometimes yield to the same seduction. "A set is the *mental construct* obtained by regarding several discrete things as constituting a single whole. Forming a set is thus a *mental* act: the human mind arbitrarily brings together certain things and regards the collection itself as a new kind of thing. This new thing is an artificial entity, in the sense that the unity lies entirely in the concept and not in the things themselves." [8]

So "the human mind" can annihilate space and time in an extraordinarily productive way. Apparently, it succeeds in "arbitrarily bringing together" entities, such as 7, 11, and 13, say, which remain *three* distinct and separate numbers, in order to produce *one* thing of "a new kind." How is this feat accomplished? We are left wholly in the dark as to what this new thing might be. We could, of course, add the numbers together and get 31, or multiply them and get 1001, but that is certainly not intended. But so long as no sense has been supplied for "assembling" or "collecting," the expression "assembling *in thought*" is a *flatus vocis:* what makes no sense in reality, the mysterious conversion of several things into one, makes no sense "in thought" either.

Given all this, there seems hardly any point in adding that appeal to the generative powers of "thought" would make mathematics a branch of psychology, and would limit the stock of sets to the finite number of "mental acts" exercised in "creating" them. Or can the mind go in for wholesale creation?

Russell, after considering at length the question "Is a class which has many terms to be regarded as itself one or many?" has nothing better to say by way of an answer than: [there is] "an ultimate distinction between a class as many and a class as one, [so that we must] hold that the many are only many, and are not also one. The class as one may be identified with the whole composed of the terms of the class, i.e., in the case of men, the class as one will be the human race." [9] When in perplexity, invoke an "ultimate distinction."

Isn't this an example of what Berkeley called the "darkness and confusion of mathematics"?

The truth is that once the elements of a set have been identified, *nothing* need or can be done to produce the corresponding set. (The "numerical conjunction" of which Russell once spoke, the bringing together of objects by means of "and" [10] is a mere chimera.) A quadrilateral is sometimes regarded as a set of four lines; but once I have drawn four mutually intersecting lines in a plane, no more needs to be done in order to produce the quadrilateral. It would be ridiculous to require in addition some—no doubt immaterial—curve encircling the quadrilateral.

Yet the picture behind all this talk about assembling, gathering, and collecting is indeed that of tying things in a bundle—that famous "unity" composed of its "elements." But it is essential for this picture's effect that the string be of that kind—useful for doing in thought what cannot be done in reality—that is invisible and intangible. The "line" around the elements must be as "imaginary" as the equator.

In the thought of those who follow Cantor, the notion of "assembly" or "collection" is treated in a figurative way that voids it of application by stretching it to absurdity. (One is reminded of the celebrated Euclidean "definition" of a point as something having position but no magnitude—a body so small that it doesn't exist at all.) If so, Cantor's formula is useless: it cannot begin to help somebody to understand what a set is. The learner asks for clarification, but receives obfuscation.

If the notion of "collection," as unified assembly without displacement is mysterious, how much more so are the notions of "collecting" a single thing to produce *another* entity (the corresponding unit-set) and collecting nothing at all to produce a unique object (the "null set"). From the standpoint of Cantor's formula, this is mystification on stilts.

Sets as Abstract Entities

The apparently insuperable difficulties in regarding sets as collections, in Cantor's style, are responsible for such remarks as the following:

> A class is an abstract entity, a universal, even if it happens to be a class of concrete things. . . . Indeed, there is no call even to distinguish attributes from classes, unless it is on this one technical score: classes are identified when they coincide in point of members, whereas it may be held that attributes sometimes differ though they are attributes of just the same things.
>
> (W. V. Quine, in P. Schilpp, ed., *The Philosophy of Alfred North Whitehead*, Evanston, Northwestern University, 1941, p. 147.)

> Sets, as ordinarily understood, are what philosophers call universals.
> (A. A. Fraenkel and Y. Bar-Hillel, *Foundations of Set Theory*, Amsterdam, 1958, p. 333.)

> Classes (or sets) may be viewed as properties of a special kind. . . . On this view there corresponds to each property f a special property called the *extension of f* or *the class defined by f*.
> (R. Feys and F. B. Fitch, *Dictionary of Symbols of Mathematical Logic*, Amsterdam, 1969, p. 13.)

Such uses of "universal" and "property" are careless at best: a man is not properly called an *instance* or a *case of* the class of men; and it would be ungrammatical to say that a man had the property of *the class of men.* But the grammatical solecisms can perhaps be patched up: behind them, there is an imposing alternative to the "collection" view which merits close examination.

Frege, to take the ablest of the sets-as-abstract-entities theorists, conceives of what would nowadays be called a set as the "extension" of *a concept* (or, as we might say, of a property or attribute). This approach would make a set a feature of something itself abstract, a property. And the way is then open to recognize that properties having no instances, or only single instances, have extensions in the *same sense* in which all properties do. So now "set" must be taken to be short for "set connected with a certain property, of which it is the extension;" and where there is no such property there can be no question of a corresponding set.

The term "extension" suggests something like the reach, or scope, or incidence of the corresponding property. Which now makes sense of that perplexing notion of "assembly"; what, figuratively speaking, "assembles" the members of the extension is just their *having* the associated property. And that is why the "assembly" is of the peculiar sort that in no way changes the things "assembled." To locate the members of a set is the very same thing as giving a necessary and sufficient condition for membership in the set, i.e., formulating a suitable property having the extension in question.

Those who invite us to consider a set as a property "of a special kind" (as in the above quotation from Feys and Fitch, for example) sometimes have the following conception in mind: Consider a case where a number of properties, P_1, P_2, P_3, . . . are "co-extensional," i.e., apply to the very same objects a, b, . . . , k, so that these objects, and no others, all have each of the properties P_1. Then conceive of these same objects as all having some *more abstract property*, say Q, by virtue of which they and they alone also have each of the properties P_1. Q is to be considered as the sophisticated replacement for the crude notion of the "collection" composed of a, b, . . . , k.

The following might serve as an analogy. A number of material bodies might have in common such properties as *acquiring the same acceleration under equal impressed forces, balancing the same bodies in scales,* and so on. Now these observable properties, shared in common by the bodies, are commonly ascribed to the joint possession of a more abstract, theoretical property, that of *having a certain mass.* Similarly, in the more general case, the "more abstract" property, Q (or "the extension" of any of the P_1), is to be construed as a certain property, co-extensive with each of the P_1, but not identical with any of them. On this view, a formula of the form $a \in Q$ is just another way of writing the more familiar formula $Q(a)$; and "is a member of" is just a somewhat unhappy way of saying "has" or, "has the property of."

But then, what is Q? One might be tempted to identify it with the disjunction $P_1 \vee P_2 \vee P_3 \vee \ldots$ of all the P's, except that we are not told yet how to complete the disjunction, i.e., how to eliminate the terminal dots.

A more natural answer would run somewhat like: *the property of being just the "members" in question and no others.* In symbols: $Q(x) \;.=.\; x = a \vee x = b \vee \ldots$ But this conception will satisfy only those willing to accept identity with a given object ($x = a$) as an intelligible "property." (And it will break down, of course, for infinite sets.)

Suppose one of the P's is the property of being a prime number ($Px = x$ has no factors except 1 and x). Then the Q-property would presumably have to be identified by some such phrase as: "the property of being just *those things that are prime numbers.*" But here, the italicized phrase is just an example of ordinary language's primitive way of designating a set. This becomes clear if we try to symbolize the supposed definition of Q as:

$$Qx = x \in \hat{x}(Px)$$

in which the role of 'Q' is that of '$\in \hat{x}(Px)$'. So it looks as if our conception of Q presupposes our prior possession of the concept of a set. In any case, the suggested conception of the Q-property will fail for null-extensions: the form of words, "the property of being just those things and no others" reduces to vacuity when "those things" are *no* things at all.

Sets as Arising from Equivalence Relations between Properties

Frege thought he could by-pass these difficulties by applying a technique for "introducing abstract entities" already familiar in geometry and elsewhere.

Consider the problem of defining a "direction" in geometry (Frege's own example in *Foundations of Arithmetic,* secs. 64–67). The relation of parallelism between lines (say Pxy), assumed to be given, is plainly reflexive, symmetrical, and transitive. Therefore, the set of all straight lines can be split, by reference to P, into a set of mutually exclusive subsets, each composed of mutually parallel lines. With the idea at the back of our minds of eventually identifying "directions" with the subsets previously mentioned, we can begin to explain the notion of "direction" by introducing Dx(= 'x is a direction') by the formula

$$(1) \qquad Dx = Dy \;.\equiv.\; Pxy$$

That is to say, we count lines as having the same direction if and only if they are parallel.

Now if this procedure is respectable, as it certainly is, why should we not similarly hope to "introduce" the "abstract property" Q by the analogous formula

$$(2) \qquad Q(X) = Q(Y) \;.\equiv.\; (u)\,(Xu \equiv Yu)$$

where X and Y now stand for properties, whose coextensionality plays a role like that of parallelism between lines in (1)?

But Frege is clear about the limitations of such a formula as (1), which provides a means of translating identities between "directions" into intelligible assertions of parallelisms. For (1) does not yet tell us what a direction *is,* does not say whether a direction is to be regarded as a privileged line, some property of lines or, perhaps, if we want to be fanciful, even some number or other. We know, from (1), the necessary and sufficient conditions for directions to be identical, but everything else about directions is left completely open.

This lack of definition of a notion such as direction is easily overcome in geometry by taking a given direction as the *class* or *set* of all lines parallel with a selected line, e:

$$D(e) = x\,(Pxe)$$

But this presupposes that the notion of a class or set is already available, and obviously fails for our attempt to introduce Q by an abstractive definition. The analogue there would be to say that the extension of a given property, P, is the *class* or *set* of all properties co-extensional with P. As an explanation of 'extension' this is clearly circular, since 'class,' on the view here under discussion is a mere synonym for 'extension.' [11]

How did Frege cope with this problem in his crowning masterpiece, the great *Grundgesetze?* The crucial context is that in which he tries to pass from assertions of co-extensionality [in modern notation, $(x)\,(P_1x \equiv P_2x)$] to the admission of certain associated *objects,* which we may regard as "extensions." (Frege actually speaks of '*Werthverläuf*' or "courses of values," that is to say, something like a class of pairs of ordered entities corresponding to a given function, the first member of each pair being a value of the independent variable and the second the corresponding value of the function. But this difference between an "extension" and a "course-of-values" is unimportant for the present discussion.)

Frege's defense of this procedure, which is an essential step in his attempt to reduce arithmetic to logic, consists of the following two assertions: (i) the step from the *equivalence* of a certain sort is a "logical principle" (*ein logisches Grundgesetz*)[12], indeed one that mathematicians and logicians have in fact universally accepted in talking about 'classes,' 'sets,' 'multiplicities,' and the like.[13] For instance, the whole of the Leibniz-Boole calculus of classes depends upon it.[14] (ii) This step from properties to certain associated objects, their extensions, is indispensible, in order to define integers in such a way that generalization over integers and so the assertion of arithmetical *laws* is to be possible. (Any attempt to construe extensions as "improper objects" (*uneigentliche Gegenstände*) and their purported names as mere "apparent names" (*Scheineigennamen*), Frege rejects for this reason.)[15]

What this amounts to is that we need extensions, in order to construe arithmetic as Frege wishes (as a series of analytical principles); and that, in any case, their existence is, in fact, generally accepted. So far as I know, this is the only place where Frege ever rests his case upon the authority of established practice. Both of these reasons, as Frege later came to see, are swept aside by Russell's discovery that unrestricted assignment of extensions to predicates results in contradictions.

Of course, Frege had no illusions about what he was doing in introducing extensions as objects. With his unremitting hostility to so-called "creative definitions," he was (unlike Weyl) armored against supposing that the mere introduction of new signs can somehow conjure entities into existence. He is at pains to point this out in connection with his own acceptance of extensions,[16] and, indeed, makes it perfectly clear that extensions in his system are *not* defined, since they are introduced only by means of identities in which extension signs stand on both sides. (He tries to remedy the lack of reference thus produced by stipulating the truth-values of certain combinations into which extensions enter, but that is a matter we need not pursue here.)

If we think of the situation quite naively, allowing ourselves to use the term 'set' as if we really understood it, we might say that the abstraction procedure that purports to lead us from equivalent properties to certain abstract entities, leaves much that we need to know still unspecified. If P_1 and P_2 are held to have *something* in common by virtue of the fact that they apply to exactly the same objects, we might think of that common feature, indifferently, as the set of all properties equivalent to P_1 (or, of course, to P_2, which would come to the same thing), as the disjunction of P_1 and all properties equivalent to P_1 as some *second-order* property of all the properties equivalent to P_1, and so on indefinitely. Without a supplementary determination of the meaning of 'set' or some synonym thereof, we are unable to choose between these alternatives and are left in the dark as to what extensions are really intended to be. Whatever emerges will be so unlike what laymen, and mathematicians too, think of as "sets" that it might be advisable to have some different technical label to mark the difference—say "the theory of extensions" or even "the theory of functional spreads."

The formula, "a set is something shared in common by coextensive properties," is somewhat better than "a set is anything you like." But this formula still leaves the concept of a set much darker than that of a number. If *that* is what a set really is, it would be hard to see how beginners could learn what it is "from their own experience"—or, indeed learn it at all. Getting the idea of "coextensive properties" across would be hard enough. But suppose the teacher managed to produce two such properties, P_1 and P_2; consider the difficulty of communicating the notion of "something that P_1 and P_2 have in common" to which the teacher (no wiser surely than Frege) *can attach no firm sense.* Would it be a *stupid* child who asked, "What do P_1 and P_2 have in common?" (Is it fair to feed caviar to the young—and to pretend that it is breakfast food?)

Sets in Ordinary Language

We are not forced to choose between thinking of sets as mysterious aggregations of distinct things into "unified" wholes and thinking of them as unknown things shared in common by certain coextensive properties. For, in ordinary language, we do manage to identify sets and reason about them, without the inconveniences of a superfluous mythology. If we can become sufficiently clear about how we *talk* about sets, we shall have all the clarity about "the concept of set" that we need for a start.

It is easy to forget while wrestling with the mystifying explanations that master logicians have offered, that the word 'set' *does* belong to ordinary language. If the word is used so naturally in such expressions as 'my set of chessmen' or 'that set of books,' the task of exhibiting the underlying rules of use should not be insuperably difficult. Let us then consider how we manage to talk about *several things at once.*

Plural reference. The most obvious ways of referring to a single thing are by using a name or a definite description: 'Aristotle' or 'the president of the United States.' Equally familiar, although strangely overlooked by logicians and philosophers, are devices for referring to several things *together:* 'Berkeley and Hume' or 'the brothers of Napoleon.' Here, *lists* of names (usually, but not necessarily, coupled by occurrences of 'and') and what might be called "plural descriptions" (phrases of the form 'the-so-and-so's' in certain uses) play something like the same role that names and singular descriptions do. Just as 'Nixon' identifies *one* man for attention in the context of some statement, the list 'Johnson and Kennedy' identifies two men at once, in a context in which something is considered that involves both of them at once. And just as 'the President of the United States' succeeds in identifying one man by description, so the phrase 'the American presidents since Lincoln' succeeds in identifying several, in a way that allows something to be said that involves all of them at once.

The notion of 'reference' here invoked is only a slight sophistication of an ordinary language concept. In ordinary life, the question, "To whom are you referring?" is usually in place only when there is some *prima facie* possibility of doubt as to the person in question. (The question, "To whom were you referring when you said 'Aristotle' just now?" would be met with a blank stare.) But suppressing this pragmatic condition gives us just what we need: *E* is about *P* if and only if the proposition in which *E* is being used attributes something to *P*. We can here ignore the difficulties in making this kind of formula sufficiently accurate to cope with some obvious objections. What we are talking about is sufficiently perspicuous at the level of intelligibility that we are aiming at.

The notion of "plural" or simultaneous reference to several things at once is really not at all mysterious. Just as I can point to a single thing, I can point to two things at once—using two hands, if necessary; pointing to two things at once need be no more perplexing than touching two things at once. Of course, it would be a mistake to think that the rules for "multiple pointing" follow automatically from the rules for pointing proper; but the requisite conventions are almost too obvious to need specification. The rules for "plural reference" are no harder to elaborate.

A rough test of the occurrence of a singular referring expression is provided by considerations of identity. If I say, referentially, "The president of the U.S. is a Republican" and if the president is Nixon, then my assertion is true if and only if Nixon is a Republican. (If I *think* that Nixon is a Republican, I must consider myself committed to the second statement when I make the first.) A similar test involving identity will certify plural expressions as occurring with genuinely multiple reference. If I say, referentially, "All Napoleon's brothers were Corsicans" and if Napoleon's brothers were Tom, Dick, and Harry, then my original statement is true if and only if Tom, Dick, and Harry were Corsicans. (If I think the multiple identity is true, I must consider myself committed to the second statement if I make the first one.)

The main point of using "plural referring expressions," such as lists and plural descriptions, is obvious enough. It is typically to permit concise statements about several things at once. A further important benefit in using plural descriptive phrases is that of being able to talk about several things at once, in cases where the precise identity of the things in question is unknown or irrelevant. Consider, for instance, the announcement, "All arriving passengers proceed to customs." The announcer, in this case, does not know and need not care *who* the arriving passengers are, but his announcement is intended to refer to all of them. This second use, but not the first, parallels an obvious advantage in using singular definite descriptions.

In an elaborated description of the rules governing the uses of plural referring expressions, we might find it convenient to speak about their *identity conditions, membership conditions,* and *retrieval conditions.* The first kind of condition determines when two plural referring expressions count as referring to the same things and are then mutually substitutable *salva veritate;* the second kind of condition (a special case of the first) determines *which* things are referred to by a given plural referring expression. Roughly speaking, both types of condition reduce identities involving plural referring expressions (names or definite descriptions). The detailed specification of conditions of these types (which are within the competence of all users of ordinary language) would be tedious, but not difficult. More interesting are the routes by which we can—at least in principle—eliminate plural referring expressions from certain basic statement-contexts in which they occur. In trivial cases ("perfect distribution"), assertions about plural subjects, identified by lists or plural descriptions, are immediately convertible into conjunctions of separate statements about each of them; in the typical and more interesting cases ("imperfect distribution"), the original predicates also need transformation, in ways not readily reducible to simple formulas. (Cf. "Tom and Harry were young" with "Most of the Beatles were ready to part company.")

Suppose that π is some plural referring expression, occurring in the context $f(\pi)$ and referring there simultaneously to the objects $a_1, a_2, \ldots, a_n$. Then, in order to understand $f(\pi)$, I must know the *retrieval conditions* for referential descent, for the use of π in this context. That is to say, I must know how to pass from the assertion f to the assertions g, that are explicitly about the *a*'s. (To take a fairly awkward case: In order to understand "His brothers are two in numbers" I must be able to pass from that assertion to "Tom (say) is one of his brothers, and Dick is another, and that is all of them.") Anybody who has sufficient mastery of the retrieval conditions for certain plural referring expressions, and of the identity conditions for those expressions already has *some* mastery of the concept of a set. But so far there has been no need to introduce "the general notion of a set." Let us consider how that word is used.

One primitive use of the word 'set' is as a stand-in for plural referring expressions of the kinds discussed above. If I say "A certain set of men are running for office" and am asked to be more specific, then I might say, "To wit, Tom, Dick, and Harry"—or, in the absence of knowledge of their names, I might abide by my original assertion. One might therefore regard the *word* 'set,' in its most basic use, as an indefinite surrogate for lists and plural descriptions. To know how to use the word 'set' correctly at this level is just to know the linguistic connections between such uses of 'set' and the uses of more definite multiple-referring devises.

Ostensibly Singular Plural Referring Expressions

It is a peculiarity of English and many other languages to admit collective expressions such as 'the Hungarian Quartet' and 'the Cabinet' which are allowed to behave, at least part of the time, as if they were singular names or descriptions. (And this fact is perhaps one of the main sources of the inclination to regard a set as some peculiar entity constituted by, but not identical with, its elements.) This ambivalent role of such expressions is betrayed in the absence of firm rules for their grammatical "number"—thus Fowler says that an expression like 'the Cabinet' can be indifferently followed by a verb in the singular or the corresponding plural.

Expressions that look superficially like singular descriptions but really serve to refer to several things at once can fit quite comfortably into our program. The role previously played by "plural identities" (of the form 'the *P*'s are $a_1, a_2, a_3, \ldots, a_n$') is now taken over by statements of the form 'the *G* *is composed of* $a_1, a_2, \ldots, a_n$' (where 'the *G*' is now an ostensibly singular referring expression). What was said earlier about the need for "identity conditions" and "retrieval conditions" still applies. We must, for instance, know when we should identify, say, the Cabinet and the Smith brothers; and how, in principle at least, we should proceed from statements about the Cabinet to statements about Tom, Dick, and Harry, if they are, indeed, all the members of the Cabinet.

It is not at all clear that we need "ostensibly singular referring expressions," or that we would be seriously inconvenienced by their absence. (It is worth recalling that we have regular ways for converting them into overtly plural referring expressions, as when we pass from 'the Cabinet' to 'the members of the Cabinet.' The losses of nuances of meaning that result from such transformations are not irreparable).

Sets of sets in ordinary language. Set talk (the use of plural referring expressions) is especially convenient when we cannot or need not identify the corresponding memberships—as when we are interested only in the number of members and in nothing more. But these and other considerations that give talk about sets its rationale can also lead naturally to forming lists of sets (second-order lists). If a number of committees are to be formed from a certain group of men, and each committee is to have a separate secretary or separate meeting place, we shall be interested only in the number of the sets or their diverse memberships. And then there will naturally emerge lists in which plural referring expressions follow one another ('the finance committee, the membership committee, and the rules committee'). Hence, also, we might find it useful to introduce second-order expressions such as 'the set of committees.' (Or we might raise such a question as "How many different committees could we form from this set of men?") I do not think this interesting and undoubtedly useful extension of primitive set talk offers any serious obstacles for our program (although the step is certainly of crucial importance). We need only be sure that we have at our disposal adequate devices for connecting such "second-order" discourse with the lower level discourse already discussed, by means of "retrieval conditions." In short, we need to know how we could, if necessary, convert the more abstract talk about "sets of sets" into assertions about sets simpliciter (sets composed of persons or other things that are not sets).

The way is now open for mathematicians to introduce sets of sets of sets, and so on. The necessary restriction upon this kind of reiteration in order to prevent inconsistency and paradox do not concern us here. Nor are we at present interested in ticklish questions about "existence axioms" for sets or the "richness" of the requisite ontology.

Of course, any transition from colloquial set talk to the idealised and sophisticated notion of making sense of a "null set" and of a "unit set" (regarded as distinct from its sole member) will cause trouble. From the standpoint of ordinary usage, such sets can hardly be regarded as anything else than convenient fictions (like the zero exponent in X^0) useful for rounding off and simplifying a mathematical set theory. But they represent a significant extension of ordinary use; and nothing but muddle will result from ignoring this point.

A Retrospect, and Some Objections

The program here recommended, for building the idealised set talk of mathematicians upon the rough but serviceable uses in ordinary language of "plural referring expressions" has the merit of being pedagogically feasible. Beginners, who will be already competent to handle plural referring expressions—though, of course, not under that title—can be readily shown the connections between the uses of such expressions, the uses of "ostensibly singular expressions," and the intended use of the more abstract word, 'set.' Collective terms such as 'herd,' or 'team,' or 'orchestra,' etc. have the sense of *a number of things* (animals, persons, etc.) *such that so-and-so.* We need only drop the such-that clause, and the specific information it conveys, to get the abstract notion of a set as *a number of things considered together* (Cf. the French name for a set, *ensemble*) or *several things referred to at once.* This is the nearest we need to approach a formal definition and if it helps to elucidate the technical notion—as I think it does—that will suffice. (We shall, of course, eschew mystifying formulas about "assembly" and "unification.") And difficulties presented by the null set and by unit-sets will have to be faced head-on, in full recognition of the sophisticated conceptual manoeuvre involved. (Whether it is wise to introduce mathematical dodges at the very outset of an introduction to mathematics need not be discussed here.)

Now it may be objected that, at best, what is here being advocated is a theory of sets as collections or aggregates and not—as would be desirable—of sets as certain "abstract entities." To which the following replies may be made: (1) Mathematicians do in fact think of sets as collections and the heuristic utility of the notion of a set depends upon their so thinking. (2) There seems no ready way of connecting sets as "abstract entities" with ordinary language or informal thought. If set-theory is to be identified with the theory of *such* entities, then beginners will, in fact, be taught something else and there is a risk of a fraud being perpetrated upon them. (3) Nobody knows what the supposedly "abstract entities" are intended to be (as I have argued earlier in this essay) so it is not an exaggeration to say that the intended "theory" has no firm or intelligible subject-matter.

Escaping from Mythology

Now the questions that invite a mythology of sets can perhaps be side-stepped.

(1) "Is a set a thing 'in its own right?' " "Well, it is clearly nonsensical to identify 'Tom, Dick, and Harry' with either Tom, or Dick, or Harry or anyone else. And if that is what is meant by saying that a set is a thing in its own right, I can agree."

(2) "But what *kind of a thing* is a set?" "This is a question that deserves another. What sort of answer do you want? I might reply uniformatively that a set—is a set. But that wouldn't satisfy you. But why should there be *any* answer? (Call a set *sui generis* if that helps—but it really shouldn't.)"

(3) "What is it that 'unites' the elements of a set into the set?" Nothing at all. Nothing happens to Tom, Dick, and Harry when you refer to them all at once as 'Tom, Dick, and Harry.' But you *can* refer to them *at the same time,* in the context of a *single* statement."

(4) "But then does the set exist?" "Well, if this is the question whether Tom, Dick, and Harry exist, the answer is, of course, Yes—and similarly for other cases. But what more do you want?"

(5) "But are all the sets waiting to be discovered, like stars before the astronomers see them? Or do they come into existence when one of your 'plural referring expressions' is used? And what about sets that never have been and never will be thus identified? An unseen star's a star for all that—but is an unidentified 'set,' on your conception, *anything* at all?" "You presuppose an obfuscating analogy. Comparing sets with stars is as helpful as comparing mustard with four o'clock. (Constellations would have been another story.) Were the lines of longitude 'waiting to be discovered' before we talked about them? Set talk is a verbal pattern projected on the universe, and set-boundaries are as 'real' or 'imaginary' as territorial boundaries. But statements *in* set talk may be as true as those of cartography. And isn't that enough?"

The foregoing sketch, crude as it may be, perhaps sufficiently illustrates how we might envisage the introduction of sets (or rather, the introduction of set talk) without recourse to mystifying explanations that neither explain nor illuminate. Would it not be enough if we could *talk* in full awareness "about sets?" Do we need to know anything more? Is there anything more to be known?

Notes

[1] *Set Theory and the Continuum Hypothesis*, New York, 1966, p. 50.

[2] *Synopses for Modern School Mathematics*, Paris, O.E.E.C., 1961, emphasis added.

[3] P. Geach and M. Black, eds., *Translations from the Philosophical Writings of Gottlob Frege*, 2nd ed., Oxford, 1961, p. 106.

[4] *Gesammelte Abhandlungen*, Hildesheim, 1962, p. 282, translated.

[5] *Op. cit.*, p. 379, emphasis in original.

[6] R. G. D. Allen, *Basic Mathematics*, London, 1962, p. 88.

[7] *Set Theory and Its Logic*, Cambridge, Mass., 1963.

[8] W. L. Schaaf, *Basic Concepts of Elementary Mathematics*, New York, 1960, p. 11.

[9] *Principles of Mathematics*, p. 76.

[10] *Op. cit.*, pp. 57, 67.

[11] Cf. Charles D. Parsons in M. Black, ed., *Philosophy in America*, London, 1965, pp. 184–185.

[12] *Grundgesetze, II*, p. 147.

[13] *Loc. cit.*, p. 148.

[14] *Op. cit.*, I, p. 14.

[15] *Op. cit.*, II, p. 255.

[16] *Op. cit.*, II, p. 147.

Bibliographies

As indicated in the Preface, relatively little work has been done to date on the history of the philosophy of mathematics, and even less has been done in exploring the relations between pre-twentieth century philosophies of mathematics and present-day work in this area. The following bibliography has been compiled to assist anyone who is interested in examining recent work in the philosophy of mathematics or the history of the philosophy of mathematics. This bibliography is not complete: it includes mainly basic works which themselves contain useful bibliographies on these topics, and items (mostly recent) which are not included in these other bibliographies.

A basic reference work which contains much good material on all aspects of the philosophy of mathematics and its history is *The Encyclopedia of Philosophy*, Paul Edwards, Editor in Chief (Macmillan-Free Press, 1967). This eight-volume work contains separate essays on each of the main figures in the history of Western philosophy, although only a few of these essays treat questions of the philosophy of mathematics directly. The articles on Logic, History of, and Mathematics, Foundations of, contain discussions of selected elements of the history of the philosophy of mathematics. Each article in the *Encyclopedia* also contains a bibliography of standard works on the various topics.

Plato

Brown, M. S., "Plato Disapproves of the Slave-Boy's Answer." *Review of Metaphysics*, vol. XXI, no. 1 (January, 1970).

Cherniss, H., "Plato as Mathematician." *Review of Metaphysics*, vol. IV, no. 3 (March, 1951).

Fleming, N., and Sesonske, A., eds., *Meno: Text & Criticism*. Belmont, Calif.: Wadsworth Publishing Co., 1965.

Gibson, A. B., "Plato's Mathematical Imagination." *Review of Metaphysics*, vol. IX, no. 1 (September, 1955).

Lusserre, F., *The Birth of Mathematics in the Age of Plato*. Cleveland: World Publishing Co., 1966.

Morrow, G. R., "Plato and the Mathematicians." *Philosophical Review*, vol. 79 (July, 1970).

Rose, Lynn, "Plato's Divided Line." *Review of Metaphysics*, vol. XVII, no. 3 (March, 1964).

Taylor, C. C. W., "Plato and the Mathematicians." *Philosophical Quarterly*, vol. 17, no. 68 (July, 1967).

Aristotle

Apostle, H., *Aristotle's Philosophy of Mathematics.* Chicago: University of Chicago Press, 1955.

Brumbaugh, R., "Aristotle as a Mathematician." *Review of Metaphysics*, vol. 8 (March, 1955).

Greenwood, T., "Plato and Aristotle: A Contrast between their Mathematical Outlooks." *The New Scholasticism*, vol. 18 (1944).

Hintikka, J., "Aristotelean Infinity." *Philosophical Review*, vol. LXXV (April, 1966).

Ross, W. D., "Some Thoughts on Aristotle's Logic." *Proceedings of the Aristotelian Society*, vol. XL (1939–40).

Slakey, T. J., "Aristotle on Sense Perception." *Philosophical Review*, vol. 70 (October, 1961).

Descartes

Boyer, C. B., "Cartesian Geometry from Fermat to Lacroix." *Scripta Mathematica*, vol. 13 (September, 1947).

Boyer, C. B., "Review of J. Vuillemin, 'Mathématique et Métaphysique chez Descartes.'" *Isis*, vol. 53 (July, 1962).

Cronin, T. J., "Eternal Truths in the Thought of Descartes." *Journal of the History of Ideas*, vol. 21 (October, 1960).

Doney, W., "The Cartesian Circle." *Journal of the History of Ideas*, vol. 16 (June, 1955).

Frankfurt, H. G., "Memory and the Cartesian Circle." *Philosophical Review*, vol. 71 (October, 1962).

Gewirth, A., "The Cartesian Circle Reconsidered." *Journal of Philosophy*, vol. 67 (October 8, 1970).

Kenny, A., "The Cartesian Circle and the Eternal Truths." *Journal of Philosophy*, vol. 67 (October 8, 1970).

Miller, L. G., "Descartes, Mathematics and God." *Philosophical Review*, vol. 66 (October, 1966).

Salmon, E. G., "The Mathematical Roots of Cartesian Metaphysics." *The New Scholasticism*, vol. 39 (1965).

Sebla, G., *Bibliographia Cartesiana: A Critical Guide to the Descartes Literature, 1800–1960.* The Hague: Martinus Nijhoff, 1960.

Hobbes

Watkins, J. W. N., *Hobbes' System of Ideas*, London: Hutchinson, 1965.

Newton

Koyrè, A., *Newtonian Studies.* Chicago: University of Chicago Press, 1968.

Perl, M. R., "Physics and Metaphysics in Newton, Leibniz and Clarke." *Journal of the History of Ideas*, vol. 30 (October, 1969).

Power, J. E., "Henry More and Isaac Newton on Absolute Space." *Journal of the History of Ideas*, vol. 31 (April, 1970).

Strong, E. W., "Newton's Mathematical Way." *Journal of the History of Ideas*, vol. 12 (January, 1951).

Suchting, W. A., "Berkeley's Criticisms of Newton on Space and Motion." *Isis*, vol. 58 (July, 1967).

Toulmin, S., "Criticism in the History of Science: Newton on Absolute Space, Time, and Motion." *Philosophical Review*, vol. 58 (January-April, 1959).

Whiteside, D. T., "Sources and Strengths of Newton's Early Mathematical Thought." *Texas Quarterly*, vol. 10 (Autumn, 1967).

Locke

Aaron, R., and Walters, P., "Locke and the Intuitionist Theory of Number." *Philosophy*, vol. XL (July 1965).

Armstrong, R. L., "The Cambridge Platonists and Locke on Innate Ideas." *Journal of the History of Ideas*, vol. 30 (April, 1969).

Margolis, J., "Locke and Scientific Realism." *Review of Metaphysics*, vol. 22 (December, 1968).

Perry, D. L., "Locke on Mixed Modes, Relations and Knowledge." *Journal of the History of Philosophy*, vol. V (July, 1967).

Wilson, M. W., "Leibniz and Locke on First Truths." *Journal of the History of Ideas*, vol. 28 (July, 1967).

Leibniz

Ballard, K. E., "Leibniz's Theory of Space and Time." *Journal of the History of Ideas*, vol. 21 (January, 1960).

Heineman, F. H., "Truths of Reason and Truths of Facts." *Philosophical Review*, vol. 57 (September, 1948).

McRae, R., "The Unity of the Sciences: Bacon, Descartes and Leibniz." *Journal of the History of Ideas*, vol. 18 (January, 1957).

Russell, B., *A Critical Exposition of the Philosophy of Leibniz*. Cambridge: Cambridge University Press, 1900.

Wells, R., "Leibniz Today: Work on Leibniz, 1946–56." *Review of Metaphysics*, vol. 10 (December, 1956-March, 1957).

Wilson, M. D., "Leibniz and Locke on First Truths." *Journal of the History of Ideas*, vol. 28 (July, 1967).

Berkeley

Baum, R. J., "The Instrumentalist and Formalist Elements of Berkeley's Philosophy of Mathematics." *Studies in History and Philosophy of Science*, vol. 3 (August, 1972).

Strong, E. W., "Mathematical Reasoning and Its Object," in *University of California Publications in Philosophy*. Berkeley & Los Angeles: University of California Press, 1957.

Turbayne, C. M., and Ware, R., "Bibliography of George Berkeley, 1933–1962." *Journal of Philosophy*, vol. 60 (February 14, 1963).

Hume

Atkinson, R. F., "Hume on Mathematics." *The Philosophical Quarterly*, vol. 10 (April, 1960).

Flew, Antony, "Did Hume Distinguish Pure from Applied Geometry?" *Ratio*, vol. VIII (June, 1966).

Grossman, R., "Two Unpublished Essays on Mathematics in the Hume Papers." *Journal of the History of Ideas*, vol. 21 (July, 1960).

Hausman, Alan, "Hume's Theory of Relations." *Nous*, vol. 1 (August, 1967).

Ushenko, A., "Hume's Theory of General Ideas." *Review of Metaphysics*, vol. 9 (December, 1955).

Wolff, R. P., "Hume's Theory of Mental Activity." *Philosophical Review*, vol. 69 (July, 1960).

Zabeeh, Farhang, "Hume on Pure and Applied Geometry." *Ratio*, vol. VI (December, 1964).

Kant

Broad, C. D., "Kant's Theory of Mathematical and Philosophical Reasoning." *Proceedings of the Aristotelian Society*, vol. XLII (1941–42).

Hintikka, J., "Kant on the Mathematical Method." *Monist*, vol. LI (July, 1967).

Parsons, C. "Infinity and Kant's Conception of the Possibility of Experience." *Philosophical Review*, vol. 73 (April, 1964).

Parsons, C., "Kant's Philosophy of Arithmetic," in Morgenbesser, S., Suppes, P., and White, M., *Philosophy, Science and Method.* New York: St. Martin's Press, 1969.

Toohey, J. J., "Kant on the Propositions of Pure Mathematics." *The New Scholasticism*, vol. II (1937).

Vuillemin, J., "La Théorie Kantienne de l'Espace à la Lumière des Groupes de Transformations." *Monist*, vol. LI (July, 1967).

Mill

Britton, K., "The Nature of Arithmetic. A Reconsideration of Mill's Views." *Proceedings of the Aristotelian Society*, vol. XLVIII (1947–48).

Thomas, M. H., "J. S. Mill's Theory of Truth: A Study in Metaphysics and Logic." *Philosophical Review*, vol. 56 (May, 1947).

Whitmore, C., "Mill and Mathematics: An Historical Note." *Journal of the History of Ideas*, vol. 6 (January, 1945).

General—Anthologies

Benacerraf, Paul, and Putnam, Hilary, eds., *Philosophy of Mathematics: Selected Readings.* Englewood Cliffs, N.J.: Prentice-Hall, 1964.

Hintikka, J., ed., *Philosophy of Mathematics*, London: Oxford University Press, 1969.

Klibansky, R., ed., *Contemporary Philosophy—vol. 1: Logic and Foundations of Mathematics.* Florence: La Nuova Italia Editrice, 1968.

Lakatos, I., ed., *Problems in the Philosophy of Mathematics.* Amsterdam: North Holland Publishing Co., 1967.

Van Heijenoort, J., ed., *A Source Book in Mathematical Logic, 1879–1931.* Cambridge, Mass.: Harvard University Press, 1967.

General—Recent Books

Barker, S. F., *Philosophy of Mathematics.* Englewood Cliffs, N.J.: Prentice-Hall, 1964.

Benardete, J. A., *Infinity: An Essay in Metaphysics.* Oxford: Clarendon Press, 1964.

Bochner, S., *The Role of Mathematics in the Rise of Science.* Princeton: Princeton University Press, 1966.

Boyer, C. B., *The History of the Calculus and Its Conceptual Development.* New York: Dover Publications, Inc., 1959.

Carruccio, E., *Mathematics and Logic in History and in Contemporary Thought.* Chicago: Aldine Publishing Corp., 1964.

Goodstein, R. L., *Essays in the Philosophy of Mathematics.* Leicester: Leicester University Press, 1965.

Kielkopf, C., *Strict Finitism.* New York: Humanities Press, 1970.

Kleene, S. C., and Vesley, R. E., *The Foundations of Intuitionistic Mathematics.* Amsterdam: North Holland Publishing Co., 1965.

Kneale, W. and Kneale, M., *The Development of Logic.* Oxford: Clarendon Press, 1962.

Mostowski, A., *Thirty Years of Foundational Studies.* New York: Barnes & Noble, Inc., 1966.

Nidditch, P. H., *The Development of Mathematical Logic.* New York: Dover Publications, Inc.

Poincare, H., *Mathematics and Science: Last Essays*, trans. by J. W. Bolduc. New York: Dover Publications, Inc., 1963.

General—Recent Papers

Benacerraf, P., "What Numbers Could Not Be." *Philosophical Review*, vol. 74 (January, 1965).

Bird, O., "The History of Logic." *Review of Metaphysics*, vol. XVI, no. 3 (March, 1963).

Braithewaite, R. B., Thompson, J. F., and Warnock, G. J., "Reducibility." *Proceedings of the Aristotelian Society*, Supplementary vol. XXVI (1952).

Chihara, C. S., "On the Possibility of Completing an Infinite Process." *Philosophical Review*, vol. 74 (January, 1965).

Dretske, F. I., "Counting to Infinity." *Analysis*, vol. xxv.

Engel, S. M., "Wittgenstein's Foundations and Its Reception." *American Philosophical Quarterly*, vol. IV, no. 4 (October, 1967).

Fisher, C. S., "The Death of a Mathematical Theory: A Study in the Sociology of Knowledge." *Archive for History of the Exact Sciences*, vol. 3 (1966–67).

Goodstein, R. L., "Pure and Applied Mathematics." *Ratio*, vol. VI, no. 1 (June, 1964).

Grünbaum, A., "Some Recent Writings in the Philosophy of Mathematics." *Review of Metaphysics*, vol. V, no. 2 (December, 1951).

Hochberg, H., "Peano, Russell and Logicism." *Analysis*, vol. xvi.

Keene, G. B., "Analytical Statements and Mathematical Truth." *Analysis*, vol. xvi.

Pollock, J. L., "Mathematical Proof." *American Philosophical Quarterly*, vol. IV, no. 3 (July, 1967).

Putnam, H., "Mathematics Without Foundations." *Journal of Philosophy*, vol. LXIV, no. 1 (January 19, 1967).

Sellars, W., "Classes as Abstract Entities and the Russell Paradox." *Review of Metaphysics*, vol. XVII, no. 1 (September, 1963).

Whiteside, D. T., "Patterns of Mathematical Thought in the Later Seventeenth Century." *Archive for History of the Exact Sciences*, vol. I (1960–62).

Index

a priori-a posteriori distinction, 212 ff., 216 ff., 273 ff.
a priori knowledge, 32, 58, 216 ff., 273 ff.
abstract entities, 52, 298 ff.
abstract principles, 2, 40
abstraction, 40, 54 ff., 66, 74, 108, 117, 125 ff., 136, 176 ff., 185, 197 ff., 241, 251, 255, 266 ff.
aesthetics, 11, 16
algebra, 183, 186, 247, 258 ff.
analysis, 136, 148, 158
analytic vs. synthetic knowledge, 218, 272 ff. (see *truth*)
Aristotle, 39–77, 78, 117, 156, 270, 286
Aristoteleanism, 78 ff., 116, 135
arithmetic, 30, 164, 186, 219, 244
  objects of, 60, 183 ff., 187, 245 ff., 265 ff.
  value of, 30
  vs. geometry, 10, 41, 68, 74, 78, 135, 174, 206
Augustine, ix, 78, 156
Axinn, S., 42, 269, 283–291
axioms (first principles), 25, 48, 55, 58 ff., 79, 91, 114, 131, 157 ff., 163, 190, 223, 226, 233 ff., 247 ff., 253
  common, 64, 245
  created by God, 99, 169
Ayer, A. J., 273 ff.

becoming, 15
being, 20, 57, 63, 66, 72, 144, 162
  vs. becoming, 15, 26, 31
Berkeley, G., 41, 118, 172–192, 199, 235, 292
Black, M., 42, 269, 292–309
body, 7, 74, 84, 92
  vs. soul, 92, 138 ff.

calculus, 135, 149
Cantor, G., 295 ff.
Cartesianism, 149 ff.
cause, 65
  Aristotle's four kinds of, 53
certainty, 2 ff.
  absolute, 2 ff., 17, 81, 117, 227, 248 ff.
  criteria of, 12, 85, 154, 195, 237
  mathematical, 4 ff., 98, 131, 209, 282
  objective, 3, 79, 150
  subjective, 3, 79, 150
  vs. necessity, 170
class, 241, 270, 293 ff.
common sensibles, 49, 154
completeness, 11
concepts, 221 ff, 228 ff.
consistency, 11 ff.
constructivism, 117, 127, 214, 223, 225 ff., 296 ff.
continuity, 69
conventionalism, 103, 233, 275 ff.
creation vs. discovery, 9 ff., 302
  of Forms, 16
  of mathematical truths by God, 99

definition, 46 ff., 58, 65, 67, 114, 152, 157, 188, 230, 232, 245, 250 ff., 294
demonstration (proof), 46, 48, 55, 97 ff., 133 ff., 165 ff., 204, 207, 231, 234, 247, 286
Descartes, R., 2, 78–100, 101, 116, 136 ff., 140 ff., 149, 156, 173, 213
dialectic, 25
discovery vs. creation, 9, 14, 286
divisibility, infinite, 75, 143 ff., 168, 190 ff., 210 ff., 227
doubt, 79, 81 ff.
dualism, ontological, 7

Einstein, A., 172
empiricism, 116 ff., 271 ff.
epistemology, vii, 5 ff., 39, 80, 101, 149 ff., 193 ff., 269
  definition of, 5
equality, 70, 131, 202, 248, 255
error, 57, 96 ff., 204, 207, 217, 233, 240, 276 ff., 289, 290
essence, 65, 159
  perception of, 84
  real vs. nominal, 118, 132
essential attributes, 46 ff., 75

ethics, 11, 16, 17, 41, 101, 117, 155, 193, 197, 203, 214
Euclid, 39, 79, 101, 107, 147, 158, 253, 259, 286 ff., 298
evanescent quantities, 136, 146
existence, kinds of, 73
experiment, 255 ff., 283 ff.,
  crucial, 258
experimentalism, 283 ff.
extension, 140, 177, 190, 210 ff., 221

facts, 12
falsity, 50 ff.
figure, 69, 123, 141, 154
  construction of, 106
  perception of, 50, 154
formalism, 5, 41, 174, 196, 257 ff., 275 ff., 280
form—see *matter vs. form*
Forms, 15, 21, 40, 65, 67
  perception of, 22
formula, 48
free will, 96, 214
Frege, G., vii, 1, 151, 235, 263–268, 294, 299 ff.
Freud, S., 16
function, mathematical, 248

Galileo, 78, 283 ff.
geometry, 10, 26, 30, 65, 74, 81, 164, 188 ff., 206, 220, 224 ff., 250 ff., 300
  analytic, 78, 135
  non-Euclidean, 215, 235
  objects of, 60, 93 ff., 103, 177, 188, 201 ff., 214, 224 ff., 250
  pure vs. applied, 281
  (see also: *arithmetic vs. geometry*, *figure*, *line*, *point*, *space*)
God, 7, 13, 79, 87, 89, 99, 101 ff., 157, 162, 164, 169, 173, 193, 214
  proofs of the existence of, 79, 82, 86 ff., 137, 139 ff., 150
good, idea of, 23, 26, 29

Hobbes, T., 41, 101–115, 136, 149, 173, 235, 270
human nature, 196 ff.
Hume, D., 8, 41, 150, 193–211, 212, 235, 261
hypotheses, 25, 148, 249, 252 ff., 279

idealism, 8
ideas, 7, 89, 116, 119 ff., 136, 152 ff., 172 ff., 197 ff., 207 ff., 241 ff.
  abstract general, 94, 108, 125 ff., 132, 172, 176 ff., 188, 199 ff., 251, 294
  adequate, 165
  innate, 16, 79, 88 ff., 91 ff., 99, 102, 116, 119 ff., 137, 151, 193
  kinds of, 92, 116, 119 ff., 124 ff., 155
  simple vs. compound, 105, 123, 127, 197
identity, 208, 218
illusion, 82
image, 24, 52
imagination, 105, 154
immaterialist thesis, 173
impressions (vs. ideas), 197 ff., 262
induction, 56, 58, 62, 117, 136, 148, 155, 161, 217, 237, 244, 248, 250 ff., 274, 276, 282, 291
infinitesimals, 136, 167 ff., 190 ff.
infinity, 10, 75 ff., 87, 102, 129, 167 ff., 190 ff., 210, 257
  actual vs. potential, 75, 87, 136 ff., 142 ff., 167 ff., 211
  four senses of, 75
instrumentalism, 11, 174
introspection (reflection), 116, 121, 154, 157, 161, 176
intuition, rational, 56, 85, 97, 133
intuitionism, 215, 223, 227 ff., 250

Kant, I., vii, 9, 10, 41, 118, 150, 195, 212–234, 237, 262, 270 ff., 283
knower, 51 ff.
  active vs. passive, 9, 40, 213 ff.
knowledge
  degrees of, 22
  direct vs. indirect, 97, 133, 165, 216, 231, 238 ff.
  kinds of, 40, 53, 58, 65, 108, 129 ff., 152 ff., 203 ff., 222, 237
  mathematical, 9, 31 ff., 63 ff., 131, 162 ff., 227 ff., 271 ff.
  objects of, 5 ff., 9, 15, 19 ff., 44 ff., 66, 119 ff., 175 ff., 243
  perfect, 159
  pre-existent (innate), 62, 92 ff., 159 ff., 164, 216
  scientific, 46, 55, 61, 90, 108
  vs. opinion, 18 ff., 57

language, 39, 44, 101, 108 ff., 179 ff., 236, 238 ff., 257, 263, 303 ff.,
  and arithmetic, 185
Leibniz, G., vii, 9, 116, 135, 149–171, 172, 195, 212, 263, 302
line, 69, 108, 188 ff., 201
Locke, J., 41, 116–134, 172 ff., 178, 195, 213, 235
logic, deductive, vii, 11, 39, 149, 155, 197, 212, 230, 235, 238 ff., 263, 274, 293
  inductive, 236
  syllogistic, 55, 58, 60, 111, 134, 231

symbolic, vii, 1, 151, 263, 269
logicism, 151, 264, 301 ff.

materialism, 8, 102 ff.
mathematics,
certainty of, 98, 164, 195, 219, 234, 271 ff.
objects of, 51, 56, 65, 68, 73 ff., 98, 105 ff., 117, 129, 264
pure vs. applied, 11, 187, 192, 219, 225 ff., 245 ff.
vs. physics, 64, 245, 254, 283 ff.
matter, 140 ff.
perceptible vs. intelligible, 68
vs. form, 48, 53, 67, 68
material substance, 104 ff., 122, 140 ff.
meaning,
connotative, 242, 260
correspondence theory of, 6, 262
ideational theory of, 126 ff., 172, 179 ff., 193, 199 ff., 240 ff.
instrumentalist theory of, 180 ff.
theories of, 6, 117, 172, 238 ff.
memory, 54, 62, 105, 129
Meno problem, 31 ff., 58, 159
metaphor,
of the cave, 27 ff.
of the dark closet, 122
of the divided line, 24 ff.
of the marble block, 162 ff.
of the tabula rasa, 162
Mill, J. S., 41, 175, 195, 235–262, 270, 272, 277 ff.
mind, 7, 15, 22 ff., 119, 124, 194
(see also *knower* and *soul*)
monad, 150
monism, ontological, 8

names, 108 ff., 240 ff.
(see also *signs*)
natural light, 86, 90, 96, 163
necessary connections, 56
Newton, I., 135–148, 149, 172, 212
nominalism, 258
(see also *formalism*)
nothing, idea of, 112
number, 30, 69, 89, 95, 107, 123, 125, 127, 154, 183 ff., 201 ff., 205 ff., 208, 224, 243 ff., 265 ff., 295

observation—see *perception, sense*
Occam, William of, ix, 78
ontology, vii, 17, 39, 80, 101, 269, 307
definition of, 5
theories of, 6 ff.
opinion, 15
vs. knowledge—see *knowledge vs. opinion*

particulars, 56
perception, sense, 16, 40, 48 ff., 54, 61, 105 ff., 116, 120, 124, 162, 197 ff., 213, 221, 238, 260
perception,
actual vs. potential, 48
clear and distinct, 85, 154
objects of, 48, 79, 296
philosophy, nature of, 64, 227 ff.
Plato, 15–38, 39, 117, 150, 159, 217, 263
Platonism, 41, 78, 136, 264, 294, 298 ff.
pluralism, ontological, 7
point, geometrical, 108, 149, 250, 298
politics, 16, 101, 116
Pre-Socratics, 15, 39, 73, 93
Principle of Contradiction, 60, 151, 156 ff., 164, 203, 219, 230
probability, 209
proof—see *demonstration*
properties, 242, 299 (see *qualities*)
Pythagoras, 15
Pythagorean theorem, 204, 287
Pythagoreans, 67

qualities, primary vs. secondary, 90, 122 ff., 173, 184
quantity, 45, 69 ff., 107, 199 ff., 205 ff., 227 ff., 243 ff., 256
quantum, 68
Quine, W. V., 287, 294, 296, 298 ff.

realism, 296
reality, formal vs. objective, 88
reason, 29, 90
reason, intuitive, 61
recollection, 32 ff., 94, 164
reflection—see *introspection*
relation, 130 ff., 205 ff., 300 ff.
kinds of, 205 ff.
numerical, 71
of ideas, 205 ff.
religion, 196, 214
philosophy of, 11, 173, 193
Russell, B., vii, 1, 16, 151, 263, 293, 297

Saccheri, G., 290
Scholasticism, 78, 116
science, 23, 26, 48, 53, 96, 157, 181, 196, 205 ff., 220, 250, 253, 283 ff
demonstrative, 60, 250
objects of, 56
philosophy of, 11
physical, 41
scientific method, 39, 253, 283 ff.
Scotus, Duns, ix, 78
semantics, 269, 272 ff. (see *language*)
sense perception—see *perception, sense*

sets, 270, 293 ff.
signs, 101, 108 ff., 127, 174, 179 ff., 182 ff., 240 ff., 259, 264, 291, 302
  natural vs. arbitrary, 110
  uses of, 182 ff., 294 ff.
simplicity, 12 ff.
soul, 32, 62, 89, 154, 157, 163, 176
  immortality of, 18, 38
  nature of, 16, 18, 23, 26, 40, 51 ff., 92, 159
Space, 69, 107 ff., 116, 136, 138 ff., 214, 224 ff.
  absolute vs. relative, 138 ff., 166 ff., 224
Stenius, E., 269 ff.
substance, 6, 72, 79, 87, 102 ff., 116, 150, 155, 166, 169, 172, 193, 217, 236
  Aristotle's conception of, 39, 45 ff.
Sufficient Reason, Principle of, 150, 157, 164
syntax, 281
synthetic-a priori knowledge, 118, 132, see also *truth*, *synthetic-a priori*

theology, 65, 135
theory, 12 ff., 285
thing-in-itself, 222, 225
triangle, abstract general idea of, 94, 177 ff.
truth, 2, 21 ff., 29, 57 ff., 96, 103, 113 ff., 162, 195, 240, 248
  absolute, 26
  analytic, 132, 150, 156 ff., 169, 195, 203 ff., 218, 271 ff., 278, 288
  contingent, 114, 150, 156 ff., 203 ff., 212, 244
  correspondence theory of, 12, 90, 119 ff., 127 ff.
  criteria of, 12, 83, 85
  discovery of, 36
  eternal, 31, 91, 99, 114, 169, 265
  kinds of, 56, 156, 203 ff., 272 ff.
  necessary, 30, 83, 114, 150, 156, 161 ff., 170 ff., 195, 203 ff., 212, 217, 250 ff., 253, 273 ff.
  synthetic-a priori, 117 ff., 131 ff., 212 ff., 271 ff.
  universal, 40, 73, 109, 117, 189

understanding, faculty of, 85, 113, 154, 231 ff.
unity, 64, 71 ff., 127, 184, 201, 208, 215, 298
universals, 40, 54 ff., 62, 67, 102, 126, 188 ff., 201, 298 ff., 308
  (see also *abstract entities*, *Forms* and *ideas, abstract general*)

value, of mathematics, 11

Wittgenstein, L., 42, 270, 282

Zeno of Elea, 15